# Biotechnology of Edible Fungi

# Biotechnology of Edible Fungi

Editors

**Mingwen Zhao**
**Gen Zou**
**Jing Zhu**

MDPI

Basel • Beijing • Wuhan • Barcelona • Belgrade • Novi Sad • Cluj • Manchester

*Editors*

Mingwen Zhao
College of Life Sciences
Nanjing Agricultural University
Nanjing
China

Gen Zou
Institite of Edible Fungi
Shanghai Academy of Agricultural Sciences
Shanghai
China

Jing Zhu
College of Life Sciences
Nanjing Agricultural University
Nanjing
China

*Editorial Office*
MDPI
St. Alban-Anlage 66
4052 Basel, Switzerland

This is a reprint of articles from the Special Issue published online in the open access journal *Journal of Fungi* (ISSN 2309-608X) (available at: www.mdpi.com/journal/jof/special_issues/edible_fungus).

For citation purposes, cite each article independently as indicated on the article page online and as indicated below:

| |
|---|
| Lastname, A.A.; Lastname, B.B. Article Title. *Journal Name* **Year**, *Volume Number*, Page Range. |

**ISBN 978-3-7258-1110-6 (Hbk)**
**ISBN 978-3-7258-1109-0 (PDF)**
**doi.org/10.3390/books978-3-7258-1109-0**

# Contents

Journal of *Fungi*

*Editorial*

# Biotechnology of Edible Fungi

**Gen Zou [1], Jing Zhu [2] and Mingwen Zhao [2,*]**

1 Institute of Edible Fungi, Shanghai Academy of Agricultural Sciences, Shanghai 201403, China; zougen@sibs.ac.cn
2 College of Life Sciences, Nanjing Agricultural University, Nanjing 210095, China; jingzhu@njau.edu.cn
* Correspondence: mwzhao@njau.edu.cn

Edible fungi are generally defined as macrofungi with large fruiting bodies that may be consumed by humans and are commonly referred to as mushrooms. As a valuable source of proteins, fibers, minerals, vitamins, and other bioactive compounds, mushrooms have been consumed by humans for cultural, medicinal, recreational, and religious purposes, extending beyond their use as food for centuries. A typical fungal cell has precisely two genetically distinct but allelically compatible nuclei. When cell division occurs, a bulge-like hyphal outgrowth is formed over the cross wall between the two daughter cells and provides a clamp connection to maintain the binucleate state of the resulting cells. These unique properties make genetic manipulation tools unsuitable or inefficient in mushrooms. This has further caused the slow development of biotechnological research of edible fungi in the past.

Articles in this Special Issue have demonstrated a favorable transition in the situation of adapting the genetic technology to mushrooms [1,2]. The CRISPR/Cas9 genome-editing system is a revolutionary technology and a powerful tool for precision molecular breeding. CRISPR/Cas9 systems were established in some edible fungi based on in vivo expressed Cas9 and guide RNA. Liu Xiaotian et al. employed plasmids harboring the codon-optimized Cas9 and dual sgRNAs to edit *pyrG* in *Flammulina filiformis*. It was the first successful CRISPR/Cas9 genome editing system in *F. filiformis* [1]. Compared with this system, the in vitro assembled Cas9 and sgRNA ribonucleoprotein complexes (RNPs) have more advantages. Liu Jianyu et al. developed and optimized a CRISPR/Cas9 genome-editing method based on in vitro assembled RNP complexes in the same mushroom. The surfactant Triton X-100 played a critical role in this research [2]. These reports indicate that the development of edible mushroom biotechnology will accelerate along with this cutting-edge technology.

Most of the articles in this Special Issue are devoted to the bioactive components of *Ganoderma lingzhi* [3] and *G. lucidum* [4–6]. As traditional medicinal mushrooms with a history of more than 6800 years, the most concerned is the expression level and regulatory mechanism of genes related to the biosynthesis pathway of ganoderic acid [3–6]. Yan et al. investigated the metabolic flux distribution of three ganoderic acids. Intriguingly, the metabolic flux of ganoderic acids R was mostly benefited with a dramatical increase of 97.48% after adding oleic acid [6]. Mitochondria was proved that affected the content of polysaccharide and triterpenoid via starch and sucrose metabolism, steroid biosynthesis, and pentose and glucuronate interconversions [4]. The other two articles investigated the biosynthesis of these bioactive components through genome sequencing and transcriptome data [3,5].

Due to the commodity value of fruiting bodies, researchers frequently focus on morphological development and fungi–environment interactions. Guo et al. conducted the transcriptional and metabolic patterns of *Lentinula edodes* when the mycelial tissues were exposed to high temperature [7]. Yang et al. described a special link of spermidine and lysine which was probably involved in the development of mushroom fruiting body and in

check for updates

**Citation:** Zou, G.; Zhu, J.; Zhao, M. Biotechnology of Edible Fungi. *J. Fungi* **2023**, *9*, 1025. https://doi.org/10.3390/jof9101025

Received: 9 October 2023
Accepted: 16 October 2023
Published: 19 October 2023

response to the multiple environmental factors [8]. Another study corroborated the long-chain fatty acid synthesis pathway which was responsible for stipe gradient elongation in *F. filiformis* [9]. Gong et al. offered a novel strategy of *Hypsizygus marmoreus* cultivation using vernalization-like low-temperature fruiting [10]. Li et al. speculated the horizontal transfer of functional 1-aminocyclopropane-1-carboxylate oxidase via analyzing phylogenesis [11]. Finally, a novel application of mushroom was conducted to enhance the nutritional value of highland barley via solid-state fermentation with *Agaricus sinodeliciosus* [12].

The Editors of this Special Issue express their sincere gratitude to all the authors who contributed to this Special Issue, as well as to MDPI's staff for their professional help and efficient decisions.

**Author Contributions:** Conceptualization, M.Z., G.Z. and J.Z.; writing—original draft preparation, G.Z. and J.Z.; writing—review and editing, M.Z., G.Z. and J.Z. All authors have read and agreed to the published version of the manuscript.

**Conflicts of Interest:** The authors declare no conflict of interest.

## References

1. Liu, X.; Dong, J.; Liao, J.; Tian, L.; Qiu, H.; Wu, T.; Ge, F.; Zhu, J.; Shi, L.; Jiang, A.; et al. Establishment of CRISPR/Cas9 Genome-Editing System Based on Dual sgRNAs in *Flammulina filiformis*. *J. Fungi* **2022**, *8*, 693. [CrossRef] [PubMed]
2. Liu, J.; Cui, H.; Wang, R.; Xu, Z.; Yu, H.; Song, C.; Lu, H.; Li, Q.; Xing, D.; Tan, Q.; et al. A Simple and Efficient CRISPR/Cas9 System Using a Ribonucleoprotein Method for *Flammulina filiformis*. *J. Fungi* **2022**, *8*, 1000. [CrossRef] [PubMed]
3. Wu, Q.; Liu, H.; Shi, Y.X.; Li, W.T.; Huang, J.; Xue, F.F.; Liu, Y.N.; Liu, G.Q. Characteristics of the Genome, Transcriptome and Ganoderic Acid of the Medicinal Fungus *Ganoderma lingzhi*. *J. Fungi* **2022**, *8*, 1257. [CrossRef] [PubMed]
4. Ye, L.Y.; He, X.F.; Su, C.B.; Feng, H.Y.; Meng, G.L.; Chen, B.Z.; Wu, X.P. The Effect of Mitochondria on *Ganoderma lucidum* Growth and Bioactive Components Based on Transcriptomics. *J. Fungi* **2022**, *8*, 1182. [CrossRef] [PubMed]
5. Wu, T.H.; Cai, M.J.; Hu, H.P.; Jiao, C.W.; Zhang, Z.; Liu, Y.C.; Chen, J.; Xiao, C.; Li, X.M.; Gao, X.; et al. Whole-Genome Sequencing and Transcriptome Analysis of *Ganoderma lucidum* Strain Yw-1-5 Provides New Insights into the Enhanced Effect of Tween80 on Exopolysaccharide Production. *J. Fungi* **2022**, *8*, 1081. [CrossRef] [PubMed]
6. Yan, M.Q.; Su, X.W.; Liu, Y.F.; Tang, C.H.; Tang, Q.J.; Zhou, S.; Tan, Y.; Liu, L.P.; Zhang, J.S.; Feng, J. Effects of Oleic Acid Addition Methods on the Metabolic Flux Distribution of Ganoderic Acids R, S and T's Biosynthesis. *J. Fungi* **2022**, *8*, 615. [CrossRef] [PubMed]
7. Guo, Y.; Gao, Q.; Fan, Y.Y.; Song, S.; Yan, D.; Zhao, J.; Chen, Y.L.; Liu, Y.; Wang, S.X. Two Strains of Lentinula edodes Differ in Their Transcriptional and Metabolic Patterns and Respond Differently to Thermostress. *J. Fungi* **2023**, *9*, 179. [CrossRef] [PubMed]
8. Yang, Y.Y.; Shi, L.; Xu, X.Y.; Wen, J.; Xie, T.Y.; Li, H.; Li, X.Y.; Chen, M.Y.; Dou, X.Y.; Yuan, C.J.; et al. Spermidine Synthase and Saccharopine Reductase Have Co-Expression Patterns Both in Basidiomycetes with Fusion Form and Ascomycetes with Separate Form. *J. Fungi* **2023**, *9*, 352. [CrossRef] [PubMed]
9. Yan, J.J.; Tong, Z.J.; Han, X.; Gan, Y.; Liu, Y.Y.; Chen, J.; Duan, X.L.; Lin, J.B.; Gan, B.C.; Xie, B.G. Transcriptome Profiling Reveals Candidate Genes Related to Stipe Gradient Elongation of *Flammulina filiformis*. *J. Fungi* **2023**, *9*, 64. [CrossRef] [PubMed]
10. Gong, M.; Huang, T.; Li, Y.; Li, J.; Tang, L.; Su, E.; Zou, G.; Bao, D. Multi-Omics Analysis of Low-Temperature Fruiting Highlights the Promising Cultivation Application of the Nutrients Accumulation in *Hypsizygus marmoreus*. *J. Fungi* **2022**, *8*, 695. [CrossRef] [PubMed]
11. Li, Y.A.; Qi, M.; Zhang, Q.; Xu, Z.X.; Zhang, Y.; Gao, Y.Q.; Qi, Y.C.; Qiu, L.Y.; Wang, M.D. Phylogenesis of the Functional 1-Aminocyclopropane-1-Carboxylate Oxidase of Fungi and Plants. *J. Fungi* **2023**, *9*, 55. [CrossRef] [PubMed]
12. Liu, B.; Lu, H.Y.; Shu, Q.; Chen, Q.H.; Wang, J.L. The Influence of Different Pretreatment Methods of Highland Barley by Solid-State Fermentation with *Agaricus sinodeliciosus* var. Chaidam ZJU-TP-08 on Its Nutrient Content, Functional Properties and Physicochemical Characteristics. *J. Fungi* **2022**, *8*, 940. [CrossRef] [PubMed]

Journal of *Fungi*

MDPI

*Communication*

# Establishment of CRISPR/Cas9 Genome-Editing System Based on Dual sgRNAs in *Flammulina filiformis*

**Xiaotian Liu [1,†], Jianghan Dong [1,†], Jian Liao [1], Li Tian [1], Hao Qiu [1], Tao Wu [1], Feng Ge [1], Jing Zhu [1], Liang Shi [1], Ailiang Jiang [1], Hanshou Yu [1], Mingwen Zhao [1] and Ang Ren [1,2,3,*]**

1 Key Laboratory of Microbiological Engineering of Agricultural Environment, Ministry of Agriculture, Department of Microbiology, College of Life Sciences, Nanjing Agricultural University, Nanjing 210095, China; sheltonliu@foxmail.com (X.L.); 10318127@njau.edu.cn (J.D.); 10317128@njau.edu.cn (J.L.); 2020116058@stu.njau.edu.cn (L.T.); 2020816130@stu.njau.edu.cn (H.Q.); wt1170450810@hotmail.com (T.W.); 2021816129@stu.njau.edu.cn (F.G.); jingzhu@njau.edu.cn (J.Z.); shiliang@njau.edu.cn (L.S.); aljiang@njau.edu.cn (A.J.); yuhans@njau.edu.cn (H.Y.); mwzhao@njau.edu.cn (M.Z.)
2 Sanya Institute of Nanjing Agricultural University, Sanya 572025, China
3 Institute of Biology, Guizhou Academy of Sciences, Guiyang 550009, China
* Correspondence: angren@njau.edu.cn; Tel./Fax: +86-25-84395602
† These authors contributed equally to this work.

**Abstract:** *Flammulina filiformis*, previously known as Asian *Flammulina velutipes*, is one of the most commercially important edible fungi, with nutritional value and medicinal properties worldwide. However, precision genome editing using CRISPR/Cas9, which is a revolutionary technology and provides a powerful tool for molecular breeding, has not been established in *F. filiformis*. Here, plasmids harboring expression cassettes of Basidiomycete codon-optimized Cas9 and dual sgRNAs targeting *pyrG* under the control of the *gpd* promoter and FfU6 promoter, respectively, were delivered into protoplasts of *F. filiformis* Dan3 strain through PEG-mediated transformation. The results showed that an efficient native U6 promoter of *F. filiformis* was identified, and ultimately several *pyrG* mutants exhibiting 5-fluorooric acid (5-FOA) resistance were obtained. Additionally, diagnostic PCR followed by Sanger sequencing revealed that fragment deletion between the two sgRNA target sites or small insertions and deletions (indels) were introduced in these *pyrG* mutants through the nonhomologous end joining (NHEJ) pathway, resulting in heritable changes in genomic information. Taken together, this is the first report in which a successful CRISPR/Cas9 genome-editing system based on dual sgRNAs was established in *F. filiformis*, which broadens the application of this advanced tool in Basidiomycetes.

**Keywords:** *Flammulina filiformis*; CRISPR/Cas9; gene editing; dual sgRNAs; U6 promoter; *pyrG*



**Citation:** Liu, X.; Dong, J.; Liao, J.; Tian, L.; Qiu, H.; Wu, T.; Ge, F.; Zhu, J.; Shi, L.; Jiang, A.; et al. Establishment of CRISPR/Cas9 Genome-Editing System Based on Dual sgRNAs in *Flammulina filiformis*. *J. Fungi* **2022**, *8*, 693. https://doi.org/10.3390/jof8070693

Academic Editor: Sotiris Amillis

Received: 27 May 2022
Accepted: 27 June 2022
Published: 30 June 2022



## 1. Introduction

As one of the most common edible mushrooms worldwide, *Flammulina filiformis*, also named winter mushroom or enokitake, and previously known as *Flammulina velutipes*, has been renamed according to a recent report [1]. Based on phylogenetic analyses, genetic structure analyses, and haplotype network analysis, Wang et al. proposed that *F. filiformis* cultured in East Asia should be treated as a separate species that is different from European *F. velutipes*. It is widely recognized that *F. filiformis* is rich in amino acids, vitamins, minerals, and unsaturated fatty acids, which are essential and beneficial to human health [2]. In addition to its nutritional value, an increasing amount of research has focused on the medicinal properties of *F. filiformis*, such as immune-modulatory, anti-inflammatory, antioxidant, and intestinal flora activities [3–6]. With the development of the *F. filiformis* industry, generating highly valuable cultivars and improving the quality have become two important issues faced in *F. filiformis* industrial production. Unfortunately, classical breeding is time-consuming and laborious due to the tedious crossing steps [7]. In contrast, molecular breeding has attracted increasing attention, largely due to its flexibility, accuracy, and high efficiency [8–10].

CRISPR/Cas9-mediated genome editing is a revolutionary technology that has been widely applied in filamentous fungi, including Basidiomycetes [11–13]. However, due to the limited genetic manipulation and transformation methods, there have been relatively few reports describing successful CRISPR systems in edible fungi. For the convenience of screening and verification, most studies attempting to establish CRISPR systems in Basidiomycetes firstly preferred to choose genes that encoded clear morphological phenotypes or physiological properties for editing, such as *pyrG* in *Pleurotus eryngii* [14] and *ura3* in *Ganoderma lucidum* [15]. Because *pyrG/ura3* encodes orotidine-5′-monophosphate decarboxylase, which is not only involved in the key pathway of uracil synthesis, but also can convert 5-fluorooric acid (5-FOA) into the toxic substance 5-fluorouridine, *pyrG/ura3* mutants survive and wild-type (WT) die on a medium supplied with uracil and 5-FOA.

Efficient expression of Cas9 and sgRNA (single-guide RNA) guarantees successful CRISPR genome editing. DNA double-strand breaks (DSBs) caused by Cas9 can be repaired mainly via nonhomologous end joining (NHEJ) or homology-directed repair (HDR) mechanisms [16], which alter genetic information near the target site permanently. Theoretically, once the genomic information of CRISPR-targeted recipient cells has been changed, there will be no need for the existence of Cas9 and sgRNA. However, given that there are few reports on inducible promoters in Basidiomycetes, the conditional expression of Cas9 and sgRNA has not been feasible yet. Thus, the RNA polymerase (Pol) II constitutive promoters are adopted to express Cas9, while Pol III promoters such as U6 are suitable for expression of sgRNA, owing to the fact that mRNA processing events such as splicing, 5′ capping, and 3′ poly(A)-tail addition are not involved. Up to now, the *gpd* (glyceraldehyde-3-phosphate dehydrogenase) or *ef3* (elongation factor 3) promoter (Pol II) and U6 promoter (Pol III) have been successfully applied in some edible fungi, including *P. eryngii* [14], *Pleurotus ostreatus* [17], and *Lentinula edodes* [18].

In essence, checking the integrity of genomic information is the core work of screening and verification of mutants after gene editing. Several strategies have been developed to track changes in sequence information, such as Sanger sequencing and next-generation sequencing (NGS) [19], despite the fact that they are time-consuming and expensive. The PCR/restriction enzyme (PCR/RE) method [20], which is a sequencing-free manner, depends on electrophoresis by means of visualizing the differences in PCR products between the WT and mutants. Unfortunately, small insertions and deletions (indels) are the most common mutation types based on a single sgRNA in the CRISPR/Cas9 system through NHEJ [21], which means sequence differences cannot be distinguished using electrophoresis. The dual-sgRNA-mediated fragment-deletion strategy has attracted increasing interest [22–25], probably due to its high efficiency and divergence visualization. In this method, two independent DSBs are simultaneously generated at the two sgRNA-targeted loci, and then the DSBs rejoin through NHEJ, resulting in the generation of chromosomal segment deletion between the two cutting sites of Cas9.

Although many attempts have been made [26,27], so far there is no report on successful CRISPR-mediated gene editing in *Flammulina.* In the present study, by identifying U6 promoters in the *F. filiformis* genome, we obtained *pyrG* mutants with fragment or base deletion profiting from expression of Basidiomycete codon-optimized Cas9 and dual sgRNAs under the control of the *gpd* promoter and FfU6 promoter, respectively.

## 2. Materials and Methods

### 2.1. Strains and Culture Conditions

The *F. filiformis* strain Dan3, obtained from Shanghai Academy of Agricultural Sciences (SAAS), was used as the host for gene disruption. The *F. filiformis* was cultured in CYM medium (10 g/L maltose, 20 g/L glucose, 2 g/L tryptone, 2 g/L yeast extract, 0.5 g/L $MgSO_4 \cdot 7H_2O$, and 4.6 g/L $KH_2PO_4$) at 25 °C. Rescreening of possible 5-FOA-resistance transformants was performed through culturing on MM medium (20 g/L glucose, 2 g/L L-asparagine, 0.5 g/L $MgSO_4 \cdot 7H_2O$, 0.46 g/L $KH_2PO_4$, and 1 g/L $K_2HPO_4$) containing 600 μg/mL 5-FOA (Sangon Biotech, Shanghai, China) and 100 μg/mL uracil (Sangon Biotech). The *Escherichia coli* strain DH5α was used for vector construction.

### 2.2. Selection of sgRNAs of *pyrG* and U6 Promoters of *F. filiformis*

To design the sgRNA sequence used in gene editing, the nucleotide sequence of *pyrG* was uploaded to the online design tool EuPaGD (http://grna.ctegd.uga.edu (accessed on 9 November 2021)). Only those with high scores and "G" at 5′ were chosen as candidate sgRNAs.

In order to identify the FfU6 promoters, we performed a BLAST search of the *F. filiformis* genome using the nucleotide sequence of human U6 snRNA (NCBI accession: NR_004394.1) as a query to search for its homologs by means of BioEdit software, and then selected 400 bp upstream of the transcription start site (TSS) as the predicted FfU6 promoters.

### 2.3. In Vitro Cas9 Cleavage Assay

The primer pairs FfpyrG-vitro-F1/FfpyrG-vitro-R1 and FfpyrG-vitro-F2/FfpyrG-vitro-R2 (Table 1) were used to amplify the *pyrG* fragments containing the respective target site. First, 5 μL Cas9 nuclease (500 ng/μL) and 10 μL sgRNA (50 ng/μL) purchased from GenScript (Nanjing, China) were incubated for 5 min at 37 °C to form a ribonucleoprotein (RNP) complex. Subsequently, the RNP complex was incubated with the PCR-amplified fragment (100 ng) in Cas9 reaction buffer for 1 h at 37 °C. Finally, the mixture was visualized using 1.5% agarose gel electrophoresis.

**Table 1.** List of oligonucleotides used in this study.

| Name | Sequence (5′–3′) | Descriptions |
|---|---|---|
| FfpyrG-vitro-F1 | ATGGCACATATCCTCAACTTTG | Amplification of *pyrG* for in vitro cleavage assay |
| FfpyrG-vitro-R1 | TCATGCTGTTCTCTCCAAGTATG | |
| FfpyrG-vitro-F2 | ACGCCGCTCGCGCCTCAAAACATTC | Amplification of *pyrG* for in vitro cleavage assay |
| FfpyrG-vitro-R2 | TGGCCCATTCCATCGCCCTTGAC | |
| FfU6-1-F1 | TCGGGAAGAGCAGAGCGGGCAAGTTATCAACAAGCGTG | Amplification of FfU6-1 promoter |
| FfU6-1-R1 | TTGACATCCAGTCCCACTCCATTAAACGATGAAAAGGAGACATC | |
| pyrG-sgRNA1-F | GGAGTGGGACTGGATGTCAAGTTTTAGAGCTAGAAATAGCAAG | Amplification of sgRNA1 |
| pyrG-sgRNA1-R | TCTAAAACAAAAAAGCACCGACTCGGTG | |
| FfU6-1-F2 | GTCGGTGCTTTTTTGTTTTAGAGGGCAAGTTATCAACAAGCGTG | Amplification of FfU6-1 promoter |
| FfU6-1-R2 | ACCCGGTTCCAGGTCCATCCATTAAACGATGAAAAGGAGACATC | |
| pyrG-sgRNA2-F | GGATGGACCTGGAACCGGGTGTTTTAGAGCTAGAAATAGCAAG | Amplification of sgRNA2 |
| pyrG-sgRNA2-R | GGATCCTCTAGAGATGCGGCCGCTCTAAAACAAAAAAGCACCGAC | |
| FfU6-3-F1 | TCGGGAAGAGCAGAGCGTATACGGCTTAATAGCCCTCTT | Amplification of FfU6-3 promoter |
| FfU6-3-R1 | TTGACATCCAGTCCCACTCCATAAGCCTGTGGAAGAGAGGC | |
| FfU6-3-F2 | GTCGGTGCTTTTTTGTTTTAGACGTATACGGCTTAATAGCCCTCTT | Amplification of FfU6-3 promoter |
| FfU6-3-R2 | ACCCGGTTCCAGGTCCATCCATAAGCCTGTGGAAGAGAGGC | |
| FfpyrG-chk-F | CAACTTTGCGAGCATAGACCC | Amplification of *pyrG* for sequencing |
| FfpyrG-chk-R | AATAATCCGTCCCATACACCC | |

### 2.4. Construction of CRISPR Plasmids

Plasmids used for *pyrG* editing were derived from the vector pXT-300, which successfully enabled genome editing in *G. lucidum* (data not shown), harboring the *gpd* promoter and Basidiomycete codon-optimized Cas9. Cas9 was optimized according to humanized *Streptococcus pyogenes* Cas9 (SpCas9) open reading frame (ORF) in the commercial vector pX330 (addgene #42230), with a SV40 NLS (nuclear localization signal of simian virus 40) and a nucleoplasmin NLS located at 5′ and 3′, respectively. For the construction of the sgRNA expression cassette, the primer pair FfU6-1-F1/FfU6-1-R1 (Table 1) was used to amplify the FfU6-1-1 promoter from the *F. filiformis* genomic DNA. Using the primer pair pyrG-sgRNA1-F/pyrG-sgRNA1-R, the fragment sgRNA1 containing FfpyrG-sgRNA1, backbone region of gRNA scaffold, and a "TTTTTT" transcription terminator were amplified from pX330. Similarly, fragments FfU6-1-2 and sgRNA2 were amplified by the primer pairs FfU6-1-F2/FfU6-1-R2 and pyrG-sgRNA2-F/pyrG-sgRNA2-R, respectively. By means of a ClonExpress Ultra One Step Cloning Kit (Vazyme, Nanjing, China), the four purified fragments—FfU6-1-1, sgRNA1, FfU6-1-2, and sgRNA2—were sequentially ligated into the *Not* I digested vector pXT-300, yielding plasmid pXT-302-FfpyrG-1. The plasmid-containing FfU6-3 promoter, named pXT-302-FfpyrG-3, was constructed through the same steps as described above, except for the primers used. Briefly, the primer pairs FfU6-3-F1/FfU6-3-R1 and FfU6-3-F2/FfU6-3-R2 were used to amplify the fragments FfU6-3-1 and FfU6-3-2, respectively. Subsequently, FfU6-3-1, sgRNA1, FfU6-3-2, and sgRNA2 were ligated into the *Not* I digested vector pXT-300, yielding plasmid pXT-302-FfpyrG-3.

### 2.5. PEG-Mediated Transformation of Protoplasts

Protoplast preparation and transformation were performed as described previously [15] with minor modifications. The mycelia were collected and washed with 0.6 M mannitol (Sangon Biotech) and digested with 2% (*w*/*v*) lywallzyme (Guangdong Institute of Microbiology, Guangzhou, China) for 3 h at 30 °C. After filtering insufficiently digested mycelia, the protoplasts were resuspended in MTC buffer (0.6 M mannitol, 100 mM $CaCl_2$, and 100 mM Tris-HCl at pH 7.5) and adjusted to a final concentration of approximately $10^7$ cells/mL. Then, the protoplasts in 160 μL MTC buffer were transformed with 20 μg linearized pXT-302-FfpyrG-1/3 plasmid, 10 μL 20 mM aurintricarboxylic acid (ATA) (Sigma-Aldrich, Shanghai, China), 5 μL 50 mM spermidine (Sigma-Aldrich), 2 μL 50 mg/mL heparin (Sangon Biotech), and 60 μL PTC buffer (40% polyethylene glycol (PEG) 3350, 100 mM $CaCl_2$, and 10 mM Tris-HCl at pH 7.5). Subsequently, the mixture was incubated on ice for 30 min, then 1 mL PTC buffer was added and incubated for another 30 min at 25 °C. After centrifugation at 4000 rpm for 10 min, protoplasts were resuspended in 1 mL liquid CYM medium containing 0.6 M mannitol for 24–30 h at 25 °C.

### 2.6. Screening and Verification of Transformants

Protoplasts resuspended in 1 mL liquid CYM medium were poured onto CYM selective regeneration medium containing 0.6 M mannitol and 600 μg/mL 5-FOA. A total of 28 days after the PEG-mediated protoplast transformation, all transformants present on CYM selective regeneration medium were transferred to MM medium containing 600 μg/mL 5-FOA and 100 μg/mL uracil. After 14 days of culture, with the use of MightyAmp$^{TM}$ DNA Polymerase Ver.3 (Takara, Beijing, China), all isolates were verified using diagnostic PCR or sequencing analysis using mycelia as the template. With the primers FfpyrG-chk-F/FfpyrG-chk-R, PCR amplification of *pyrG* fragments of all transformants that contained the target sites were compared with the WT *pyrG* sequence using Sanger sequencing to determine whether indels occurred at or near the expected sites.

## 3. Results

### 3.1. In Vitro Cas9 Cleavage Assay

Successful genome editing was ensured by high-efficiency sgRNA. To verify whether the *pyrG* target sites could be recognized and cleaved by Cas9 endonuclease under the guidance of designed sgRNAs (Figure 1A, sgRNA1: 5′-GGAGTGGGACTGGATGTCAA-3′; sgRNA2: 5′-GGATGGACCTGGAACCGGGT-3′), we performed a cleavage validation experiment in vitro. Using genomic DNA of *F. filiformis* as the template, the *pyrG* fragments containing target regions with sizes of 714 bp and 726 bp were amplified through the primers FfpyrG-vitro-F1/FfpyrG-vitro-R1 and FfpyrG-vitro-F2/FfpyrG-vitro-R2, respectively. Theoretically, cleavage of the 714 bp fragment caused by RNP complex would yield two small bands of 527 bp and 187 bp, while cleavage of the 726 bp fragment would yield two small bands of 446 bp and 280 bp (Figure 1B). As expected, under the circumstance that only Cas9 or sgRNA was present in the reaction system, there was no observed difference in fragments compared with the groups to which only *pyrG* PCR fragments were added. On the contrary, the 714 bp and 726 bp PCR fragments were almost completely digested and two respective small bands appeared, provided both Cas9 and sgRNA existed in the reaction system (Figure 1C). The results were consistent with the theory, which indicated that the targeting efficiency of the two sgRNAs was high enough, and they could be utilized in subsequent experiments.

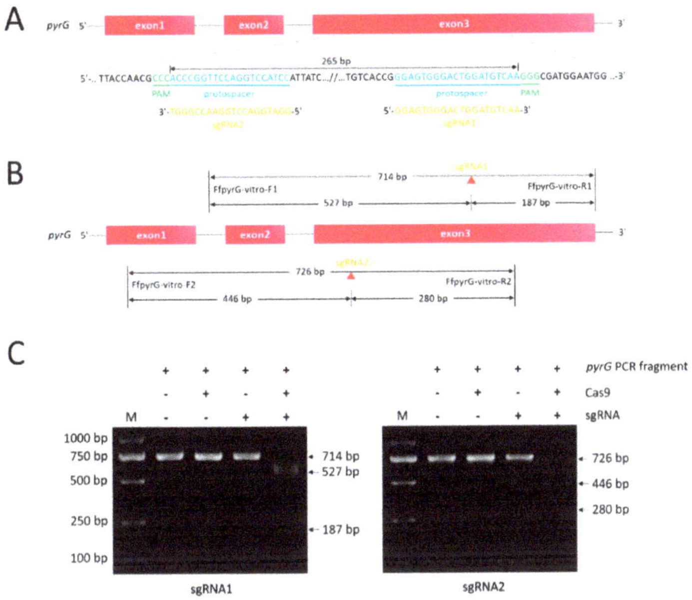


**Figure 1. Schematic illustration of sequence information and in vitro Cas9 cleavage assay.** (**A**) Sequences of the two sgRNAs targeting *pyrG* are shown in yellow font, both of which were located at exon3. Sequence directions are all 5′–3′ as shown. Schematic representation of exons was not drawn to scale. (**B**) Fragments required for in vitro Cas9 cleavage assay were amplified with primer pairs FfpyrG-vitro-F1/FfpyrG-vitro-R1 and FfpyrG-vitro-F2/FfpyrG-vitro-R2, respectively. Target regions that could be recognized by sgRNA1 or sgRNA2 were cleaved by Cas9 (red triangles), yielding two small fragments. (**C**) Visualization of the in vitro Cas9 cleavage assay through agarose gel electrophoresis. M, marker.

### 3.2. Four U6 Promoters of F. filiformis Were Identified Based on Homologous Search

Four candidate genes, termed FfU6-1, FfU6-2, FfU6-3, and FfU6-4, were identified through a homologous search in the *F. filiformis* genome with the human U6 snRNA gene, and all of them shared a conserved region with RNU6-1. Since nucleotide "G" was the TSS of U6 promoters [28], we selected 400 bp upstream of the TSS as the predicted FfU6 promoters. An analysis of multiple sequence alignments between FfU6 promoters and *Homo sapiens* U6 promoter showed that the promoter regions of these U6 varied greatly, including the TATA-like box, which was critical for transcription in the Pol III promoter, according to previous reports [29] (Figure 2). Interestingly, the four FfU6 promoters lacked a typical TATA box but contained a "GGTAATGCAAAACT" motif, which could be beneficial to effective transcription. In view of this, FfU6-1 and FfU6-3 promoters were randomly selected and adopted for the respective expression of sgRNA.

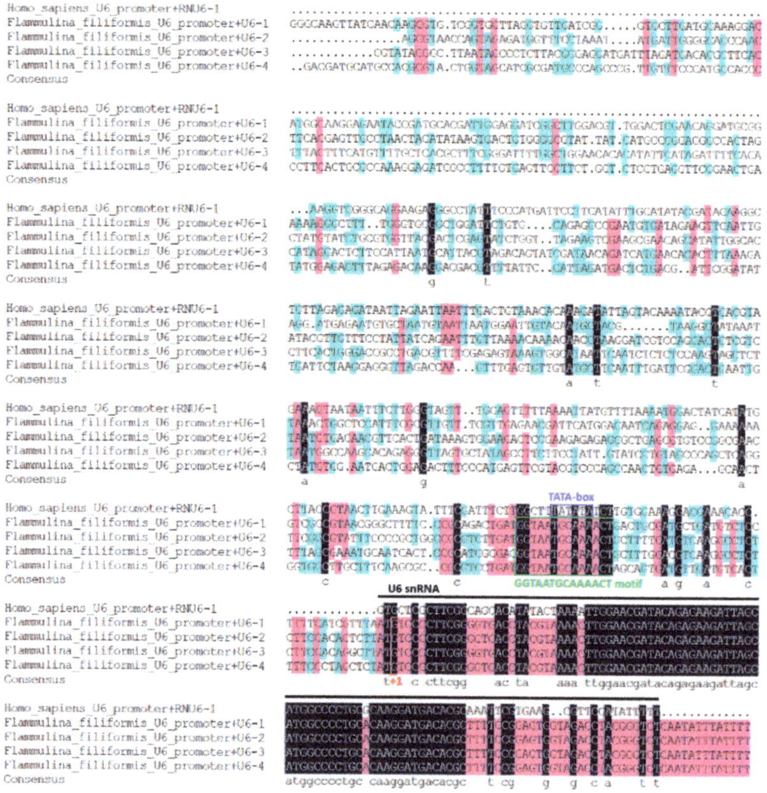


**Figure 2. Analysis of *F. filiformis* U6 and promoter sequences.** Multiple alignments of *F. filiformis* and human U6 snRNA are illustrated. Black lines indicate human U6 snRNA sequence. The conserved elements TATA box and "GGTAATGCAAAACT" motif are indicated by blue and green frames, respectively. The nucleotides "G" that could be recognized by FfU6 promoters for transcription initiation (+1) are labeled with a red box. Different colors denote different levels of sequence identity. The strictly conserved nucleotides (100% identity) are highlighted in black, while those with ≥75% and ≥50% identity are highlighted in red and blue, respectively. The multiple sequence alignments were performed using DNAMAN software.

### *3.3. Transformants Harboring pyrG Fragment Deletion or Small Indels Were Obtained*

Twenty-eight days after the PEG-mediated protoplast transformation (Figure 3A,B), colonies with white and dense hyphae were selected from 5-FOA-containing CYM regeneration medium and transferred to MM medium supplied with 5-FOA and uracil. A total of six transformants that exhibited resistance to 5-FOA (Figure 3C) were obtained from several replicates, and all of them were subjected to PCR to amplify the *pyrG* fragment using mycelia as the template with the primer pair FfpyrG-chk-F/FfpyrG-chk-R (Figure 3D). The agarose gel electrophoresis result showed the presence of variant bands that were about 250 bp smaller than that of WT (positive amplification control, Figure 3E). After gel electrophoresis purification and recovery, PCR products were subjected to Sanger sequencing with the primer FfpyrG-chk-F or cloned into the pMD19-T vector for further sequencing with the universal primer M13F-47. The sequencing results showed that a total length of 259 bp in *F. filiformis pyrG* locus was deleted, which was the interval between dual sgRNAs; more specifically, a region starting from 3 bp upstream of the PAM (the reverse complement of CCC; i.e., GGG) of sgRNA2 and ending at 3 bp upstream of the PAM (GGG) of sgRNA1 (Figure 3F). However, the molecular mechanism by which two DSBs generated by Cas9 directly rejoin together to generate chromosomal fragment deletion in vivo remains unclear [24]. In addition, one mutant harboring small indels ("CG" deletion and "T" insertion), whose size of PCR product was almost the same as that of WT, was determined (Figure 3F). Notably, all mutants obtained in the present study benefited from FfU6-1 under current experimental conditions. Taken together, our results demonstrated that the dual-sgRNA system could be applied as an efficient tool to introduce fragment deletion or indels for CRISPR/Cas9-mediated genome editing in *F. filiformis*.

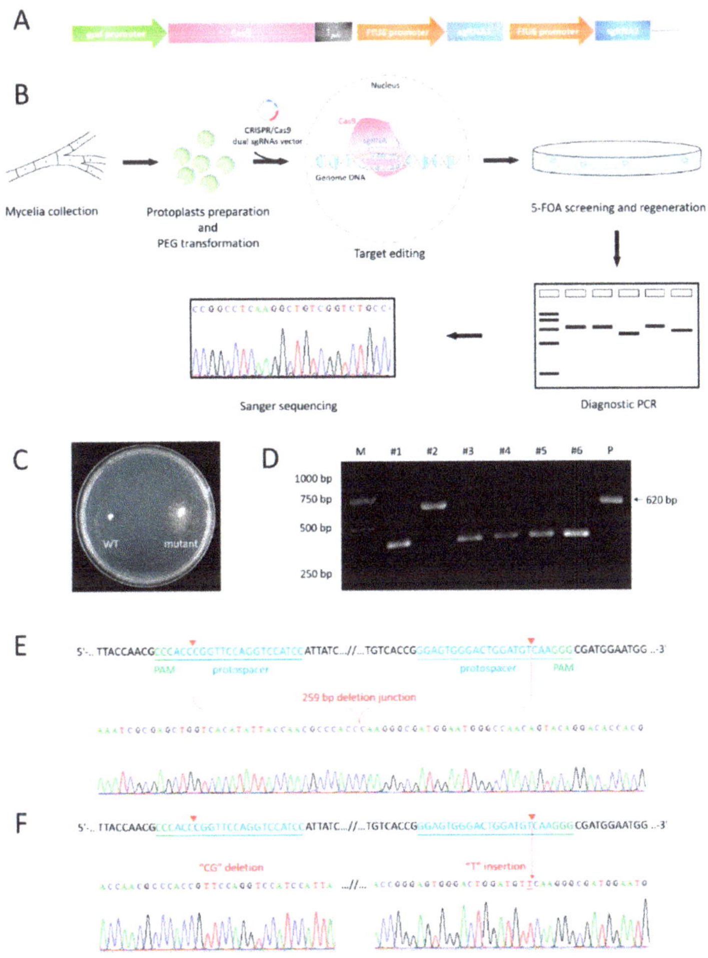


**Figure 3. Screening and confirmation of *pyrG* mutants of *F. filiformis*.** (**A**) Schematic diagram of vectors used in mutation of *pyrG*. (**B**) PEG-mediated protoplast transformation and screening process. (**C**) Rescreening on MM medium containing 5-FOA and uracil. (**D**) Diagnostic PCR of *pyrG* fragment harboring target sites. "P" indicates positive amplification control amplified from WT, with a size of 620 bp. (**E**) Sequencing result confirming the *pyrG* mutants harboring fragment deletion. (**F**) Sequencing result of the *pyrG* mutant harboring small indels.

## 4. Discussion

Since the first application in eukaryotic cells about a decade ago [16,28], CRISPR/Cas9 has exhibited its huge potential with great success in a tremendous number of species, strongly promoting the development of biotechnology, agriculture, and medicine [30].

However, in view of numerous species of Basidiomycetes, establishment of CRISPR/Cas9 system has just begun and remains scarce, especially in mushrooms, notwithstanding versatile and convenient application. Generally, determining appropriate target sites could be helpful for evaluation of editing efficiency, such as *pksP* and *lae1* in *Aspergillus fumigatus* [31] and *Trichoderma reesei* [32], respectively. Pigment production is blocked due to inactivation of these genes, which makes it easy to distinguish mutants from WT. As a negative selection marker, mutation of *pyrG/ura3* confers 5-FOA resistance in a host; meanwhile, the uracil synthesis pathway is interrupted. Here, we achieved the goal of fragment or base deletion of *pyrG* in *F. filiformis* by means of endogenous expression of Cas9 and dual sgRNAs.

According to the mechanism of CRISPR/Cas9, just a single sgRNA would be competent for gene disruption, despite the fact that dual sgRNAs were applied in this study. In practice, adoption of a dual-sgRNA strategy has obvious merits. First, a low mutation efficiency caused by a single sgRNA itself can be avoided as far as possible. Second, fragment deletion mediated by dual sgRNAs can be visualized directly via electrophoresis, which greatly improves the efficiency of mutant screening. Third, due to fragment deletion of the target region between the two sgRNAs, it is clear that a more complete deletion in terms of the primary structure is obtained in this manner, which is especially helpful in research on noncoding RNAs (ncRNAs) or other gene structures with biological functions [33,34]. Previous studies have reported several strategies for the expression of multiple sgRNAs; for instance, Pol III promoter::sgRNA tandem repeats [35], the endogenous tRNA-processing system [24], Csy4 ribonuclease [36], RZ ribozyme [37], hepatitis delta virus (HDV) and hammerhead (HH) ribozymes [38], and so on. An off-target effect that accompanies more sgRNAs, nevertheless, should be noticed.

CRISPR-related components and their delivery strategy will depend on the assay and cell type. In general, direct delivery of an in vitro Cas9/sgRNA complex (namely ribonucleoprotein (RNP)) or vectors containing Cas9-encoding and sgRNA-encoding information are the major methods [33]. Even though the Pol II promoter and Cas9 applied in this study were derived from *G. lucidum*, we still gained expected *pyrG* mutants. However, only the U6 promoter of *F. filiformis* could successfully screen mutants, while the use of the U6 promoter of *G. lucidum* failed (data not shown), probably due to unsuccessful transcription of sgRNA. Although a few articles using heterologous promoters reported successful CRISPR/Cas9 genome editing [17], it seems that most studies regarding CRISPR/Cas9 in Basidiomycete fungi, especially edible and medicinal fungi, tended to adopt their native promoters [14,39,40]. Furthermore, we noticed that the U6 promoter of *F. filiformis* did not contain a TATA box, and a similar phenomenon was also found in *G. lucidum* [39], despite the fact that it was vital for U6 transcription in many species [29]. We tried to analyze the U6 promoters in *P. eryngii*, and the same result was found (data not shown). Whether the lack of a TATA box is a common feature in edible and medicinal fungi remains to be further investigated. As classic genetic-manipulation methods, PEG-mediated transformation and *Agrobacterium tumefaciens*-mediated transformation were widely applied in filamentous fungi, generally, resulting in homologous or heterologous integration of CRISPR-related elements into the host genome. In this study, although we obtained *pyrG* mutants with fragment deletion or small indels via PEG-mediated transformation, the screening efficiency was relatively low. One reason could be that the vectors containing CRISPR-related elements were not delivered efficiently. There are few reports on autonomous replicators, except for AMA1 [41,42], which is only found in Aspergillus. Hence, CRISPR-related components will duplicate along with genome duplication. That is to say, the integration in genome is permanent, which does not cater to agriculture and food safety. On the contrary, an RNP complex formed by Cas9 and sgRNA, which are both transcribed and assembled in vitro; i.e., without restrictions on species, is an ideal strategy for genome editing due to its short half-life [32,43,44].

In conclusion, this was the first report to successfully establish a CRISPR/Cas9 genome-editing system through identifying native U6 promoters and applying a dual-sgRNA strategy in the edible mushroom *F. filiformis*, which broadened the application of this powerful tool in Basidiomycetes. Further research should be focused on increasing editing efficiency and targeting other genomic regions such as genes and functional elements.

**Author Contributions:** Conceptualization, X.L., J.D., M.Z. and A.R.; Methodology, X.L., J.D., J.L., J.Z., M.Z. and A.R.; Investigation, X.L., J.D. and J.L.; Formal analysis, X.L., J.D. and J.L.; Visualization, X.L. and J.D.; Writing—original draft, all authors; Writing—review and editing, M.Z. and A.R.; Resources, L.S., A.J., H.Y., M.Z. and A.R.; Funding acquisition, M.Z. and A.R.; Project administration, M.Z. and A.R. All authors have read and agreed to the published version of the manuscript.

**Funding:** This work was supported by the National Natural Science Foundation of China (project no. 81773839), the Guidance Foundation, the Sanya Institute of Nanjing Agricultural University (project no. NAUSY-MS17), and the China Agriculture Research System of MOF and MARA (no. CARS20).

**Institutional Review Board Statement:** Not applicable.

**Informed Consent Statement:** Not applicable.

**Data Availability Statement:** Not applicable.

**Conflicts of Interest:** The authors declare no conflict of interest.

## References

1. Wang, P.M.; Liu, X.B.; Dai, Y.C.; Horak, E.; Steffen, K.; Yang, Z.L. Phylogeny and species delimitation of *Flammulina*: Taxonomic status of winter mushroom in East Asia and a new European species identified using an integrated approach. *Mycol. Prog.* **2018**, *17*, 1013–1030. [CrossRef]
2. Tang, C.; Hoo, P.C.-X.; Tan, L.T.-H.; Pusparajah, P.; Khan, T.M.; Lee, L.-H.; Goh, B.-H.; Chan, K.-G. Golden needle mushroom: A culinary medicine with evidenced-based biological activities and health promoting properties. *Front. Pharmacol.* **2016**, *7*, 474. [CrossRef] [PubMed]
3. Liang, Q.; Zhao, Q.; Hao, X.; Wang, J.; Ma, C.; Xi, X.; Kang, W. The effect of *Flammulina velutipes* polysaccharide on immunization analyzed by intestinal flora and proteomics. *Front. Nutr.* **2022**, *9*, 841230. [CrossRef] [PubMed]
4. Zhao, R.; Hu, Q.; Ma, G.; Su, A.; Xie, M.; Li, X.; Chen, G.; Zhao, L. Effects of *Flammulina velutipes* polysaccharide on immune response and intestinal microbiota in mice. *J. Funct. Foods* **2019**, *56*, 255–264. [CrossRef]
5. Yuan, F.; Gao, Z.; Liu, W.; Li, H.; Zhang, Y.; Feng, Y.; Song, X.; Wang, W.; Zhang, J.; Huang, C.; et al. Characterization, antioxidant, anti-aging and organ protective effects of sulfated polysaccharides from *Flammulina velutipes*. *Molecules* **2019**, *24*, 3517. [CrossRef]
6. Wu, D.; Duan, W.; Liu, Y.; Cen, Y. Anti-inflammatory effect of the polysaccharides of golden needle mushroom in burned rats. *Int. J. Biol. Macromol.* **2010**, *46*, 100–103. [CrossRef]
7. Sonnenberg, A.S.M.; Baars, J.J.P.; Gao, W.; Visser, R.G.F. Developments in breeding of *Agaricus bisporus* var. *bisporus*: Progress made and technical and legal hurdles to take. *Appl. Microbiol. Biotechnol.* **2017**, *101*, 1819–1829. [CrossRef]
8. Tsukihara, T.; Honda, Y.; Watanabe, T.; Watanabe, T. Molecular breeding of white rot fungus *Pleurotus ostreatus* by homologous expression of its versatile peroxidase MnP2. *Appl. Microbiol. Biotechnol.* **2006**, *71*, 114–120. [CrossRef]
9. Tu, J.-L.; Bai, X.-Y.; Xu, Y.-L.; Li, N.; Xu, J.-W. Targeted gene insertion and replacement in the basidiomycete *Ganoderma lucidum* by inactivation of nonhomologous end joining using CRISPR/Cas9. *Appl. Environ. Microbiol.* **2021**, *87*, e0151021. [CrossRef]
10. Boontawon, T.; Nakazawa, T.; Horii, M.; Tsuzuki, M.; Kawauchi, M.; Sakamoto, M.; Honda, Y. Functional analyses of *Pleurotus ostreatus pcc1* and *clp1* using CRISPR/Cas9. *Fungal Genet. Biol.* **2021**, *154*, 103599. [CrossRef]
11. Jin, F.-J.; Wang, B.-T.; Wang, Z.-D.; Jin, L.; Han, P. CRISPR/Cas9-based genome editing and its application in *Aspergillus* species. *J. Fungi* **2022**, *8*, 467. [CrossRef] [PubMed]
12. Shi, T.-Q.; Liu, G.-N.; Ji, R.-Y.; Shi, K.; Song, P.; Ren, L.-J.; Huang, H.; Ji, X.-J. CRISPR/Cas9-based genome editing of the filamentous fungi: The state of the art. *Appl. Microbiol. Biotechnol.* **2017**, *101*, 7435–7443. [CrossRef]
13. Song, R.; Zhai, Q.; Sun, L.; Huang, E.; Zhang, Y.; Zhu, Y.; Guo, Q.; Tian, Y.; Zhao, B.; Lu, H. CRISPR/Cas9 genome editing technology in filamentous fungi: Progress and perspective. *Appl. Microbiol. Biotechnol.* **2019**, *103*, 6919–6932. [CrossRef] [PubMed]
14. Wang, T.; Yue, S.; Jin, Y.; Wei, H.; Lu, L. Advances allowing feasible *pyrG* gene editing by a CRISPR-Cas9 system for the edible mushroom *Pleurotus eryngii*. *Fungal Genet. Biol.* **2021**, *147*, 103509. [CrossRef] [PubMed]
15. Qin, H.; Xiao, H.; Zou, G.; Zhou, Z.; Zhong, J.-J. CRISPR-Cas9 assisted gene disruption in the higher fungus *Ganoderma* species. *Process Biochem.* **2017**, *56*, 57–61. [CrossRef]
16. Cong, L.; Ran, F.A.; Cox, D.; Lin, S.; Barretto, R.; Habib, N.; Hsu, P.D.; Wu, X.; Jiang, W.; Marraffini, L.A.; et al. Multiplex genome engineering using CRISPR/Cas systems. *Science* **2013**, *339*, 819–823. [CrossRef]
17. Boontawon, T.; Nakazawa, T.; Inoue, C.; Osakabe, K.; Kawauchi, M.; Sakamoto, M.; Honda, Y. Efficient genome editing with CRISPR/Cas9 in *Pleurotus ostreatus*. *AMB Express* **2021**, *11*, 30. [CrossRef]
18. Moon, S.; An, J.Y.; Choi, Y.-J.; Oh, Y.-L.; Ro, H.-S.; Ryu, H. Construction of a CRISPR/Cas9-mediated genome editing system in *Lentinula edodes*. *Mycobiology* **2021**, *49*, 599–603. [CrossRef]

19. Xue, L.-J.; Tsai, C.-J. AGEseq: Analysis of genome editing by sequencing. *Mol. Plant* **2015**, *8*, 1428–1430. [CrossRef]
20. Lloyd, A.; Plaisier, C.L.; Carroll, D.; Drews, G.N. Targeted mutagenesis using zinc-finger nucleases in *Arabidopsis*. *Proc. Natl. Acad. Sci. USA* **2005**, *102*, 2232–2237. [CrossRef]
21. Zhang, H.; Zhang, J.; Wei, P.; Zhang, B.; Gou, F.; Feng, Z.; Mao, Y.; Yang, L.; Zhang, H.; Xu, N.; et al. The CRISPR/Cas9 system produces specific and homozygous targeted gene editing in rice in one generation. *Plant Biotechnol. J.* **2014**, *12*, 797–807. [CrossRef] [PubMed]
22. Zhou, H.; Liu, B.; Weeks, D.P.; Spalding, M.H.; Yang, B. Large chromosomal deletions and heritable small genetic changes induced by CRISPR/Cas9 in rice. *Nucleic Acids Res.* **2014**, *42*, 10903–10914. [CrossRef]
23. Kraft, K.; Geuer, S.; Will, A.J.; Chan, W.L.; Paliou, C.; Borschiwer, M.; Harabula, I.; Wittler, L.; Franke, M.; Ibrahim, D.M.; et al. Deletions, inversions, duplications: Engineering of structural variants using CRISPR/Cas in mice. *Cell Rep.* **2015**, *10*, 833–839. [CrossRef] [PubMed]
24. Xie, K.; Minkenberg, B.; Yang, Y. Boosting CRISPR/Cas9 multiplex editing capability with the endogenous tRNA-processing system. *Proc. Natl. Acad. Sci. USA* **2015**, *112*, 3570–3575. [CrossRef] [PubMed]
25. Liu, K.; Sun, B.; You, H.; Tu, J.; Yu, X.; Zhao, P.; Xu, J. Dual sgRNA-directed gene deletion in basidiomycete *Ganoderma lucidum* using the CRISPR/Cas9 system. *Microb. Biotechnol.* **2020**, *13*, 386–396. [CrossRef] [PubMed]
26. Luo, R.; Lin, J.-F.; Guo, L.-Q.; Ye, Z.-W.; Guo, T.-F.; Yun, F. Construction of *Flammulina velutipes* genome editing vector by using CRISPR/Cas9 system. *Sci. Technol. Food Ind.* **2016**, *37*, 230–234. [CrossRef]
27. Lin, J.-D.; Yang, X.-Q.; Wei, T.; Guo, L.-Q.; Lin, J.-F.; Chen, Y.-S.; Huang, S.-S. Construction and transformation of CRISPR/Cas9 genome editing vector of *Flammulina filiformis* G protein-coupled receptor gene. *Mycosystema* **2019**, *38*, 349–361. [CrossRef]
28. Mali, P.; Yang, L.; Esvelt, K.M.; Aach, J.; Guell, M.; DiCarlo, J.E.; Norville, J.E.; Church, G.M. RNA-guided human genome engineering via Cas9. *Science* **2013**, *339*, 823–826. [CrossRef]
29. Gao, Z.; Herrera-Carrillo, E.; Berkhout, B. RNA polymerase II activity of type 3 Pol III promoters. *Mol. Ther.-Nucleic Acids* **2018**, *12*, 135–145. [CrossRef]
30. Adli, M. The CRISPR tool kit for genome editing and beyond. *Nat. Commun.* **2018**, *9*, 1911. [CrossRef]
31. Zhang, C.; Meng, X.; Wei, X.; Lu, L. Highly efficient CRISPR mutagenesis by microhomology-mediated end joining in *Aspergillus fumigatus*. *Fungal Genet. Biol.* **2016**, *86*, 47–57. [CrossRef] [PubMed]
32. Zou, G.; Xiao, M.; Chai, S.; Zhu, Z.; Wang, Y.; Zhou, Z. Efficient genome editing in filamentous fungi via an improved CRISPR-Cas9 ribonucleoprotein method facilitated by chemical reagents. *Microb. Biotechnol.* **2021**, *14*, 2343–2355. [CrossRef] [PubMed]
33. Wang, H.; La Russa, M.; Qi, L.S. CRISPR/Cas9 in genome editing and beyond. *Annu. Rev. Biochem.* **2016**, *85*, 227–264. [CrossRef] [PubMed]
34. Terns, M.P. CRISPR-based technologies: Impact of RNA-targeting systems. *Mol. Cell* **2018**, *72*, 404–412. [CrossRef] [PubMed]
35. Ma, X.; Zhang, Q.; Zhu, Q.; Liu, W.; Chen, Y.; Qiu, R.; Wang, B.; Yang, Z.; Li, H.; Lin, Y.; et al. A robust CRISPR/Cas9 system for convenient, high-efficiency multiplex genome editing in monocot and dicot plants. *Mol. Plant* **2015**, *8*, 1274–1284. [CrossRef]
36. Čermák, T.; Curtin, S.J.; Gil-Humanes, J.; Čegan, R.; Kono, T.J.Y.; Konečná, E.; Belanto, J.J.; Starker, C.G.; Mathre, J.W.; Greenstein, R.L.; et al. A multipurpose toolkit to enable advanced genome engineering in plants. *Plant Cell* **2017**, *29*, 1196–1217. [CrossRef]
37. Tang, X.; Zheng, X.; Qi, Y.; Zhang, D.; Cheng, Y.; Tang, A.; Voytas, D.F.; Zhang, Y. A single transcript CRISPR-Cas9 system for efficient genome editing in plants. *Mol. Plant* **2016**, *9*, 1088–1091. [CrossRef]
38. Xu, L.; Zhao, L.; Gao, Y.; Xu, J.; Han, R. Empower multiplex cell and tissue-specific CRISPR-mediated gene manipulation with self-cleaving ribozymes and tRNA. *Nucleic Acids Res.* **2016**, *45*, e28. [CrossRef]
39. Wang, P.-A.; Xiao, H.; Zhong, J.-J. CRISPR-Cas9 assisted functional gene editing in the mushroom *Ganoderma lucidum*. *Appl. Microbiol. Biotechnol.* **2020**, *104*, 1661–1671. [CrossRef]
40. Sugano, S.S.; Suzuki, H.; Shimokita, E.; Chiba, H.; Noji, S.; Osakabe, Y.; Osakabe, K. Genome editing in the mushroom-forming basidiomycete *Coprinopsis cinerea*, optimized by a high-throughput transformation system. *Sci. Rep.* **2017**, *7*, 1260. [CrossRef]
41. Liu, W.; An, C.; Shu, X.; Meng, X.; Yao, Y.; Zhang, J.; Chen, F.; Xiang, H.; Yang, S.; Gao, X.; et al. A dual-plasmid CRISPR/Cas system for mycotoxin elimination in polykaryotic industrial fungi. *ACS Synth. Biol.* **2020**, *9*, 2087–2095. [CrossRef] [PubMed]
42. Maruyama, J. Genome editing technology and its application potentials in the industrial filamentous fungus *Aspergillus oryzae*. *J. Fungi* **2021**, *7*, 638. [CrossRef] [PubMed]
43. Boontawon, T.; Nakazawa, T.; Xu, H.; Kawauchi, M.; Sakamoto, M.; Honda, Y. Gene targeting using pre-assembled Cas9 ribonucleoprotein and split-marker recombination in *Pleurotus ostreatus*. *FEMS Microbiol. Lett.* **2021**, *368*, fnab080. [CrossRef]
44. Vonk, P.J.; Escobar, N.; Wösten, H.A.B.; Lugones, L.G.; Ohm, R.A. High-throughput targeted gene deletion in the model mushroom *Schizophyllum commune* using pre-assembled Cas9 ribonucleoproteins. *Sci. Rep.* **2019**, *9*, 7632. [CrossRef] [PubMed]

Journal of
*Fungi*

MDPI

*Article*

# A Simple and Efficient CRISPR/Cas9 System Using a Ribonucleoprotein Method for *Flammulina filiformis*

**Jianyu Liu [1,†], Haiyang Cui [2,†], Ruijuan Wang [1], Zhen Xu [1], Hailong Yu [1], Chunyan Song [1], Huan Lu [1], Qiaozhen Li [1], Danrun Xing [2], Qi Tan [1], Weiming Sun [2,*], Gen Zou [1,*] and Xiaodong Shang [1,*]**

1 National Engineering Research Center of Edible Fungi, Institute of Edible Fungi, Shanghai Academy of Agricultural Sciences, Shanghai 201403, China
2 College of Marine Resources and Environment, Hebei Normal University of Science & Technology, Qinghuangdao 066004, China
* Correspondence: swimming122@aliyun.com (W.S.); zougen@sibs.ac.cn (G.Z.); shangxiaodong@saas.sh.cn (X.S.); Tel.: +86-335-8058992 (W.S.); +86-13671512909 (G.Z.); +86-21-62209760 (X.S.)
† These authors contributed equally to this work.

**Abstract:** CRISPR/Cas9 systems were established in some edible fungi based on in vivo expressed Cas9 and guide RNA. Compared with those systems, the in vitro assembled Cas9 and sgRNA ribonucleoprotein complexes (RNPs) have more advantages, but only a few examples were reported, and the editing efficiency is relatively low. In this study, we developed and optimized a CRISPR/Cas9 genome-editing method based on in vitro assembled ribonucleoprotein complexes in the mushroom *Flammulina filiformis*. The surfactant Triton X-100 played a critical role in the optimal method, and the targeting efficiency of the genomic editing reached 100% on a selective medium containing 5-FOA. This study is the first to use an RNP complex delivery to establish a CRISPR/Cas9 genome-editing system in *F. filiformis*. Moreover, compared with other methods, this method avoids the use of any foreign DNA, thus saving time and labor when it comes to plasmid construction.

**Keywords:** *Flammulina filiformis*; CRISPR/Cas9; genomic editing; RNPs; *pyrG*



**Citation:** Liu, J.; Cui, H.; Wang, R.; Xu, Z.; Yu, H.; Song, C.; Lu, H.; Li, Q.; Xing, D.; Tan, Q.; et al. A Simple and Efficient CRISPR/Cas9 System Using a Ribonucleoprotein Method for *Flammulina filiformis*. *J. Fungi* **2022**, *8*, 1000. https://doi.org/10.3390/jof8101000

Academic Editor: Anna Muszewska

Received: 29 July 2022
Accepted: 19 September 2022
Published: 23 September 2022



## 1. Introduction

*Flammulina filiformis* from East Asia (previously referred to as *F. velutipes* or *F. velutipes* var. *filiformis*) [1] is a commercially valuable and edible fungus. In recent years, with the development of the *F. filiformis* industry and increased market demand, generating cultivars with high-yield and improved quality has caused important production issues [2]. However, the lack of efficient genetic engineering tools makes it difficult to improve the physiological characteristics of this species [3]. Therefore, the development of new strategic approaches, such as genome editing, are being used to overcome this hurdle [4–6].

The CRISPR/Cas9 genome-editing system is a revolutionary technology and a powerful tool for precision molecular breeding [7]. A typical system comprises nuclease (Cas9), mature CRISPR RNA (crRNA), and trans-activating crRNA (tracrRNA). The crRNA can combine with the tracrRNA to generate a single-guide RNA (sgRNA) [8,9], which can effectively induce the Cas9 nuclease to cleave the target sequences. DNA double-strand breaks (DSBs) are formed when the sequence is cleaved. Then, the genomic DNA initiates the repair process. In eukaryotes, there are two DNA self-repair mechanisms: the non-homologous end-joining (NHEJ) and homologous directed repair (HDR) pathways. The NHEJ, as the dominant repair pathway, can lead to genomic alteration by causing random insertion, deletion, or replacement at DSB locations [10–12].

CRISPR/Cas9 is now becoming the standard methodology for improving genome-editing efficiency in fungi. Typically, the implementation of CRISPR/Cas9 systems in fungi is based on in vivo expression of Cas9 and sgRNA. Although plasmid construction is becoming straightforward, it takes time for plasmid propagation and extraction. In addition, the most notable merit is that the RNP-based CRISPR-Cas9 is a fast, easy, and accurate strategy in gene editing while avoiding transgenes in many organisms [13]. Thus, implementing Cas9 and the sgRNA components as an in vitro assembled ribonucleoprotein (RNP) complex in transformation may be a viable alternative.

Genetic manipulation of basidiomycete mushrooms is challenging [14]. To facilitate early phenotypic screening of mutants, most studies attempting to establish CRISPR/Cas9 systems have chosen target genes that encode clear morphological phenotypes or physiological properties for editing, such as *pyrG* (encoding orotidine 5′-phosphate decarboxylase) in *Pleurotus eryngii* [15] and *ura3* (syn. *pyrG*) in *Ganoderma lucidum* [16]. As *pyrG* and *ura3* have negative selection effects, which can convert 5-fluorooric acid (5-FOA) into the toxic substance 5-fluorouridine, *pyrG/ura3* mutants survive, and the wild-type (WT) die on a medium supplied with uracil and 5-FOA [9].

Although there have been some reports of the establishment of gene editing systems for *F. filiformis* in recent years [17–19], these technologies are all based on genetic modification (GM). However, public attitudes towards GM-based agricultural products are still conservative. Therefore, the establishment of editing technology independent from GM or exogenous DNA is beneficial to evade regulation of policy. Moreover, the reported editing efficiency is still very low [2]. In this study, we used an RNP delivery strategy to develop a CRISPR/Cas9 transformation method in *F. filiformis*, in which the editing efficiency on the *pyrG* gene was 100%.

## 2. Materials and Methods

### 2.1. Strains and Culture Conditions

*Flammulina filiformis* homokaryon strain Dan3, used in this study, was stored at the Shanghai Key Laboratory of Agricultural Genetics and Breeding. *Flammulina filiformis* was cultured on PDA medium (200 g/L potato starch, 20.0 g/L dextrose, and 20.0 g/L agar) at 25 °C. The uracil auxotrophic mutants were grown on PDA containing 100 μg/mL uracil (Sangon Biotech, Shanghai, China).

### 2.2. Screening for Optimum Concentration of Triton X-100 Reagent

Different gradients (0, 0.005%, 0.01%, 0.15%, 0.2%, and 0.3% [*w*/*v*]) of Triton X-100 were prepared to screen its optimum concentration. Triton X-100 is a chemical reagent that can improve cell membrane permeability to ensure RNPs cross the fungal cell membrane and nuclear membrane successfully [20,21].

### 2.3. Preparingation of sgRNA

The protospacer sequence was designed and analyzed online (https://www.idtdna.com/site/order/designtool/index/CRISPR_PREDESIGN, accessed on 12 September 2020). The crRNA containing the 20 bp protospacer sequence for *pyrG* was chemically synthesized together with a 36 bp consensus sequence provided by Integrated DNA Technologies (Coralville, IA, USA). The equimolar concentrations of crRNA and tracrRNA (purchased from IDT, Coralville, IA, USA) were mixed in equal proportions and annealed at 95 °C for 5 min and then left at room temperature (5 °C/min to 25 °C) to form sgRNA. The integrity of the hybridized products was visualized using agarose gel electrophoresis. The sequences of crRNA and tracrRNA are shown in Table 1.

**Table 1.** RNA sequence information.

| Name | Sequence (5′ to 3′) |
|---|---|
| crRNA (56 bp) | <u>GCGTACACAAAATCGCGAGC</u>CUCCUUCACCUCCUCUCAU CGUUUUAGAGCUAUGCU [1] |
| tracrRNA (67 bp) | AAAUAGCAAGUUAAAAUAAGGCUAGUCCGUUAUCAACU UGAAAAAGUGGCACCGAGUCGGUGCUUUU |

[1] The underline indicates the protospacer sequences.

### 2.4. Preparation of RNP Complex

The RNP complex was assembled in a 33 μL volume reaction containing a mixture of 1.6 μL of commercial Cas9 nuclease (62 μM, purchased from IDT), 8.3 μL of sgRNA (10 μM), 3.3 μL of 10 × Cas9 nuclease reaction buffer (20 mM HEPES, 100 mM NaCl, 5 mM MgCl2, 0.1 mM EDTA; pH 6.5), and 19.8 μL of RNase free water. The mixture was incubated for 20 min at room temperature (18–25 °C) to form RNPs.

### 2.5. In Vitro Cas9 Cleavage Assay

To determine the activity of sgRNA, an in vitro cleavage assay was performed. The primers FfpyrG-1F/1R and FfpyrG-4F/4R (Table 2) were used to amplify the *pyrG* fragments containing the target site. One microliter of RNP complex (3000 nM), 2 μL of PCR-amplified fragment (100 ng/μL), 1 μL of 10 × Cas9 nuclease reaction buffer, and 6 μL of RNase-free water were combined and mixed. The mixture was incubated for 1.5 h at 37 °C. Then, 1 μL of protease K (20 mg/mL) was added to terminate the reaction. Finally, the mixture was visualized using 2% agarose gel electrophoresis.

**Table 2.** List of oligonucleotides used in this study.

| Name | Fragment Length (bp) | Sequence (5′ to 3′) |
|---|---|---|
| *pyrG*-1F | 578 | GAGACTATGGAACGCAAAA |
| *pyrG*-1R | | CCTCTGAGCGATGAAGC |
| *pyrG*-4F | 902 | ATGCAGTCCTACGCCGCTCG |
| *pyrG*-4R | | TCATGCTGTTCTCTCCAAGT |

### 2.6. Preparation of Protoplasts

Protoplast preparation and transformation were performed as described previously [16] with minor modifications. The mycelia were collected, washed with 0.6 M mannitol (Sangon Biotech), and digested with 2% (*w*/*v*) lywallzyme (Guangdong Institute of Microbiology, Guangzhou, China) for 90 min. After filtering off the insufficiently digested mycelia, the protoplasts were re-suspended in MTC buffer (0.6 M mannitol, 100 mM $CaCl_2$, and 100 mM Tris-HCl; pH 7.5) and adjusted to a range of concentrations: $10^4$, $10^5$, $10^6$, and $10^7$ cells $mL^{-1}$.

### 2.7. PEG-Mediated Transformation of Protoplasts

The resuspended protoplasts were transformed with RNP complex (0, 90, 170, 250, 300, or 400 nM), 1 μL Triton X-100 [0.01% (*w*/*v*) final concentration in transformation reaction], 10 μL 10 × Cas9 nuclease reaction buffer, 31.5 μL 2 × MTC buffer, and 12.5 μL PTC buffer [60% polyethylene glycol (PEG) 4000, 100 mM CaCl2, and 10 mM Tris-HCl; pH 7.5]. The mixture was chilled on ice for 20 min and then 500 μL PTC buffer was added and incubated for another 70 min at 20 °C. Then, 1 mL MTC buffer and 2 mL resuscitation medium PDMU (200 g/L potato starch, 20.0 g/L dextrose, 109.3 g/L mannitol, and 100 μg/mL uracil) were added, and the aliquot was incubated for 24 h at 20 °C. Finally, 3.6 mL low melting top-agar medium PDLMUF (200 g/L potato starch, 20.0 g/L dextrose, 20.0 g/L low melting-point agarose, 109.3 g/L mannitol, 100 μg/mL uracil, and 0.1 mg/mL 5-FOA) was added to the aliquot and poured onto the bottom-medium PDAMUF (200 g/L potato starch, 20.0 g/L dextrose, 20.0 g/L agarose, 109.3 g/L mannitol, 100 μg/mL uracil, and 0.1 mg/mL 5-FOA) selective regeneration medium.

### 2.8. Screening and Verification of Transformants

At 11 d after the PEG-mediated protoplast transformation, all transformants present on PDAMUF selective regeneration medium were transferred to PDAUF medium (200 g/L potato starch, 20.0 g/L dextrose, 20.0 g/L agarose, 100 μg/mL uracil, and 0.1 mg/mL 5-FOA). After 10 d of culture, all isolates were examined using diagnostic PCR or sequencing analysis utilizing mycelia as the template. With either the primers FfpyrG-1F/1R or

FfpyrG-4F/4R, PCR amplification of the *pyrG* fragments from all the transformants that contained the target sites was compared with the WT *pyrG* sequence. Sanger sequencing was used to determine whether indels occurred at, or near, the expected sites. Randomly selected transformants were transferred to PDA plates containing uracil and subcultured 2–4 times to ensure hereditary stability.

## 3. Results

### 3.1. Effects of Different Concentrations of Triton X-100 on Protoplast Regeneration

The surfactant Triton X-100 can be used to improve cell membrane permeability and ensure that RNPs successfully cross the fungal cell membrane and nuclear membrane during PEG-mediated protoplast transformation [20–23]. However, overloading Triton X-100 causes cell death due to the complete destruction of the cell membrane. Therefore, we first analyzed the optimal concentration of Triton X-100 for *F. filiformis*. The results showed that, as the concentration of Triton X-100 increased, the number of regenerated protoplasts showed an obvious reduction. Although the number of protoplasts (123 colonies in Figure 1C) obtained at the concentration of 0.01% Triton X-100 has decreased, it is still sufficient for the next step, prompting our choice of 0.01% as the optimum concentration (Figure 1).

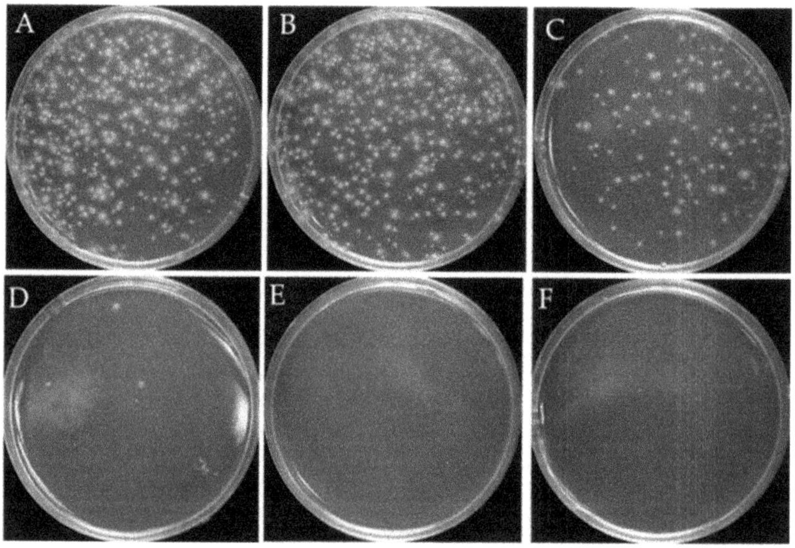


**Figure 1.** Effect of the Triton X-100 on the protoplast regeneration. (**A–F**): The different concentrations of Triton X-100: 0, 0.005%, 0.01%, 0.15%, 0.2%, 0.3%, respectively.

### 3.2. In Vitro Cas9 Cleavage Assay

In many reported fungal CRISPR systems, *pryG* is commonly employed as a bidirectional selective marker to avoid the use of antibiotic screening markers. To verify whether the pyrG target site could be recognized and cleaved by Cas9 endonuclease under the guidance of designed sgRNA, we performed an in vitro cleavage validation experiment (Figure 2A). Using genomic DNA of *F. filiformis* as the template, the pyrG fragments containing target regions with sizes of 578 bp and 902 bp were amplified through the primers FfpyrG-1F/FfpyrG-1R and FfpyrG-4F/FfpyrG-4R, respectively. Theoretically, cleavage of the 578 bp fragment by the RNP complex would yield two small bands of 367 bp and 210 bp, while cleavage of the 902 bp fragment would yield two small bands of 427 bp and 475 bp (Figure 2B). The result of the agarose gel electrophoresis showed that the crRNA and tracrRNA annealed in vitro to form a stable sgRNA, with complete structure and no degradation (Figure 2C). As expected, the 578 bp and 902 bp PCR fragments were almost completely digested, and two respective small bands appeared when the RNP complex was present in the reaction system (Figure 2D). The results were consistent with our expectations and indicated that the targeting efficiency of the sgRNA was sufficient to be used in subsequent experiments.

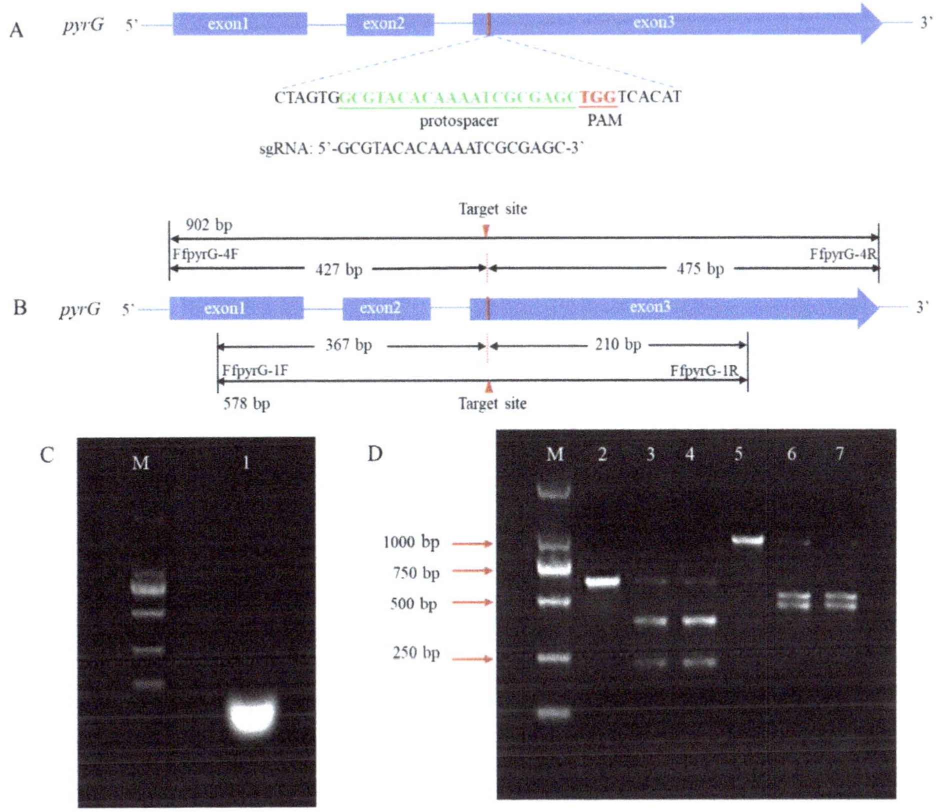


**Figure 2.** Schematic illustration of sequence information and in vitro Cas9 cleavage assay. (**A**) Sequences of the sgRNA targeting pyrG are shown in green font, which were located at exon3. Sequence direction was 5′-3′ as shown. Schematic representations of exons are drawn to scale. The protospacer adjacent motif (PAM) is shown in red. (**B**) Fragments required for the in vitro Cas9 cleavage assay were amplified with primers FfpyrG-1F/FfpyrG-1R and FfpyrG-4F/FfpyrG-4R. The red triangle represents the cleavage site of Cas9. The annealed products of crRNA and tracrRNA (**C**) and cleavage assay of Cas9 nuclease in vitro were visualized using agarose gel electrophoresis (**D**) M, DL 2000 marker; 1, sgRNA products; 2, pyrG-578 bp without RNPs; 3–4, pyrG-578 bp with RNPs; 5, pyrG-902 bp without RNPs; 6–7, pyrG-902 bp with RNPs.

### *3.3. Optimization of Polyethylene Glycol (PEG)-Mediated Protoplast Transformation of RNPs in F. filiformis*

The amounts of RNPs and protoplasts are pivotal factors that restrict the editing efficiency in an RNP-based editing strategy. To determine the effect of protoplast concentration on the targeting efficiency, four concentration gradients were set (0, $10^4$, $10^5$, $10^6$, and $10^7$ cells $mL^{-1}$). After 11 days of incubation on a regeneration medium (containing 100 µg/mL uracil and 0.1 mg/L 5-FOA), the resulting number of colony-forming units (CFUs) was determined, and the corresponding efficiency was analyzed by PCR and sequencing with the FfpyrG-1F and FfpyrG-1R primers (Table 3, Figure 3 and Figure S1). Five transformants were obtained from transformations with protoplasts at concentrations of $10^7$ cells mL$^{-1}$ when 0.01% Triton X-100 was added. However, no transformants were obtained in the control group without Triton X-100 (Table 3). This result indicates that Triton X-100 promoted the passage of RNP through the cell membrane, and gene editing ensued. Conversely, we also assessed the edited transformants grown on non-selective media. Although only 4.07% of the colonies were edited, this result demonstrated that it was feasible to obtain correct transformants independent of selective marker or transgenic system.

**Table 3.** Effect of protoplast concentration on targeting efficiency.

| Protoplasts (Cell mL$^{-1}$) | 0.01% Triton X-100 | | No Triton X-100 | |
|---|---|---|---|---|
| | CFUs | Positivity Rate (%) | CFUs | Positivity Rate (%) |
| $3.5 \times 10^4$ | 0 | - | 0 | - |
| $3.5 \times 10^5$ | 0 | - | 0 | - |
| $3.5 \times 10^6$ | 2 | 100 | 0 | - |
| $3.5 \times 10^7$ | 5 | 100 | 0 | - |

```
A
WT              AGTGGCGTACACAAAATCGCGAGCTGGTCACATATTACCAACGCCCACCCGG
(+1 nt) 10⁶-1   AGTGGCGTACACAAAATCGCGAGCTGGTCACATATTACCAACGCCCACCCGG
                                     T
(+33 nt) 10⁶-2  AGTGGCGTACACAAAATCGCGAGCTGGTCACATATTACCAACGCCCACCCGG
                CAATCTCTTATACACATCTGTGATCGTCGTGCT

B
WT              AGTGGCGTACACAAAATCGCGAGCTGGTCACATATTACCAACGCCCACCCGG
(+1 nt) 10⁷-1   AGTGGCGTACACAAAATCGCGAGCTGGTCACATATTACCAACGCCCACCCGG
                                     T
(+1 nt) 10⁷-2   AGTGGCGTACACAAAATCGCGAGCTGGTCACATATTACCAACGCCCACCCGG
                                     G
(-2 nt+60 nt) 10⁷-3 AGTGGCGTACACAAAATCG--AGCTGGTCACATATTACCAACGCCCACCCGG
                TCCATCTGTAAAACCTACTAATCATTGATTAAATATTTCTTGATTTTCAGTTAATCCTCT
(-10 nt) 10⁷-4  AGTGGCGTACACAAAATC---------CACATATTACCAACGCCCACCCGG
(-28 nt) 10⁷-5  AGTGGCGTACACAAAATCGCG------------------------CGG
```

**Figure 3.** Verifying *pyrG* mutation after RNP-based gene disruption using different protoplast concentrations. Alignment of *pyrG* sequences and corresponding mutants obtained by transformation with $10^6$ cells mL$^{-1}$ protoplasts (**A**) and $10^7$ cells mL$^{-1}$ protoplasts (**B**). WT: wild-type strain Dan3. Concentrations $10^6$-1, $10^6$-2, $10^7$-1~$10^7$-5: mutants generated by transformation of RNPs. Green marked represents spacer sequences; orange marked represents PAM.

To determine the effect of the RNP concentration on the targeting efficiency, RNPs at a range of concentrations (0, 90, 170, 250, 300, and 400 nM) were transformed into *F. filiformis* protoplasts ($10^7$ cells mL$^{-1}$) and an addition of 0.01% Triton X-100. No transformants were obtained without RNPs or with RNPs at concentrations $\leq$ 170 nM. When the RNP concentration was 300 nM, the resulting number of CFUs was at a maximum (Table 4 and Figure 4). Therefore, 300 nM is the optimal RNP concentration (Figure 5).

**Table 4.** Effect of the RNPs concentration on targeting efficiency.

| RNPs (nM) | CFUs | Positivity Rate (%) |
|---|---|---|
| 0 | 0 | - |
| 90 | 0 | - |
| 170 | 0 | - |
| 250 | 3 | 100 |
| 300 | 4 | 100 |
| 400 | 2 | 100 |

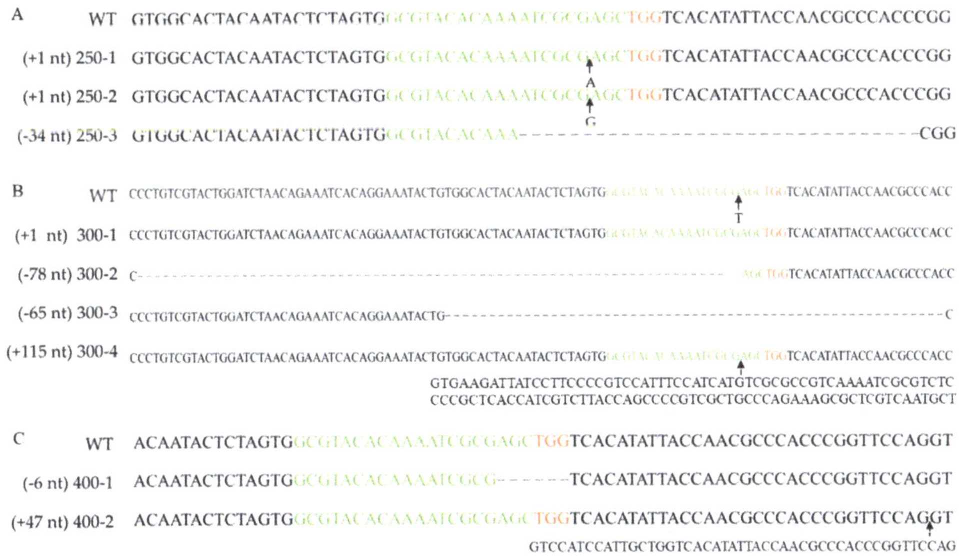

**Figure 4.** Verifying *pyrG* mutation after RNP-based gene disruption using different RNP concentrations. Alignment of *pyrG* sequences and corresponding mutants obtained by transformation with 250 nM RNPs (**A**), 300 nM RNPs (**B**) or 400 nM RNPs (**C**). WT: wild-type strain Dan3. Concentrations 250-1~250-3, 300-1~300-4, 400-1~400-2: mutants generated by transformation of RNPs. Green marked represents spacer sequences; orange marked represents PAM.

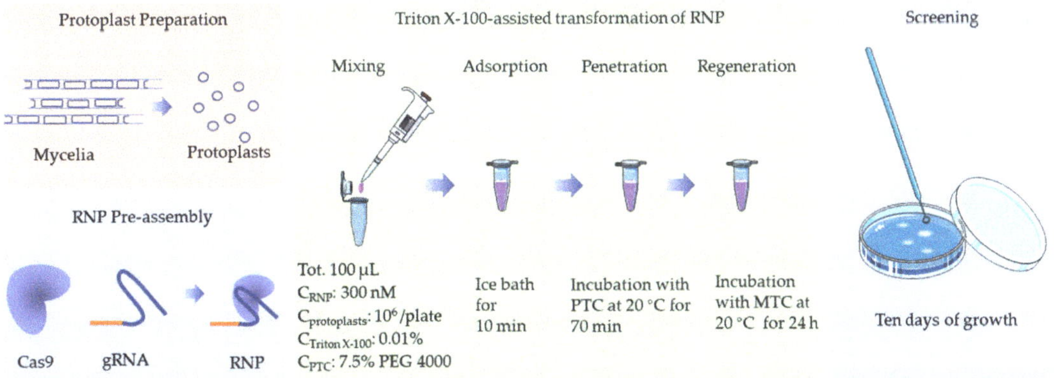


**Figure 5.** Schematic representation of the optimal protocol for CRISPR/Cas9 transformation of *F. filiformis*, which is based on an in vitro assembled RNP complex.

The *pyrG* genes of 16 *F. filiformis* transformants were sequenced, all of which exhibited expected mutations in the vicinity of the cleavage site (Figure 3, Figure 4 and Figure S1). Sequence analysis of the transformants with insertions revealed that the inserted fragments, 115 bp and 47 bp, were consistent with the genome sequences of JAJAKW010000006.1 and JAJAKW010000005.1, respectively [Fv01-10_genome GenBank assembly (GCA_022345005.1)]. The other two inserted fragments of transformant $10^6$-2 and $10^7$-3 were not able to align with the genome of *F. filirormis*.

### 3.4. Comparative Analysis of RNP-Directed Mutants and Wild-Type Strain

As a genetic manipulation toolbox, the application of such a tool should be avoided to cause phenotypic changes. To ensure that the transient introduction of RNPs would not disrupt the expression and function of other genes, the morphology and important phenotypic traits—such as growth and cellulase activity—of the selected *F. filiformis* transformants, which had undergone *pyrG* mutagenesis, were compared with those of the parental strains. The mycelial growths of five randomly selected RNP-directed mutants were observed on potato dextrose agar (PDA) medium, with or without uridine, after 7 d of incubation. On PDA plates without uridine, only the prototrophic strain Dan3 grew normally; the other five mutants showed defects in mycelial growth. By contrast, the growth of the mutants on PDA plates containing uridine was not affected (Figure 6A). No significant differences in cellulase and laccase activity were observed among the RNP-directed mutants and the wild-type Dan3 strain (Figure 6B,C).

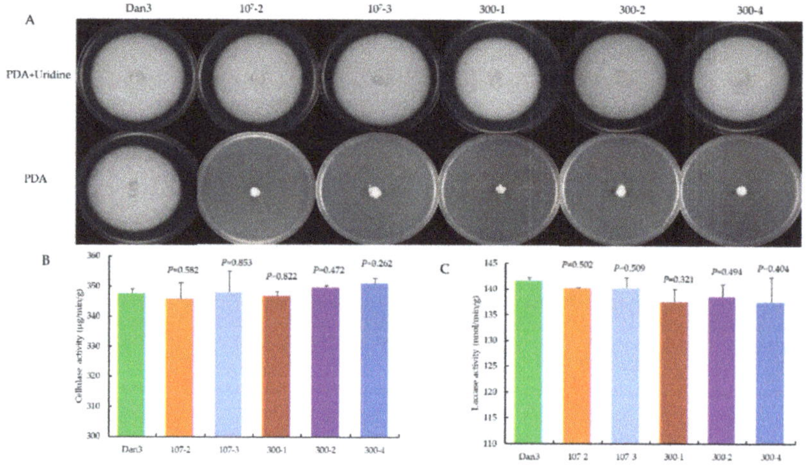


**Figure 6.** Comparative analysis of RNP-directed mutants and their parental strains. (**A**) Phenotype of Dan3 and its *pyrG* mutants on PDA plates (with/without uridine). Dan3: wild-type. $10^7$-2, $10^7$-3, 300-1, 300-2, 300-4: *pyrG* mutants edited with RNP. (**B**) Cellulase activities of Dan3 and corresponding *pyrG* mutants. (**C**) Laccase activities of Dan3 and corresponding *pyrG* mutants. Error bars indicate the standard deviation of three replicates. All *p*-values were obtained using two-tailed t-tests in Microsoft Excel.

## 4. Discussion

CRISPR/Cas9 technology has dramatically improved the efficiency of genome editing [24,25]. Determining appropriate target sites is helpful for evaluating editing efficiency [26]. As pyrG has a negative selection effect, growth inhibition can be easily observed in PDA plates containing 5-FOA (cells with wild-type pyrG converting 5-FOA into the toxic substance 5′ fluorouridine monophosphate). In this study, we produced mutations using RNP complex transformation. The indels exhibited classic mutations in the vicinity of the cleavage site. In contrast to the plasmid method, the RNP strategy is completely free of foreign DNA and can save time and labor by eliminating the tedious steps involved in plasmid construction. In the future, endogenous *pyrG* can be used to develop functional gene knockouts that do not require antibiotic markers, and the *pyrG* marker recycling system can be employed to implement multiple rounds of gene editing [27].

There are three common delivery strategies in the CRISPR-Cas9 genome-editing system: Cas9 nuclease in vivo and sgRNA in vitro (Cas9-expressing chassis with sgRNA in vitro), both Cas9 and sgRNA in vivo (all-in-one plasmid harboring Cas9-expressing cassette and sgRNA cassette), and both Cas9 and sgRNA in vitro (RNP complex) [28]. In this context, Moon et al. successfully disrupted the LeA1 locus of *Lentinula edodes* by delivering a plasmid containing the LeU6 and LeGPD promoters to express the Cas9 protein [29]. Wang et al. established a CRISPR/Cas9 genome-editing system in *G. lucidum* based on a plasmid delivery strategy, but the editing efficiency was low [14]. Boontawon et al. established an efficient CRISPR/Cas9-assisted genome-editing system based on plasmid harboring expression cassettes of Cas9 and different single-guide RNAs in *P. ostreatus* [30]. Liu et al. reported a CRISPR/Cas9 genome-editing system based on a plasmid delivery strategy in *F. filiformis* but only two mutants were obtained in their study [2]. The transformation efficiency of the plasmid method is relatively low, and there are many false positives. The RNPs strategy involves the direct delivery of an in vitro Cas9/sgRNA complex. In basidiomycetes, there was a successful transformation of RNPs, such as in *Schizophyllum commune* [31] and *P. ostreatus* [32], but only a few mutants were obtained. In our study, the Cas9/sgRNA complex was used in *F. filiformis* for direct delivery into the protoplasts through PEG-mediated transformation. In this case, the targeting efficiency of the genomic editing was 100% when the mutants were selected on a medium containing 5-FOA. The addition of the surfactant Triton X-100 may be the key to this high efficiency. Triton X-100 can improve cell membrane permeability and ensure that the RNPs cross the fungal cell membrane and nuclear membrane successfully during PEG-mediated protoplast transformation [20–22]. However, protoplasts cannot be regenerated owing to the complete disintegration of the cell membrane structure caused by excessive Triton X-100 (>0.2%), therefore, milder surfactants than Triton X-100 may be an alternative for highly efficient delivery for RNPs.

Notably, the plasmid method used with *F. filiformis* required up to 28 d for the transformants to grow after PEG-mediated transformation [2]. However, in our RNP complex method, transformant growth occurred after only 11 d. The reason for this difference is unknown. It is possible that the unwanted genomic integrations of DNA constructs expressing Cas9 and sgRNA affect the growth rate of *F. filiformis*.

In this study, the morphology and important phenotypic traits showed no significant differences between the RNP-directed mutants and the parental strain *F. filiformis* Dan3 (Figure 6). However, when mutants of *Trichoderma reesei* were obtained based on Cas9 plasmids, cellulase activities were indeed affected [23]. The authors considered that in vivo expression of Cas9 might impact ordinary physiological and biochemical processes in the transformants, which could be related to the endogenous promoter used to express the Cas9 protein. The RNP complex delivery strategy can avoid this shortcoming.

Sequence analysis of the transformants with insertions revealed that the 115 bp and 47 bp fragments were consistent with the genome sequences of JA-JAKW010000006.1 and JAJAKW010000005.1, respectively, in the *F. filiformis* genome. A large fragment insertion also occurred in the filamentous fungi *G. lucidum* [16] and *Aspergillus nidulans* [33]. Some large insertions were also identical to other genomic loci but the mechanism of capturing fragments from the whole genome to assist NHEJ remains unclear.

In conclusion, this report demonstrates a successful CRISPR/Cas9 genome-editing system through direct delivery of an RNP complex into the cultivated mushroom *F. filiformis*. This system is free from genomic integration of ectopic sequences. In the future, we will use uridine auxotrophic mutants as parental strains to study other functional genes using the HDR mechanism with donor DNA.

**Supplementary Materials:** The following are available online at https://www.mdpi.com/article/10.3390/jof8101000/s1, Figure S1: Chromatogram and alignment of pyrG sequences and corresponding mutants.

**Author Contributions:** Conceptualization, G.Z. and X.S.; Data curation, J.L. and H.C.; Formal analysis, H.Y. and Q.L.; Investigation, R.W., C.S., and H.L.; Methodology, J.L., H.C. and G.Z.; Project administration, Q.T. and X.S.; Resources, Z.X., C.S., H.L., D.X. and W.S.; Supervision, Q.T.; Writing—original draft, J.L. and H.C.; Writing—review and editing, G.Z. All authors have read and agreed to the published version of the manuscript.

**Funding:** This work was supported by "Science and Technology Innovation Action Plan", Agricultural Science and Technology Field Project, Shanghai, China (Grant No. 21N11900300), and the Shanghai Agriculture Applied Technology Development Program, China (Grant No. G20200103-Enhancement of elite germplasm and promotion of new varieties in industrial edible fungi).

**Institutional Review Board Statement:** Not applicable.

**Informed Consent Statement:** Not applicable.

**Data Availability Statement:** Not applicable.

**Conflicts of Interest:** The authors declare no conflict of interest.

## References

1. Wang, P.M.; Liu, X.B.; Dai, Y.C.; Horak, E.; Steffen, K.; Yang, Z.L. Phylogeny and species delimitation of *Flammulina*: Taxonomic status of winter mushroom in East Asia and a new European species identified using an integrated approach. *Mycol. Prog.* **2018**, *17*, 1013–1030. [CrossRef]
2. Liu, X.; Dong, J.; Liao, J.; Tian, L.; Qiu, H.; Wu, T.; Ge, F.; Zhu, J.; Shi, L.; Jiang, A.; et al. Establishment of CRISPR/Cas9 genome-editing system based on dual sgRNAs in *Flammulina filiformis*. *J. Fungi* **2022**, *8*, 693. [CrossRef] [PubMed]
3. Sonnenberg, A.S.M.; Baars, J.J.P.; Gao, W.; Visser, R.G.F. Developments in breeding of *Agaricus bisporus* var. *bisporus: Progress made and technical and legal hurdles to take. Appl. Microbiol. Biotechnol.* **2017**, *101*, 1819–1829. [PubMed]
4. Tsukihara, T.; Honda, Y.; Watanabe, T.; Watanabe, T. Molecular breeding of white rot fungus *Pleurotus ostreatus* by homologous expression of its versatile peroxidase $MnP_2$. *Appl. Microbiol. Biotechnol.* **2006**, *71*, 114–120. [CrossRef]
5. Tu, J.L.; Bai, X.Y.; Xu, Y.L.; Li, N.; Xu, J.W. Targeted gene insertion and replacement in the basidiomycete *Ganoderma lucidum* by inactivation of nonhomologous end joining using CRISPR/Cas9. *Appl. Environ. Microbiol.* **2021**, *87*, e0151021. [CrossRef]
6. Boontawon, T.; Nakazawa, T.; Horii, M.; Tsuzuki, M.; Kawauchi, M.; Sakamoto, M.; Honda, Y. Functional analyses of *Pleurotus ostreatus* pcc1 and clp1 using CRISPR/Cas9. *Fungal Genet. Biol.* **2021**, *154*, 103599. [CrossRef]
7. Chen, K.L.; Wang, Y.P.; Zhang, R.; Zhang, H.W.; Gao, C.X. CRISPR/Cas genome editing and precision plant breeding in agriculture. *Annu. Rev. Plant Biol.* **2019**, *70*, 667–697. [CrossRef]
8. Jinek, M.; Chylinski, K.; Fonfara, I.; Hauer, M.; Doudna, J.A.; Charpentier, E. A programmable dual-RNA-guided DNA endonuclease in adaptive bacterial immunity. *Science* **2012**, *337*, 816–821. [CrossRef]
9. Deltcheva, E.; Chylinski, K.; Sharma, C.M.; Gonzales, K.; Chao, Y.J.; Pirzada, Z.A.; Eckert, M.R.; Jörg Vogel, J.; Charpentie, E. CRISPR RNA maturation by *trans*-encoded small RNA and host factor RNase III. *Nature* **2011**, *471*, 602–607. [CrossRef]
10. DiCarlo, J.E.; Norville, J.E.; Mali, P.; Rios, X.; Aach, J.; Church, G.M. Genome engineering in *Saccharomyces cerevisiae* using CRISPR-Cas systems. *Nucleic Acids Res.* **2013**, *41*, 4336–4343. [CrossRef]
11. Cong, L.; Ran, F.A.; Cox, D.; Lin, S.; Barretto, R.; Habib, N.; Hsu, P.D.; Wu, X.B.; Jiang, W.Y.; Marraffini, L.A.; et al. Multiplex genome engineering using CRISPR/Cas systems. *Science* **2013**, *339*, 819–823. [CrossRef] [PubMed]
12. Jin, F.J.; Wang, B.T.; Wang, Z.D.; Jin, L.; Han, P. CRISPR/Cas9-based genome editing and its application in *Aspergillus* species. *J. Fungi* **2022**, *8*, 467. [CrossRef] [PubMed]
13. Wang, Y.; Chen, H.Y.; Ma, L.; Gong, M.; Wu, Y.Y.; Bao, D.P.; Zou, G. Use of CRISPR-Cas tools to engineer *Trichoderma* species. *Microb. Biotechnol.* **2022**, 1–12. [CrossRef]
14. Wang, P.A.; Xiao, H.; Zhong, J.J. CRISPR-Cas9 assisted functional gene editing in the mushroom *Ganoderma lucidum*. *Appl. Microbiol. Biotechnol.* **2020**, *104*, 1661–1671. [CrossRef]
15. Wang, T.L.; Yue, S.; Jin, Y.T.; Wei, H.; Lu, L. Advances allowing feasible *pyrG* gene editing by a CRISPR-Cas9 system for the edible mushroom *Pleurotus eryngii*. *Fungal Genet. Biol.* **2021**, *147*, 103509. [CrossRef] [PubMed]
16. Qin, H.; Xiao, H.; Zou, G.; Zhou, Z.; Zhong, J.J. CRISPR-Cas9 assisted gene disruption in the higher fungus *Ganoderma* species. *Process Biochem.* **2017**, *56*, 57–61. [CrossRef]
17. Luo, R.; Lin, J.F.; Guo, L.Q.; Ye, Z.W.; Guo, T.F.; Yun, F. Construction of *Flammulina velutipes* genome editing vector by using CRISPR/Cas9 system. *Sci. Technol. Food Ind.* **2016**, *37*, 230–234.
18. Lin, J.D.; Yang, X.Q.; Wei, T.; Guo, L.Q.; Lin, J.F.; Chen, Y.S.; Huang, S.S. Construction and transformation of CRISPR/Cas9 genome editing vector of *Flammulina filiformis* G protein-coupled receptor gene. *Mycosystema* **2019**, *38*, 349–361.
19. Liu, J.Y.; Liu, J.H.; Zhang, D.; Zhang, M.Y.; Xu, Z.; Wang, R.J.; Yu, H.L.; Shang, X.D. Agrobacterium-mediated gene transformation of Cas9 into *Flammulina velutipes*. *Acta Edulis Fungi* **2017**, *24*, 25–29.
20. Ahyayauch, H.; Collado, M.I.; Alonso, A.; Goni, F.M. Lipid bilayers in the gel phase become saturated by Triton X-100 at lower surfactant concentrations thanthose in the fluid phase. *Biophys. J.* **2012**, *102*, 2510–2516. [CrossRef]
21. Mattei, B.; Lira, R.B.; Perez, K.R.; Riske, K.A. Membrane permeabilization induced by Triton X-100: The role of membrane phase state and edge tension. *Chem. Phys. Lipids* **2017**, *202*, 28–37. [CrossRef] [PubMed]
22. Zou, G.; Zhou, Z. CRISPR/Cas9-mediated genome editing of *Trichoderma reesei*. In *Trichoderma reesei: Methods and Protocols, Methods in Molecular Biology*, 1st ed.; Mach-Aigner, A.R., Martzy, R., Eds.; Humana: New York, NY, USA, 2021; Volume 2234, pp. 87–98.

23. Zou, G.; Xiao, M.L.; Chai, S.X.; Zhu, Z.Z.; Wang, Y.; Zhou, Z.H. Efficient genome editing in filamentous fungi via an improved CRISPR-Cas9 ribonucleoprotein method facilitated by chemical reagents. *Microb. Biotechnol.* **2021**, *14*, 2343–2355. [CrossRef] [PubMed]
24. Montague, T.G.; Cruz, J.M.; Gagnon, J.A.; Church, G.M.; Valen, E. CHOPCHOP: A CRISPR/Cas9 and TALEN web tool for genome editing. *Nucleic Acids Res.* **2014**, *42*, W401–W407. [CrossRef]
25. Hille, F.; Richter, H.; Wong, S.P.; Bratovic, M.; Ressel, S.; Charpentier, E. The biology of CRISPR-Cas: Backward and forward. *Cell* **2018**, *172*, 1239–1259. [CrossRef] [PubMed]
26. Wang, H.F.; Russa, M.L.; Qi, L.S. CRISPR/Cas9 in genome editing and beyond. *Annu. Rev. Biochem.* **2016**, *85*, 227–264. [CrossRef] [PubMed]
27. Chai, S.X.; Zhu, Z.Z.; Tian, E.N.; Xiao, M.L.; Wang, Y.; Zou, G.; Zhou, Z.H. Building a versatile protein production platform using engineered *Trichoderma reesei*. *ACS Synth. Biol.* **2022**, *11*, 486–496. [CrossRef]
28. Kwon, M.J.; Schütze, T.; Spohner, S.; Haefner, S.; Meyer, V. Practical guidance for the implementation of the CRISPR genome editing tool in filamentous fungi. *Fungal Biol. Biotechnol.* **2019**, *6*, 15–25. [CrossRef]
29. Moon, S.; An, J.Y.; Choi, Y.J.; Oh, Y.L.; Ro, H.Y.; Ryu, H. Construction of a CRISPR/Cas9-mediated genome editing system in *Lentinula edodes*. *Mycobiology* **2021**, *49*, 599–603. [CrossRef]
30. Boontawon, T.; Nakazawa, T.; Inoue, C.; Osakabe, K.; Kawauchi, M.; Sakamoto, M.; Honda, Y. Efficient genome editing with CRISPR/Cas9 in *Pleurotus ostreatus*. *AMB Express* **2021**, *11*, 30–40. [CrossRef]
31. Vonk, P.J.; Escobar, N.; Wösten, H.A.B.; Lugones, L.G.; Ohm, R.A. High-throughput targeted gene deletion in the model mushroom *Schizophyllum commune* using pre-assembled Cas9 ribonucleoproteins. *Sci. Rep.* **2019**, *9*, 7632–7640. [CrossRef]
32. Boontawon, T.; Nakazawa, T.; Xu, H.B.; Kawauchi, M.; Sakamoto, M.; Honda, Y. Gene targeting using pre-assembled Cas9 ribonucleoprotein and split-marker recombination in *Pleurotus ostreatus*. *FEMS Microbiol. Lett.* **2021**, *368*, fnab080. [CrossRef] [PubMed]
33. Nødvig, C.S.; Nielsen, J.B.; Kogle, M.E.; Mortensen, U.H. A CRISPR-Cas9 system for genetic engineering of filamentous fungi. *PLoS ONE* **2015**, *10*, e0133085. [CrossRef] [PubMed]

Journal of *Fungi*

*Article*

# Transcriptome Profiling Reveals Candidate Genes Related to Stipe Gradient Elongation of *Flammulina filiformis*

**Junjie Yan [1,2,3], Zongjun Tong [1], Xing Han [1,3], Ying Gan [1,3], Yuanyuan Liu [2], Jie Chen [1,3], Xinlian Duan [1], Junbin Lin [1,3], Bingcheng Gan [1,3,*] and Baogui Xie [2,*]**

[1] Institute of Urban Agriculture, Chinese Academy of Agricultural Sciences, Chengdu 610000, China
[2] Mycological Research Center, College of Life Sciences, Fujian Agriculture and Forestry University, Fuzhou 350002, China
[3] Chengdu Agricultural Science and Technology Center, Chengdu 610000, China
* Correspondence: ganbingcheng@caas.cn (B.G.); fjxiebg@fafu.edu.cn (B.X.); Tel.: +86-591-83789277 (B.X.)

**Abstract:** Stipe gradient elongation is an important and remarkable feature in the development of most mushroom fruiting bodies. However, its molecular mechanism has rarely been described. Here, the decreasing trend of stipe elongation and increasing trend of cell length in a gradient from the top to the base of the stipe were determined in a model basidiomycete mushroom: *Flammulina filiformis*. According to RNA-seq results, 1409 differentially expressed genes (DEGs) were identified among elongation region (ER), transition region (TR), and stable region (SR) samples, including 26 transcription factors (TFs). Based on Short Time-series Expression Miner (STEM) clustering of DEGs, clusters 1 and 3, with obvious expression trends that were consistent with or in contrast to the elongation rate, were screened. The cluster 1 DEGs were mainly involved in the GO cellular component category and KEGG genetic information processing class; however, the cluster 3 DEGs were mainly involved in metabolic processes. Furthermore, qRT-PCR confirmed that key genes of the long-chain fatty acid synthesis pathway were involved in stipe gradient elongation and regulated by NADPH oxidase-derived ROS signaling molecules. These findings provide an essential basis for understanding the molecular mechanism of stipe gradient elongation.

**Keywords:** needle mushroom; stipe elongation; RNA-seq; ribosome; long-chain fatty acid

check for updates

**Citation:** Yan, J.; Tong, Z.; Han, X.; Gan, Y.; Liu, Y.; Chen, J.; Duan, X.; Lin, J.; Gan, B.; Xie, B. Transcriptome Profiling Reveals Candidate Genes Related to Stipe Gradient Elongation of *Flammulina filiformis*. *J. Fungi* **2023**, *9*, 64. https://doi.org/10.3390/jof9010064

Academic Editors: Mingwen Zhao, Gen Zou and Jing Zhu

Received: 24 November 2022
Revised: 28 December 2022
Accepted: 28 December 2022
Published: 31 December 2022



## 1. Introduction

The shaping of fruiting body in mushroom represents the most complex developmental process known in the fungal kingdom, and stipe elongation is one of the key features of fruiting body development [1,2]. It was demonstrated that stipe elongation is non-uniform in many mushrooms, such as *Agaricus bisporus*, *Flammulina filiformis*, and *Coprinopsis cinerea* [3–5]. The elongation rate decreases in a gradient from the apex to the base of the stipe in most basidiomycete mushrooms [2,6]. Several studies have suggested that the main driving force of stipe elongation is manifold cell elongation [7].

Two long stipe mushroom species, *C. cinerea* and *F. filiformis*, have been used as model systems to study the stipe elongation mechanism. In *C. cinerea*, the stipe can be divided into three regions: the apical 9 mm region elongates the most, followed by the middle 9 mm region with considerable elongation, and the 9 mm basal region no longer elongates during the development of the 27 mm long young fruiting body [4]. Niu et al. showed that the cell wall architecture varies in different stipe regions [8]. Several studies suggested that glucanase, chitinase, and β-glucan-degrading enzymes associated with cell wall extension are involved in stipe elongation regulation [9–12]. In addition, the genes encoding septin protein Cc.Cdc3 and MAPK homologue CcSakA were also suggested to have important roles in controlling stipe elongation in *C. cinerea* [13,14]. In *F. filiformis*, the fruiting body at a length of 8.5 cm was defined as the elongation stage; in this stage, the apical 1.5 cm showed a fast elongation rate, the 1.5–3 cm section showed a slow elongate rate, and the region below 3 cm showed no further elongation [6]. Multiple genes involved in stipe elongation regulation have been annotated based on gene expression pattern analysis, such as the genes encoding expansin-like protein [15], S-adenosylmethionine-dependent methyltransferase [16], cytochrome c peroxidase [17], and chromatin modifier protein [18]. Our recent study also demonstrated that NADPH oxidase and Mn-SOD driving ROS signaling molecule distribution have key roles in stipe gradient elongation of *F. filiformis* [6].

Transcriptomics based on RNAseq, which have been used to study the molecular mechanism of fruiting body development for several years, provide powerful tools to understand the biology of mushrooms [19]. Based on gene expression profiling, Tao et al. found that the stipe development of *Volvariella volvacea* is a fairly complex process, mainly by changes in expression levels [20]. Liu et al. identified conserved stipe-specific expression patterns during sexual development in *F. filiformis* by transcriptome data analysis [21]. Yan et al., based on comparative transcriptomics, suggested that a high $CO_2$ concentration inhibits early pileus expansion of *F. filiformis* by decreasing cell division control pathways [22]. The transcriptomics sequencing method has also been extensively used to reveal the molecular mechanism of fruiting body development in multiple mushroom species [23–26]. However, most omics studies treated the stipe as an integral part. The gene expression patterns of different stipe regions at the genome level have not been studied, and the molecular mechanism of stipe gradient elongation has yet to be elucidated.

As a commercial and factory cultivation mushroom, *F. filiformis* is widely cultivated and consumed in Asian countries due to its high nutritional value and desirable taste [27]. The long stipe and small pileus represent the typical fruiting body morphology of cultivated *F. filiformis*, making it an ideal model for investigating the stipe elongation mechanism [6]. To better understand the molecular mechanism of stipe elongation, the transcriptomes of three elongated stipe regions were analyzed and the genes correlated with the stipe elongation rate were examined. This transcriptomic information could facilitate our understanding of the genetic and molecular mechanisms of stipe gradient elongation in *F. filiformis*. Furthermore, the results are also a prerequisite for the breeding of varieties and achieving the precise control of stipe morphology.

## 2. Materials and Methods

### 2.1. Microorganism and Cultivation Conditions

*Flammulina filiformis* Fv01 (commercial white dikaryotic strain mated by 2 monokaryotic strains, Fv01-N and Fv01–10) was obtained from the Fujian Edible Fungi Germplasm Resource Collection Center of China. The fruiting bodies were cultivated with substrate medium at 10 °C, according to our previous method [6].

### 2.2. Stipe Elongation Rate Measurement and Cell Length Detection

Based on Yan et al.'s (2022) results, fruiting bodies in the elongation stage with 8.5 cm stipe length were selected based a previous study [6]. The elongation rate of different regions was measured within 12 h growth time at 10 °C. Then, 3 mm long fresh segments around the 1, 2, 4, and 6 cm stipe regions were collected and dissected into small pieces according to Shioya et al.'s (2013) methods [13], the images of stipe cells were taken under a Leica DM4B microscope (Leica Corporation, Wetzlar, Germany), and cell length between two adjacent septums were examined.

### 2.3. Sample Collection, RNA Extraction, Library Construction, and Sequencing

Different stipe regions were collected by using the paired sampling method. RNA extraction and library construction were performed according to previous methods [22]. In detail, total RNA was isolated from frozen samples using an E.Z.N.A.™ Plant RNA Kit (Omega, Stamford, CT, USA) according to the manufacturer's protocol. Then, RNA was quantified using a NanoND-1000 spectrophotometer (NanoDrop Technologies, Wilmington, DE, USA). The RNA libraries were prepared using an NEBNext Ultra RNA Library Prep Kit for Illumina (NEB, Ipswich, MA, USA) following the manufacturer's recommendations.

The complementary DNA (cDNA) libraries were sequenced on an Illumina HiSeq X platform (Illumina Inc., San Diego, CA, USA) at Novogene Co., Ltd. (Tianjin, China), and 150 bp paired-end reads were generated.

### 2.4. RNA-Sequencing Data Analysis

Clean reads were obtained after processing to remove adapters, and paired reads were removed if one of them contained more than 10% poly-Ns or more than 50% low-quality base sequences. Then, the clean reads were aligned to the *F. filiformis* strain Fv01–10 genome (accession: PRJNA769814) [6] using HISAT2 with default parameters [28]. Differential expression analysis was performed using StringTie v.1.3.3 with default parameters [29]. The gene expression levels were expressed as fragments per kilobase per million reads (FPKM). Genes with 0 FPKM in all samples were excluded from the study, and differentially expressed genes (DEGs) were identified using paired *t*-tests with a *p*-value cutoff of 0.05 and a fold-change (FC) threshold of 1.2. Gene Ontology (GO; http://www.geneontology.org, accessed on 10 August 2022) and Kyoto Encyclopedia of Genes and Genomes (KEGG; https://www.genome.jp/kegg/, accessed on 2 October 2022) were used for gene function annotation. The iTAK program (http://itak.feilab.net/cgi-bin/itak/online_itak.cgi, accessed on 2 October 2022) was used for transcription factor annotation (e-value was set as $1 \times 10^{-5}$) [30]. The bioinformatics analysis, including volcano plot, Venn diagram, cluster analysis, and heat map, was performed on BMKCloud nets (http://www.biocloud.net/). The gene expression pattern analysis was performed by Short Time-series Expression Miner (STEM) software on the OmicShare tools platform (www.omicshare.com/tools), with the threshold cluster number set to 8 and $p < 0.05$ used as the threshold for significance.

### 2.5. Gene Identification and Sequence Analysis

The gene sequences were obtained from the Fv01–10 genome and verified by Sanger sequencing; the PCR primers were designed using Primer Premier 6.0 (Table S1). Gene structure analysis was performed according to Yan et al.'s (2016) method by using strand-specific RNA-seq data of mixed samples of Fv01 mycelium and fruiting bodies [17,31]. The gene schematic diagram was drawn by the Gene Structure Display Server 2.0 (GSDS; http://gsds.cbi.pku.edu.cn/index.php, accessed on 18 August 2022) [32].

### 2.6. Quantitative Real-Time PCR (qRT-PCR)

The total RNA of samples was used for first-strand cDNA synthesis with TransScript One Step gDNA Removal and cDNA Synthesis SuperMix kits for qPCR (TransGen Biotech, Beijing, China). qRT-PCR was performed on a CFX96 multicolor real-time PCR detection system (Bio-Rad, Hercules, CA, USA) with TransStart Top Green qPCR SuperMix (TransGen Biotech, Beijing, China). Two internal control genes, glyceraldehyde-3-dehydrogenase (FfGPD) and Ras-related small GTPase (FfRAS), were used as reference genes [33]. The primers were designed using Primer Premier 6.0 (Table S2). Relative gene expression levels were calculated using $2^{-\Delta\Delta Ct}$ threshold cycle calculation [34].

### 2.7. Fatty Acid Profile Analysis

The dried samples were powdered to about 1 mm particle size for further analysis. Fatty acid isolation and derivatization were performed according to previous methods [35]. Fatty acid composition was detected by a 7890B-5977A gas chromatograph–mass spectrometer (GC/MS; Agilent Technologies, Santa Clara, CA, USA) with a DB-23 quartz capillary column (30 m × 320 um × 0.25 um; Agilent Technologies, Santa Clara, CA, USA), using helium as the carrier gas. The injector and detector temperatures were 250 and 230 °C, respectively. The oven temperature was set as follows: initiate at 50 °C and hold for 1 min, elevate to 175 °C at a rate of 25 °C/min, then to 230 °C at 4 °C/min. The injection volume was 1 μL, and duration of the analysis was 24.75 min. Detector voltage was set to 0.93 kV, and the EI ionization voltage of the metabolites was 70 eV. Mass spectra were recorded from 50 to 550 m/z.

Fatty acid identification was made by comparing the relative retention times of fatty acid methyl ester (FAME) peaks from samples with standards. The amount of FAME was quantitative by the external standard curve method, and the FAME values were converted into the respective fatty acid contents in μg per gram of DW.

### 2.8. Statistical Analysis

In this study, statistical analysis was performed by GraphPad Prism 6.0 (GraphPad Software, San Diego, CA, USA). Mean ± SEM was determined for each treatment group in the individual experiments, and significance tests were performed using Student's *t*-tests or paired *t*-test. The Pearson correlation coefficient was analyzed by SPSS Statistics v20 software with a two-tailed test.

## 3. Results

### 3.1. Stipe Gradient Elongation Features of *F. filiformis*

In order to investigate the stipe gradient elongation features of *F. filiformis*, elongation stage fruiting bodies with a stipe approximately 8.5 cm in length were selected. An elongation region of 0.6–1.5 cm (ER), a transition region of 1.5–2.4 cm (TR), and stable regions of 3–4 cm (SR′) and 5–6 cm (SR) were marked according to our previous results (Figure 1a) [6]. As shown in Figure 1b, the ER region elongated rapidly, followed by TR with much slower elongation, and SR′ and SR regions did not elongate. In addition, the cells showed differences from elongation region to stable region. In detail, the cell length was the longest in SR′ and SR regions, followed by TR region, and the cells of ER region were significantly shorter than they were in the TR region (Figure 1c).

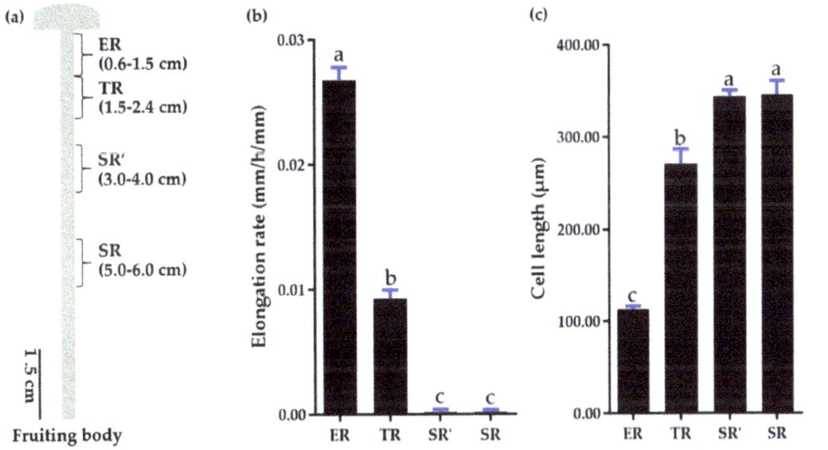


**Figure 1.** Stipe gradient elongation features of *F. filiformis*. (**a**) Stipe regions. ER, elongation region; TR, transition region; SR′ and SR, stable regions. (**b**) Elongation rate of different regions ($n$ = 15). (**c**) Cell length of different regions ($n$ = 4; value of each sample was average length of 50 cells). Different letters over columns denote significant differences ($p < 0.05$, paired $t$ test).

### 3.2. Transcriptional Changes in Different Stipe Regions of *F. filiformis*

To investigate the transcriptional change patterns during stipe elongation, three regions of ER, TR, and SR were collected for RNA-seq, applying three biological replicates for each region. As shown in Table S3, an average of 14,931,678 raw reads were generated per sample, yielding about 4.42 G clean bases on average for 9 samples (deposited in NCBI under BioprojectID PRJNA901539). For each of the 9 libraries, at least 94.03% of the clean reads had a quality score of Q30. The clean reads were compared with the genome sequence of *F. filiformis* Fv01–10 strain, and the mapping rate was higher than 95.5% for each sample.

About 13,128 genes were found to be expressed in at least one sample after the mapping of clean reads against the *F. filiformis* genome (Table S4). Pearson's correlation analysis between samples was performed (Figure 2), and the result shows r > 0.98 for replicate biological samples, indicating that the biological replicates were reliable. The SR group samples diverged from ER and TR groups, with Pearson's correlation coefficient of about 0.250–0.292 for ER and 0.356–0.459 for TR, indicating that the biological process in the SR region may be very different from that in the other two regions. However, the samples showed little difference between ER and TR groups (Pearson correlation coefficient was about 0.911–0.971), suggest the similar function between these two regions.

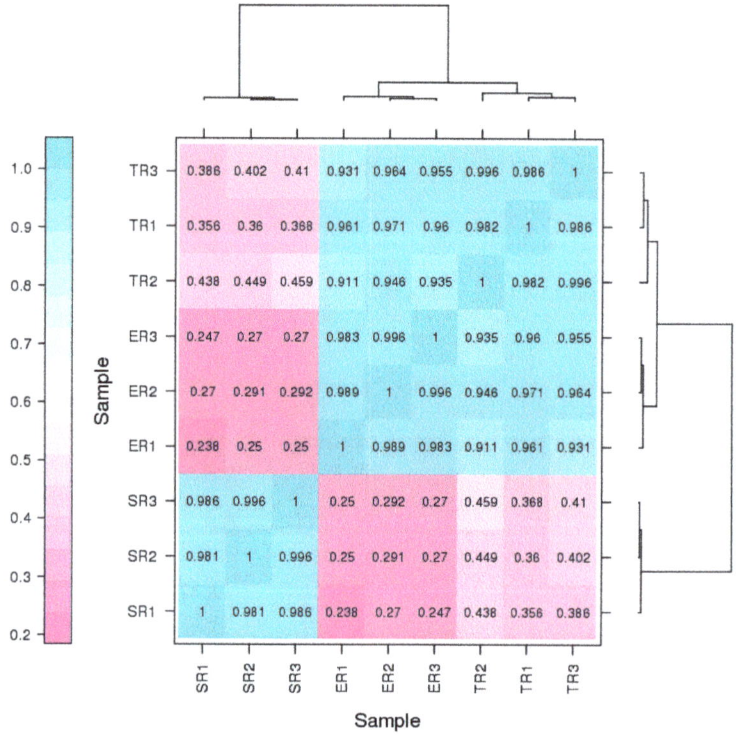


**Figure 2.** Pearson correlation coefficients for pairwise comparison of transcriptome data. ER, elongation region; TR, transition region; SR, stable region.

### *3.3. Differentially Expressed Gene (DEG) Identification*

To investigate the transcriptional divergence of the three stipe regions, DEGs were identified using the analytical thresholds of $p$-value < 0.05 and fold-change > 1.2. In ER vs. TR, about 951 upregulated and 1879 downregulated genes were detected (Figure 3a); in ER vs. SR, about 2659 upregulated and 3761 downregulated genes were detected (Figure 3b); and in TR vs. SR, about 2564 upregulated and 3018 downregulated genes were detected (Figure 3c). The results of comparative DEG analysis show that more genes were active in the high elongation rate region; however, more than 2500 upregulated genes in SR samples indicated that complex biological processes also occurred in the stable region. Venn diagram analysis showed that 1409 genes were expressed significantly differently among ER, TR, and SR groups (Figure 3d, Table S5).

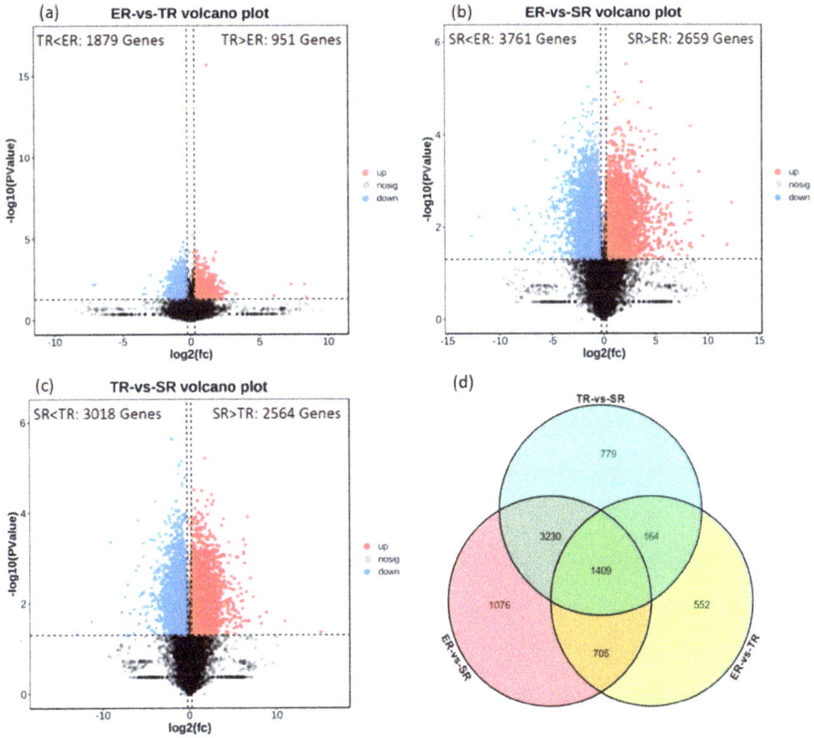


**Figure 3.** Analysis of DEGs in different samples. (**a–c**) Volcano plots illustrating DEGs between different stipe regions. (**d**) Venn diagram showing DEGs identified from comparison of ER, TR, and SR groups. ER, elongation region; TR, transition region; SR stable region.

In order to further analyze the patterns of gene expression across the three stipe regions, the expression profiles of 1409 significantly differently genes were obtained by using STEM software. Four significant ($p < 0.05$) gene expression clusters involving 1374 DEGs were screened (Figure 4a). In detail, cluster 1 (536 genes, $p$-value = $3.4 \times 10^{-154}$) exhibited a clear gene expression trend of continuing to decrease from the apex to the base, consistent with the stipe elongation rate, while in cluster 3 (285 genes, $p$-value = $4.1 \times 10^{-24}$) gene expression continued to increase from the apex to the base, in contrast to the elongation rate (Figure 4b, Table S6).

### *3.4. Transcription Factors (TFs) Annotation of DEGs*

As the master regulators control the expression of downstream genes, TFs have been demonstrated to have a critical role in mushroom fruiting body development [36]. To identify the transcription factors involved in stipe gradient elongation, about 26 TFs of 1409 DEGs were annotated, belonging to the bHLH, CSL, HMG, Homeobox, HSF, MYB, NF-YB, ZBTB, zf-C2H2, and zf-CCCH families (Figure S1). The heatmap in Figure 5 shows that all TFs were clustered into three main clades. In detail, about 18 TFs from 9 families had the highest expression in ER samples, followed by TR samples, and the lowest expression in SR samples. About 7 TFs of 3 ZBTB family genes, 1 bHLH family gene, 1 CSL family gene, 1 HMG family gene, and 1 zf-C2H2 family gene were upregulated in the stable region. Only 1 MYB family TF gene showed high expression in TR samples. These results suggest that a complicated gene network is needed to promote stipe elongation.

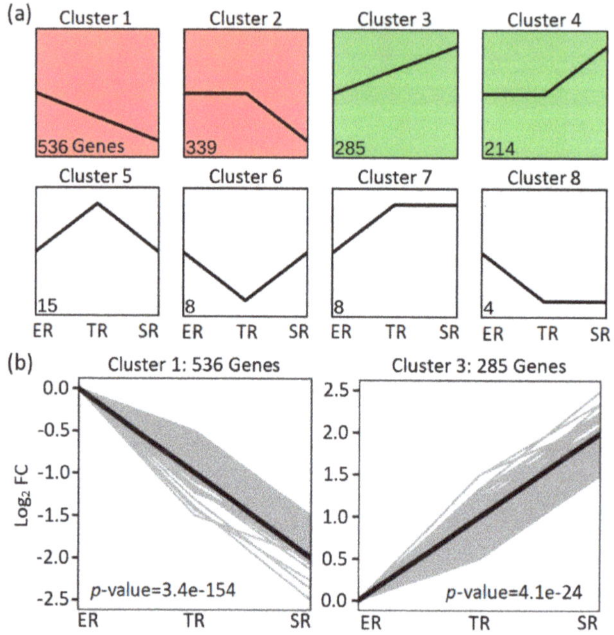


**Figure 4.** STEM analysis of 1409 DEGs based on significantly different genes in elongation region (ER), transition region (TR), and stable region (SR). (**a**) Eight expression clusters; (**b**) expression trends for genes in clusters 1 and 3.

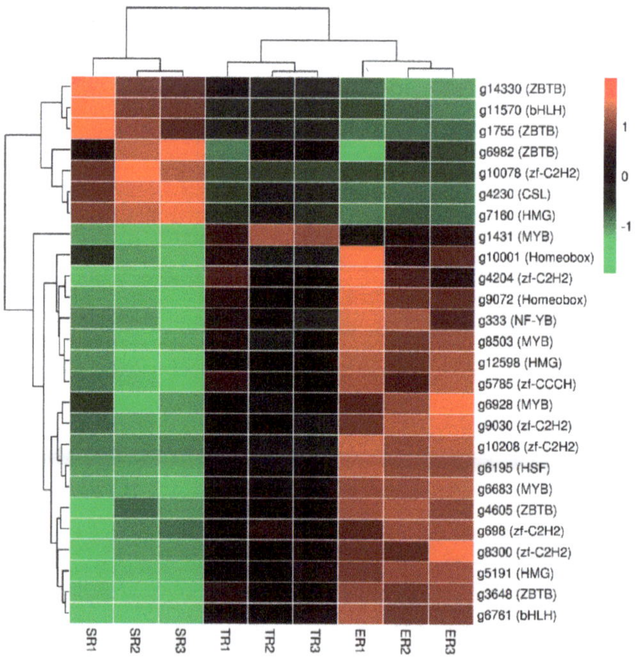


**Figure 5.** Heatmap of 26 Transcription factors in different stipe regions. Gene expression values (FPKM) were transformed to Z-score values.

### 3.5. Gene Ontology (GO) and KEGG Pathway Enrichment Analysis of DEGs

To define the functional annotation of DEGs, GO and KEGG pathway classification was carried out for the 1409 DEGs and the genes belonging to clusters 1 and 3. As shown in Figure 6 and Table S7, GO annotation was applied to 857 genes (about 60.8% of 1409 DEGs), which were assigned to 35 terms (level 2) in 3 categories. In detail, about 1429 items were enriched in the GO biological process category, mainly involved in metabolic process (430 GO items), cellular process (405 GO items), and single-organism process (306 GO items). About 1007 items were enriched in the GO molecular function category, mainly involved in binding (436 GO items), catalytic activity (391 GO items), structural molecule activity (81 GO items), and transporter activity (58 GO items). About 988 items were enriched in the GO cellular component category, mainly involved in cell (205 GO items), cell part (205 GO items), organelle (157 GO items), macromolecular complex (127 GO items), membrane (123 GO items), membrane part (83 GO items), and organelle part (68 GO items). The DEGs of cluster 1 (positively related to stipe elongation) and cluster 3 (negatively related to stipe elongation) showed different GO item enrichment. Compared with cluster 3, the genes in cluster 1 are mainly involved in macromolecular complex, organelle, cell, cell part, membrane, membrane part, organelle part, membrane-enclosed lumen, structural molecule activity and response to stimulus, cellular component organization or biogenesis terms.

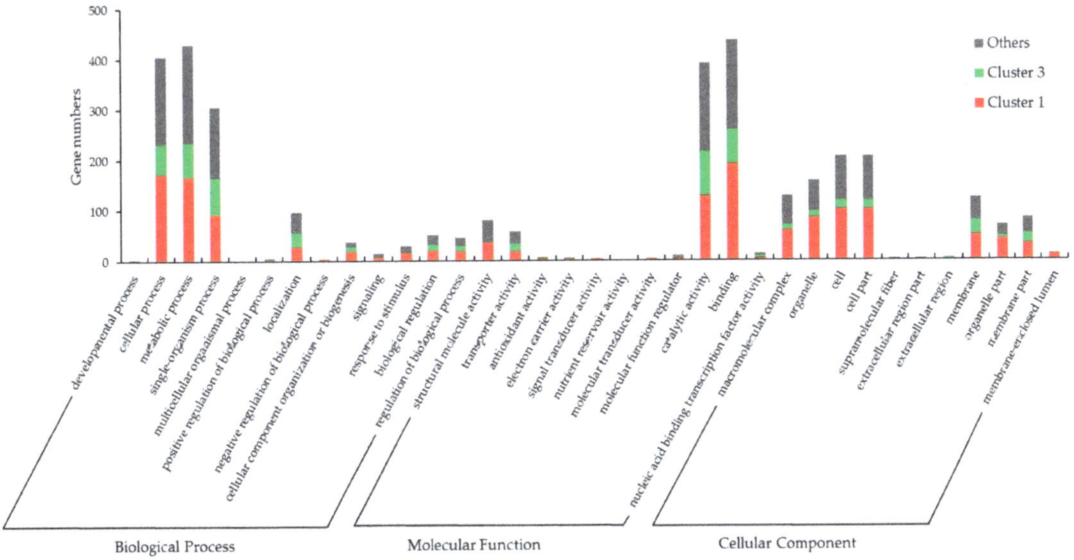


**Figure 6.** Gene Ontology (GO) classification of 1409 DEGs.

As shown in Figure 7 and Table S8, 684 genes (about 48.5% of the 1409 DEGs) mapped to 105 KEGG pathways, which can be classified into 5 groups and 21 subgroups, and are mainly enriched in the genetic information processing and metabolism classes. The genes involved in lipid metabolism, glycan biosynthesis and metabolism, translation, transcription, folding, sorting, and degradation pathways, and all genes involved in replication and repair pathway belong to cluster 1 DEGs. Cluster 3 DEGs are mainly involved in pathways of the metabolism class, including energy metabolism, carbohydrate metabolism, amino acid metabolism, and global and overview maps. The results indicate that cell replication and growth, which need DNA replication and cell component synthesis, may be the main factors promoting stipe elongation. The SR segment is the main region for nutrition and energy metabolism.

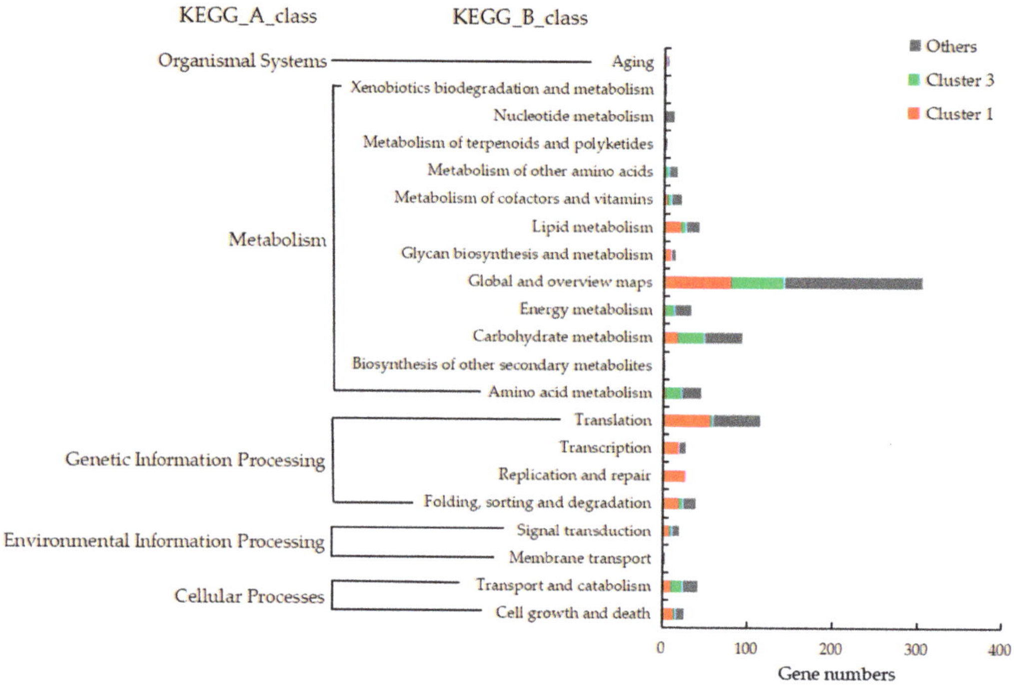


**Figure 7.** Kyoto Encyclopedia of Genes and Genomes (KEGG) classification of DEGs among different groups.

The top 10 enrichment pathways of clusters 1 and 3 and the 1409 DEGs are shown in Figure 8 and Table S8. The bubble diagrams suggest that the pathways of ribosome ($p$-value $3.15 \times 10^{-36}$, rich factor 0.73), pentose phosphate pathway ($p$-value $5.98 \times 10^{-3}$, rich factor 0.4), and fatty acid elongation ($p$-value $1.79 \times 10^{-2}$, rich factor 0.5) are significantly enriched in the 1409 DEGs. In addition, ribosome ($p$-value $4.62 \times 10^{-16}$, rich factor 0.34) and fatty acid elongation ($p$-value $2.69 \times 10^{-4}$, rich factor 0.5) are also enriched in cluster 1 genes; sulfur metabolism ($p$-value $1.26 \times 10^{-4}$, rich factor 0.26), metabolic pathways ($p$-value $4.99 \times 10^{-4}$, rich factor 0.04), pyruvate metabolism ($p$-value $4.53 \times 10^{-3}$, rich factor 0.13), galactose metabolism ($p$-value $1.70 \times 10^{-2}$, rich factor 0.15), and some other metabolism pathways are enriched in cluster 3 genes. The results indicate that the ribosome and fatty acid elongation pathways may play an important positive role related to stipe elongation.

### 3.6. *Ribosome Pathway Gene Expression Patterns Are Consistent with Stipe Elongation Rate*

Ribosomes, as important ribonucleoprotein complexes, are responsible for translating mRNA into protein and essential for cell viability [37]. Variations in the level of ribosomes can affect cell fate and developmental transitions in tissues [38]. To further investigate the ribosome pathway gene expression patterns of different stipe regions, all 13,128 expressed genes were mapped onto the ribosome pathway. As shown in Figure 9a and Table S9, 98 genes were mapped to 97 proteins in the Ribosome pathway. The heat map of gene expression patterns shows that the genes were clustered into two main clades (Figure 9b). The bigger clade included 96 genes (33 genes belonging to cluster 1 DEGs), and the gene expression was highest in ER samples, followed by TR samples, and lowest in SR samples. Our results provide evidence that ribosome gene expression is consistent with the elongation rate of different stipe regions, which may be involved in stipe gradient elongation.

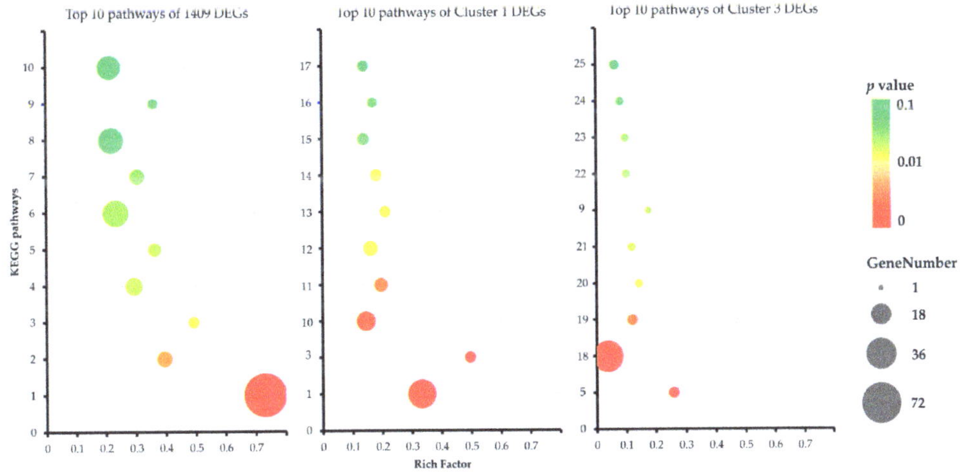


**Figure 8.** Bubble diagrams of top 10 enriched KEGG pathways of among different groups. Numbers on Y-axis represent KEGG pathways: 1. ribosome; 2. pentose phosphate pathway; 3. fatty acid elongation; 4. glycolysis/gluconeogenesis; 5. sulfur metabolism; 6. carbon metabolism; 7. citrate cycle (TCA cycle); 8. biosynthesis of amino acids; 9. ascorbate and aldarate metabolism; 10. spliceosome; 11. nucleotide excision repair; 12. mRNA surveillance pathway; 13. mismatch repair; 14. base excision repair; 15. DNA replication; 16. unsaturated fatty acid biosynthesis; 17. 2-oxocarboxylic acid metabolism; 18. metabolic pathways; 19. pyruvate metabolism; 20. galactose metabolism; 21. nicotinate and nicotinamide metabolism; 22. alanine, aspartate, and glutamate metabolism; 23. fructose and mannose metabolism; 24. valine, leucine, and isoleucine degradation; 25. ubiquitin-mediated proteolysis. Rich factor represents the ratio between DEGs and all annotated genes enriched in the pathway. Bubble scale represents the number of different genes, and intensity of bubble color represents adjusted $p$-value.

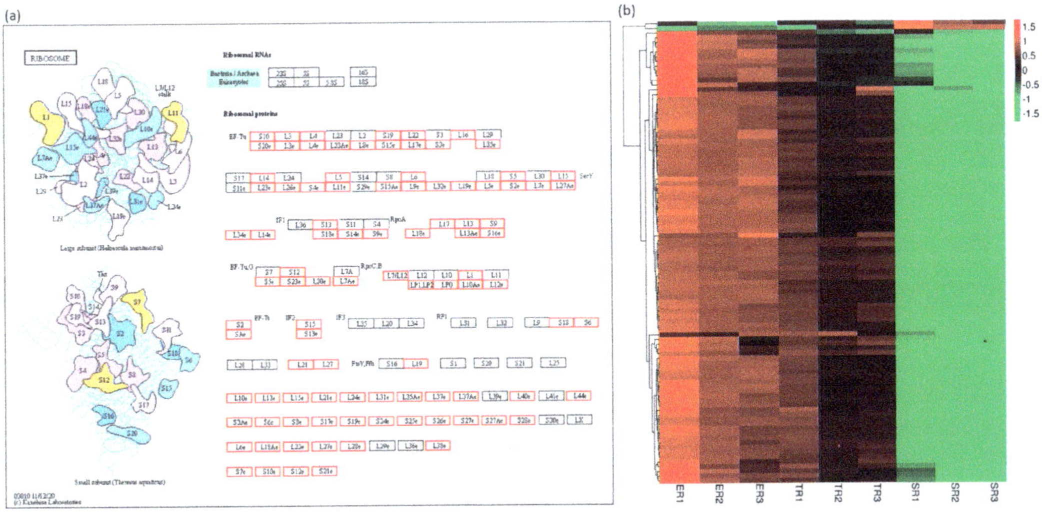


**Figure 9.** Differentially expressed genes of ribosome pathway. (**a**) Putative ribosome pathway in *F. filiformis* generated by KEGG analysis. Boxes with a black border mean no genes were annotated, boxes with a red border mean at least one gene was annotated. (**b**) Heat map showing expression profiles of ribosome genes. Gene expression values (FPKM) were transformed to Z-score values.

### 3.7. Long-Chain Fatty Acid Synthesis Pathway Involved in Stipe Gradient Elongation and Regulated by NADPH Oxidase-Derived ROS Signaling Molecules

Fatty acids are essential for membrane biosynthesis in all organisms. It was reported in plants that saturated very-long-chain fatty acids could promote cotton fiber and *Arabidopsis* cell elongation [39]. The KEGG pathway enrichment result showed that all genes mapped to long-chain fatty acid synthesis of the fatty acid elongation pathway (PATH: map00062, Module: M00415) belonged to cluster 1 DEGs (Table S8). For further investigation of long-chain fatty acid synthesis pathway gene expression patterns, five key enzymes encoding genes of *FfHSF17B12*, *FfPHS1*, *FfTER*, *FfTHEM4*, and *FfELO2* were annotated, and the sequences were submitted to NCBI after confirmation by Sanger sequencing (GenBank ID numbers OP822038-OP822042). The gene structures and FPKM values of ER, TR, and SR samples are shown in Figure 10. All five genes showed the same expression patterns, which was highest in the ER group and lowest in the SR group. To further confirm the relationship between the gene expression of the long-chain fatty acid synthesis pathway and the stipe elongation rate, the related expression of the five genes was detected by qRT-PCR using samples of elongation region (ER), transition region (TR), and two stable regions (SR' and SR). As shown in Figure 11, *FfHSF17B12*, *FfPHS1*, *FfTER*, *FfTHEM4*, and *FfELO2* genes had the same expression pattern in four stipe regions; all genes showed the highest expression in ER samples, followed by TR samples, and the lowest in SR' and SR samples. Pearson's correlation coefficient analysis showed that the expression patterns of *FfHSF17B12*, *FfPHS1*, *FfTER*, *FfTHEM4*, and *FfELO2* genes were significantly positively correlated with stipe elongation rates (Table S9).

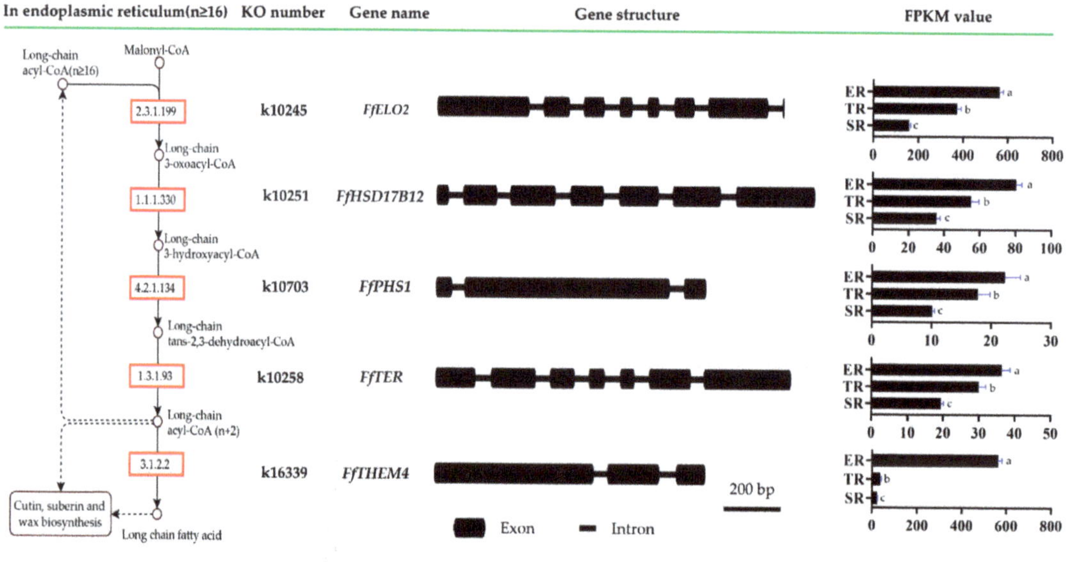


**Figure 10.** Long-chain fatty acid synthesis of fatty acid elongation pathway (PATH: map00062, Module: M00415) gene structure and expression pattern based on FPKM values of RNA-seq data. Letters over columns denote significant differences ($p < 0.05$, $n = 3$, paired $t$-test).

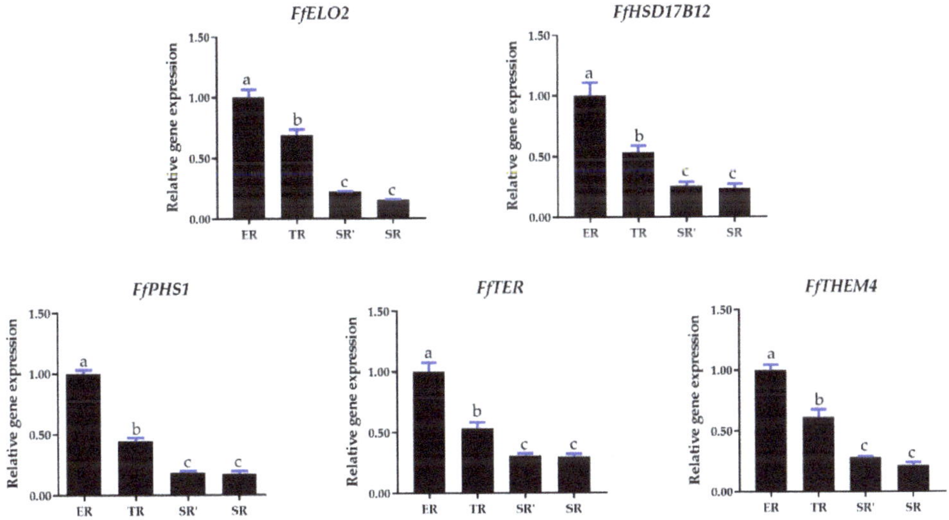


**Figure 11.** Expression of long-chain fatty acid synthesis of fatty acid elongation pathway (PATH: map00062, Module: M00415) genes in elongation region (ER), transition region (TR), and two stable regions (SR′ and SR). Letters over columns denote significant differences ($p < 0.05$, $n = 4$, paired $t$ test).

The fatty acid composition of elongation and stable regions is shown in Figure 12. Short-chain (C8 to C12), long-chain (C14 to C20), and very-long-chain (C20 to C24) fatty acid were distributed in both elongation and stable regions of the stipe. However, the concentrations were significantly different. Specifically, the saturated fatty acids C10:0, C11:0, and C12:0 and unsaturated fatty acids C16:1, C18:1n9c, C20:1, and C24:1 were significantly higher in ER than in SR; on the contrary, the saturated fatty acid C15:0, C16:0, C20:2, and C23:0 and unsaturated fatty acids C18:2n6c and C18:3n3 were significantly lower in ER than SR. Notably, the concentration of oleic acid (C18:1n9c) in ER was about 2.8 times that in SR. In *Caenorhabditis elegans*, it was demonstrated that the *ELO2* gene acts as a key enzyme during C18:1n9c synthesis from C16:0 [40]. Here, we found that the concentration of C16:0 was lower and C18:1n9c was higher in the elongation region, which was consistent with the expression pattern of the *FfELO2* gene. All of the above results suggest that the upregulation of long-chain fatty acid synthesis pathway genes, including *FfELO2* in the elongation region, may be involved in C18:1n9c synthesis and have a positive relationship with the stipe elongation rate.

Our previous results demonstrated that ROS generated by NADPH oxidase act as key signaling molecules in regulating stipe gradient elongation [6]. To study the relationship between NADPH oxidase-derived ROS signaling and long-chain fatty acid synthesis, the gene expression patterns were detected in samples treated with diphenyleneiodonium chloride (DPI, an NADPH oxidase inhibitor) [41]. As shown in Figure 13, DPI treatment downregulated the gene expression of the long-chain fatty acid synthesis pathway, and the levels of the *FfHSF17B12*, *FfPHS1*, *FfTER*, and *FfELO2* genes were significantly inhibited. In addition, the overexpression and RNA interference mutants of *FfNoxA* (membrane-bound catalytic subunit of NADPH oxidase encoding gene) generated by our previous study were also used for detecting the expression of five genes in the long-chain fatty acid synthesis pathway [6]. The results in Figure 14 show that all five genes were upregulated in *FfNoxA* overexpressing strains, and four of them were significantly downregulated in *FfNoxA* RNAi strains (except for *FfTHEM4*). These results provide evidence that the long-chain fatty acid synthesis pathway may be regulated by NADPH oxidase-derived ROS signaling.

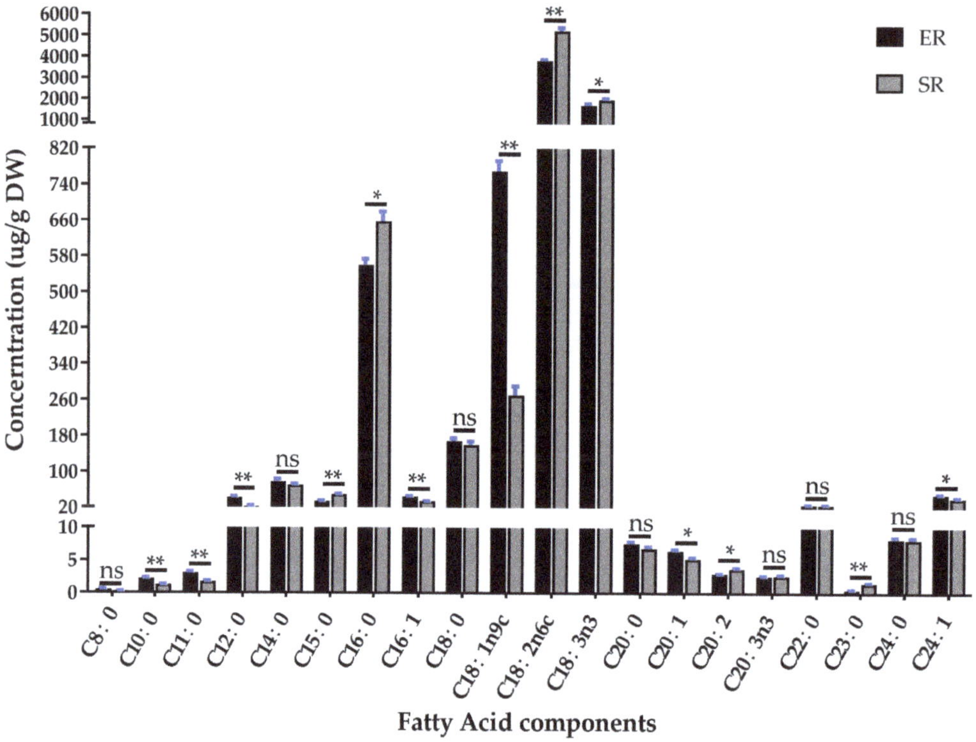


**Figure 12.** Fatty acid components of elongation region (ER) and stable region (SR). Significance levels were calculated by paired *t*-test ($n = 4$), * $p < 0.05$, ** $p < 0.01$, ns: no significance.

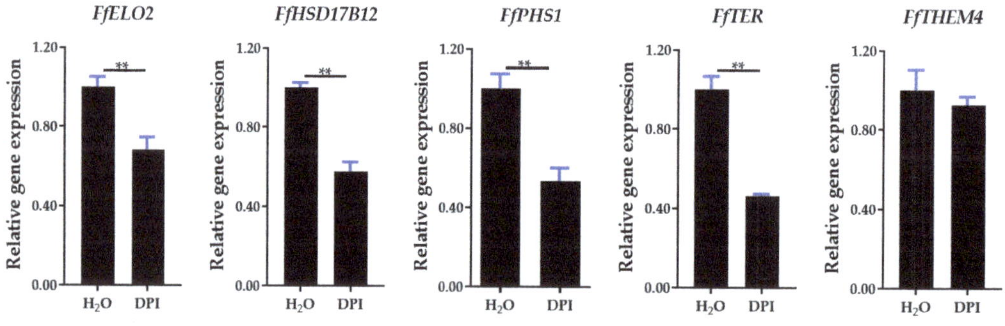


**Figure 13.** Expression of long-chain fatty acid synthesis of fatty acid elongation pathway (PATH: map00062, Module: M00415) genes in DPI and water treatment samples. Significance levels were calculated by *t*-test ($n = 4$), ** $p < 0.01$.

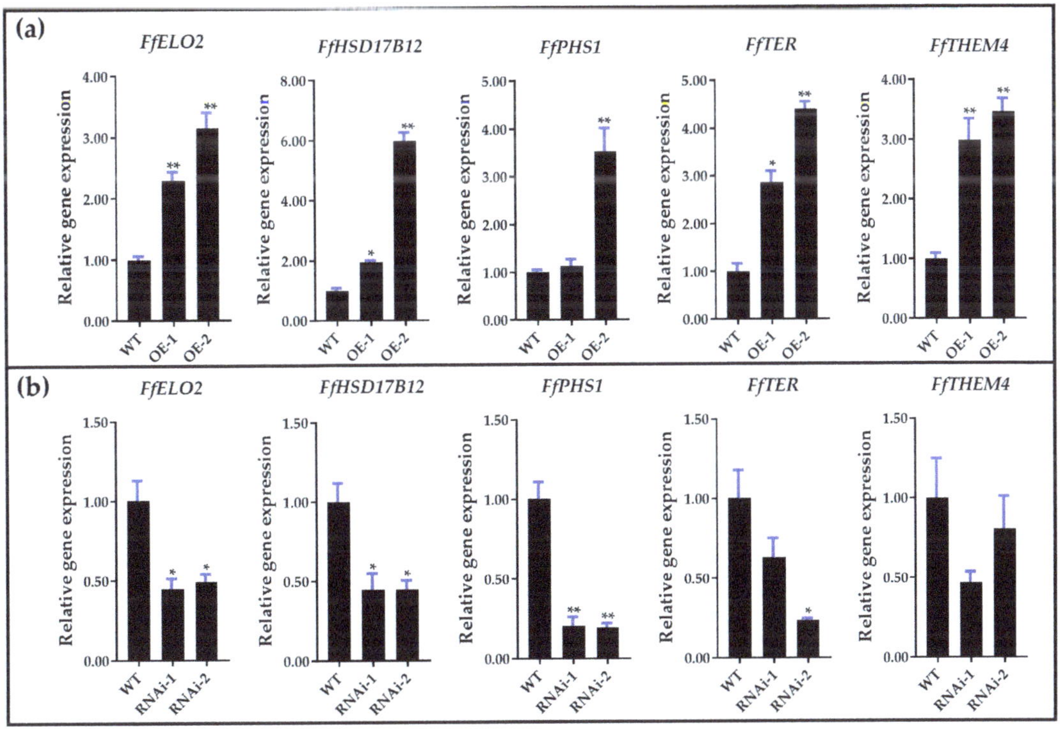


**Figure 14.** Expression of long-chain fatty acid synthesis of fatty acid elongation pathway (PATH: map00062, Module: M00415) genes in wild-type (WT) and *FfNoxA*: (**a**) overexpression lines and (**b**) RNAi lines. Significance levels were calculated by paired *t*-test compared with WT sample ($n = 3$), * $p < 0.05$, ** $p < 0.01$.

## 4. Discussion

Stipe gradient elongation is a common phenomenon in many basidiomycete fungi, such as *F. velutipes*, *C. cinerea*, and *A. bisporus* [3–5]. It is commonly accepted that stipe elongation growth is mainly attributed to manifold cell elongation rather than increased cell numbers [2]. Our study also shows that the stipe cell length significantly increased from the elongation region to the stable region, but was similar in two segments of the stable region (Figure 1c). Niu et al. suggested that the stipe cell wall architecture varies in elongation and non-elongation regions, and the thickness of stipe cell wall is inversely proportional to the cell elongation rate [8]. It was also demonstrated that several proteins and signaling pathways could mediate stipe cell extension and are involved in stipe elongation [5,6,10,13,14]. However, the stipe is usually treated as an entire organization in most transcriptome studies [42,43], and the transcriptional landscape of stipe gradient elongation remains to be clarified.

In this study, we provide a global view of stipe gradient elongation at the transcriptional level based on the RNA-seq method. We found that gene expression showed a high correlation between ER and TR samples (Figure 2), although the stipe elongation rate and cell length were significantly different (Figure 1b,c). The result suggests that the ER and TR may have a similar bio-process except for a different elongation ability. On the contrary, the gene expression in SR samples showed high variety in ER and TR samples (Figure 2). The STEM analysis of 1409 DEGs also showed that more than 1300 genes (including 857 downregulated and 553 upregulated genes) in clusters 1 to 4 had significantly different expression in SR samples compared with ER and TR samples (Figure 4a). These results indicate that bio-processes occurring in the stable region are different from those in ER and TR segments.

Some research has suggested that genes related to the cell wall synthesis are very important for the cell growth and thus cell elongation [9–12]. According to our transcriptome data, 7 glucanase genes, 14 chitinase genes, 2 glucan synthase genes, and 11 chitin synthase genes were annotated, and the FPKM value of these genes were shown in Table S10. For the most part, genes of glucanase, glucan synthase and chitin synthase showed the lowest expressed in SR segment, but the difference expression patterns were found in different in chitinase genes. These results suggested that the genes of cell wall synthesis, especially glucanase, glucan synthase, and chitin synthase genes, are involved in stipe elongated regulation.

Kamada summarized the early studies and suggested that stipe elongation is not a simple expansion of water intake, but a process involving the transmission of nutrients and accumulation of dry matter [44]. Here, we found that about 50% of GO items in the cellular component category (Figure 6), about 60% of KEGG items in the genetic information processing class (Figure 7), and about 98% of ribosome pathway mapped genes (Figure 9) were expressed consistent with the elongation rate of different regions, which provides evidence that the stipe elongation process is not only a synthesis of cellular components, but also the transmission and replication of genetic information.

Few studies have focused on the stable (non-elongation) region of stipe, since the stable region does not elongate. According to our RNA-seq data, about 2515 DEGs had the lowest expression in SR samples but 2113 DEGs had the highest expression (Figure 15). Further analysis of the 2113 upregulated DEGs in SR showed about 2070 GO items enriched in the biological process category, 1385 GO items enriched in the molecular function category, and 736 GO items enriched in the cellular component category. About 2826 GO items were enriched in the top five level 2 GO terms of catalytic activity, binding, metabolic process, cellular process, and single-organism process (Figure S2, Table S7). The KEGG pathway annotation of the 2113 DEGs also showed about 1141 DEGs enriched in 82 pathways of the KEGG A metabolism class. About 654 genes were enriched in the top five KEGG B classes: global and overview maps (310 genes), carbohydrate metabolism (114 genes), amino acid metabolism (101 genes), transport and catabolism (76 genes), and translation (53 genes) (Figure S3, Table S8). These results suggest that the complex bio-processes of nutrient metabolism, energy metabolism, substance transportation, and amino acid biosynthesis happen in the stable region.

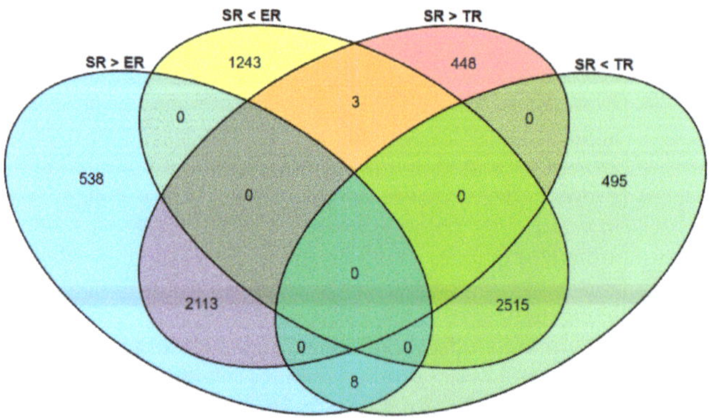


**Figure 15.** Venn diagram showing DEGs of SR samples compared with ER and TR groups. ER, elongation region; TR, transition region; SR, stable region.

Qin et al. (2007) demonstrated that saturated very-long-chain fatty acids (C24:0) could activate ethylene biosynthesis, promoting cotton fiber and *Arabidopsis* cell elongation [40]. Several studies also found that 10-oxo-trans-8-decenoic acid (ODA) synthesis from linoleic acid (C18:2n6c) could stimulate stipe elongation in *Agaricus bisporus* [45,46]. ODA was also demonstrated to stimulate the mycelial growth of Yun Chih (*Trametes versicolor*) and winter mushroom (*Flammulina filiformis*) on agar plates [47]. Our result shows that a fatty acid elongation pathway (PATH: map00062, Module: M00415) related to oleic acid (C18:1n9c) synthesis may be involved in stipe elongation and may be regulated by NADPH oxidase-derived ROS signaling molecules. Based on the similar chemical structure of oleic acid and linoleic acid, we speculate that a high concentration of oleic acid may be a precursor for some active compounds and stimulate stipe elongation. However, more studies should be carried out to determine the relationship between oleic acid and stipe elongation.

**Supplementary Materials:** The following supporting information can be downloaded at https://www.mdpi.com/article/10.3390/jof9010064/s1: Figure S1. Family classification of differently expressed Transcription factors. Figure S2. Gene ontology (GO) classification of upregulated DEGs in SR. Figure S3. Kyoto Encyclopedia of Genes and Genomes (KEGG) classification of upregulated DEGs in SR. Table S1. Primers used in PCR. Table S2. Primers used in qRT-PCR. Table S3. Quality statistics of clean sequencing data. Table S4. FPKM values of 13,128 genes. Table S5. Total of 1409 significantly different DEGs among ER, TR, and SR groups. Table S6. Mean FPKM values of clusters 1 and 3 DEGs. Table S7. GO functional classification of DEGs. Table S8. KEGG functional classification of DEGs. Table S9. Pearson's correlation coefficient analysis of expression patterns and stipe elongation rates of five genes. Table S10. FPKM values of genes related to the cell wall synthesis.

**Author Contributions:** Conceptualization, J.Y. and B.X.; methodology, J.Y.; software, Z.T., Y.L. and X.H.; validation, J.Y., X.H. and Y.L.; formal analysis, J.Y., J.L. and Z.T.; investigation, J.Y., Z.T., J.L., Y.G., J.C. and X.D.; resources, J.Y., J.L. and Y.L.; data curation, X.H., J.C. and Y.L.; writing—original draft preparation, J.Y. and Z.T.; writing—review and editing, J.Y., B.G. and B.X.; visualization, Z.T. and X.D.; supervision, B.X. and B.G.; project administration, J.Y., B.X. and B.G.; funding acquisition, J.Y., B.G. and B.X. All authors have read and agreed to the published version of the manuscript.

**Funding:** This research was funded by grants from the Agricultural Science and Technology Innovation Program (No. 34-IUA-06) and Sichuan Science and Technology Program (No. 2021YFYZ0026) to B.G., the National Natural Science Foundation of China (32102458) to J.Y., and the Seed Industry innovation and industrialization project of Fujian Province (No. zycxny2021012) to B.X.

**Institutional Review Board Statement:** Not applicable.

**Informed Consent Statement:** Not applicable.

**Data Availability Statement:** All RNA-seq data in this study are available in the National Center for Biotechnology Information Sequence Read Archive, under accession number PRJNA901539 (https://www.ncbi.nlm.nih.gov/sra/PRJNA901539, accessed on 15 November 2022). The other data are contained within the article or Supplementary Materials.

**Conflicts of Interest:** The authors declare no conflict of interest.

## References

1. Kuees, U.; Navarro-Gonzalez, M. How do Agaricomycetes shape their fruiting bodies? 1. Morphological aspects of development. *Fungal Biol. Rev.* **2015**, *29*, 63–97. [CrossRef]
2. Liu, C.; Bi, J.; Kang, L.; Zhou, J.; Liu, X.; Liu, Z.; Yuan, S. The molecular mechanism of stipe cell wall extension for mushroom stipe elongation growth. *Fungal Biol. Rev.* **2021**, *35*, 14–26. [CrossRef]
3. Craig, G.D.; Gull, K.; Wood, D.A. Stipe elongation in *Agaricus bisporus*. *Microbiology* **1977**, *102*, 337–347. [CrossRef]
4. Zhang, W.; Wu, X.; Zhou, Y.; Liu, Z.; Zhang, W.; Niu, X.; Zhao, Y.; Pei, S.; Zhao, Y.; Yuan, S. Characterization of stipe elongation of the mushroom *Coprinopsis cinerea*. *Microbiology* **2014**, *160*, 1893–1902. [CrossRef] [PubMed]
5. Fang, H.; Zhang, W.; Niu, X.; Liu, Z.; Lu, C.; Wei, H.; Yuan, S. Stipe wall extension of *Flammulina velutipes* could be induced by an expansin-like protein from *Helix aspersa*. *Fungal Biol.* **2014**, *118*, 1–11. [CrossRef]
6. Yan, J.; Chekanova, J.; Liu, Y.; Gan, B.; Long, Y.; Han, X.; Tong, Z.; Miao, J.; Lian, L.; Xie, B.; et al. Reactive Oxygen Species Distribution Involved in Stipe Gradient Elongation in the Mushroom *Flammulina filiformis*. *Cells* **2022**, *11*, 1896. [CrossRef]
7. Kües, U. Life history and developmental processes in the basidiomycete *Coprinus cinereus*. *Microbiol. Mol. Biol. Rev.* **2000**, *64*, 316–353. [CrossRef]

8. Niu, X.; Liu, Z.; Zhou, Y.; Wang, J.; Zhang, W.; Yuan, S. Stipe cell wall architecture varies with the stipe elongation of the mushroom *Coprinopsis cinerea*. *Fungal Biol.* **2015**, *119*, 946–956. [CrossRef]
9. Kang, L.; Zhou, J.; Wang, R.; Zhang, X.; Liu, C.; Liu, Z.; Yuan, S. Glucanase-Induced Stipe Wall Extension Shows Distinct Differences from Chitinase-Induced Stipe Wall Extension of *Coprinopsis cinerea*. *Appl. Env. Microbiol.* **2019**, *85*, e01345-19. [CrossRef] [PubMed]
10. Zhou, J.; Kang, L.; Liu, C.; Niu, X.; Wang, X.; Liu, H.; Zhang, W.; Liu, Z.; Latgé, J.P.; Yuan, S. Chitinases play a key role in stipe cell wall extension in the mushroom *Coprinopsis cinerea*. *Appl. Environ. Microbiol.* **2019**, *85*, e00532-19. [CrossRef]
11. Kang, L.; Zhang, X.; Liu, X.; Wang, R.; Liu, C.; Zhou, J.; Liu, Z.; Yuan, S. Comparative study of β-glucan-degrading enzymes from *Coprinopsis cinerea* for their capacities to induce stipe cell wall extension. *Int. J. Biol. Macromol.* **2020**, *152*, 516–524. [CrossRef] [PubMed]
12. Li, M.; Bi, J.; Bai, Y.; Duan, B.; Liu, Z.; Yuan, S. Accumulation and cross-linkage of β-1, 3/1, 6-glucan lead to loss of basal stipe cell wall extensibility in mushroom *Coprinopsis cinerea*. *Carbohydr. Polym.* **2021**, *259*, 117743. [CrossRef]
13. Shioya, T.; Nakamura, H.; Ishii, N.; Takahashi, N.; Sakamoto, Y.; Ozaki, N.; Kobayashi, M.; Okano, K.; Kamada, T.; Muraguchi, H. The *Coprinopsis cinerea* septin Cc.Cdc3 is involved in stipe cell elongation. *Fungal Genet. Biol.* **2013**, *58*, 80–90. [CrossRef] [PubMed]
14. Zhao, J.; Yuan, J.; Chen, Y.; Wang, Y.; Chen, J.; Bi, J.; Lyu, L.; Yu, C.; Yuan, S.; Liu, Z. MAPK CcSakA of the HOG Pathway Is Involved in Stipe Elongation during Fruiting Body Development in *Coprinopsis cinerea*. *J. Fungi* **2022**, *8*, 534. [CrossRef]
15. Huang, Q.; Han, X.; Mukhtar, I.; Gao, L.; Huang, R.; Fu, L.; Yan, J.; Tao, Y.; Chen, B.; Xie, B. Identification and expression patterns of fvexpl1, an expansin-like protein-encoding gene, suggest an auxiliary role in the stipe morphogenesis of *Flammulina velutipes*. *J. Microbiol. Biotechnol.* **2018**, *28*, 622–629. [CrossRef]
16. Huang, Q.; Mukhtar, I.; Zhang, Y.; Wei, Z.; Han, X.; Huang, R.; Yan, J.; Xie, B. Identification and characterization of two new s-adenosylmethionine-dependent methyltransferase encoding genes suggested their involvement in stipe elongation of *Flammulina velutipes*. *Mycobiology* **2019**, *47*, 441–448. [CrossRef] [PubMed]
17. Wang, R.Q.; Yan, J.J.; Li, Y.N.; Yang, H.; Ma, X.B.; Wang, M.; Tao, Y.X.; Xie, B.G. Cytochrome c peroxidase gene (ffccp) and its differential expression during stipe elongation in *Flammulina filiformis*. *Mycosystema* **2020**, *39*, 993–1005.
18. Li, J.; Shao, Y.; Yang, Y.; Xu, C.; Jing, Z.; Li, H.; Xie, B.; Tao, Y. The chromatin modifier protein FfJMHY plays an important role in regulating the rate of mycelial growth and stipe elongation in *Flammulina filiformis*. *J. Fungi* **2022**, *8*, 477. [CrossRef]
19. Nowrousian, M. Genomics and transcriptomics to study fruiting body development: An update. *Fungal Biol. Rev.* **2018**, *32*, 231–235. [CrossRef]
20. Tao, Y.; Van Peer, A.F.; Chen, B.; Chen, Z.; Zhu, J.; Deng, Y.; Jiang, Y.; Li, S.; Wu, T.; Xie, B. Gene expression profiling reveals large regulatory switches between succeeding stipe stages in *Volvariella volvacea*. *PLoS ONE* **2014**, *9*, e97789. [CrossRef]
21. Liu, X.B.; Xia, E.H.; Li, M.; Cui, Y.Y.; Wang, P.M.; Zhang, J.X.; Xie, B.G.; Xu, J.P.; Yan, J.J.; Li, J.; et al. Transcriptome data reveal conserved patterns of fruiting body development and response to heat stress in the mushroom-forming fungus *Flammulina filiformis*. *PLoS ONE* **2020**, *15*, e0239890. [CrossRef] [PubMed]
22. Yan, J.J.; Tong, Z.J.; Liu, Y.Y.; Li, Y.N.; Zhao, C.; Mukhtar, I.; Tao, Y.X.; Chen, B.Z.; Deng, Y.J.; Xie, B.G. Comparative transcriptomics of *Flammulina filiformis* suggests a high $CO_2$ concentration inhibits early pileus expansion by decreasing cell division control pathways. *Int. J. Mol. Sci.* **2019**, *20*, 5923. [CrossRef] [PubMed]
23. Tang, X.; Ding, X.; Hou, Y. Comparative analysis of transcriptomes revealed the molecular mechanism of development of *Tricholoma matsutake* at different stages of fruiting bodies. *Food Sci. Biotechnol.* **2020**, *29*, 939–951. [CrossRef] [PubMed]
24. Zhang, L.; Gong, W.; Li, C.; Shen, N.; Gui, Y.; Bian, Y.; Kwan, H.S.; Cheung, M.K.; Xiao, Y. RNA-Seq-based high-resolution linkage map reveals the genetic architecture of fruiting body development in shiitake mushroom, *Lentinula edodes*. *Comput. Struct. Biotechnol. J.* **2021**, *19*, 1641–1653. [CrossRef]
25. Orban, A.; Weber, A.; Herzog, R.; Hennicke, F.; Rühl, M. Transcriptome of different fruiting stages in the cultivated mushroom *Cyclocybe aegerita* suggests a complex regulation of fruiting and reveals enzymes putatively involved in fungal oxylipin biosynthesis. *BMC Genom.* **2021**, *22*, 324. [CrossRef]
26. Liu, D.; Sun, X.; Diao, W.; Qi, X.; Bai, Y.; Yu, X.; Li, L.; Fang, H.; Chen, Z.; Liu, Q.; et al. Comparative transcriptome analysis revealed candidate genes involved in fruiting body development and sporulation in *Ganoderma lucidum*. *Arch. Microbiol.* **2022**, *204*, 514. [CrossRef]
27. Chen, J.; Li, J.M.; Tang, Y.J.; Ma, K.; Li, B.; Zeng, X.; Liu, X.B.; Li, Y.; Yang, Z.L.; Xu, W.N.; et al. Genome-wide analysis and prediction of genes involved in the biosynthesis of polysaccharides and bioactive secondary metabolites in high-temperature-tolerant wild *Flammulina filiformis*. *BMC Genom.* **2020**, *21*, 719. [CrossRef]
28. Kim, D.; Langmead, B.; Salzberg, S.L. HISAT: A fast spliced aligner with low memory requirements. *Nat. Methods* **2015**, *12*, 357–360. [CrossRef]
29. Pertea, M.; Pertea, G.M.; Antonescu, C.M.; Chang, T.-C.; Mendell, J.T.; Salzberg, S.L. StringTie enables improved reconstruction of a transcriptome from RNA-seq reads. *Nat. Biotechnol.* **2015**, *33*, 290–295. [CrossRef]
30. Zheng, Y.; Jiao, C.; Sun, H.; Rosli, H.G.; Pombo, M.A.; Zhang, P.; Banf, M.; Dai, X.; Martin, G.B.; Giovannoni, J.J.; et al. iTAK: A program for genome-wide prediction and classification of plant transcription factors, transcriptional regulators, and protein kinases. *Mol. Plant* **2016**, *9*, 1667–1670. [CrossRef]

31. Yan, J.-J.; Zhang, L.; Wang, R.-Q.; Xie, B.; Li, X.; Chen, R.-L.; Guo, L.X.; Xie, B.G. The sequence characteristics and expression models reveal superoxide dismutase involved in cold response and fruiting body development in *Volvariella volvacea*. *Int. J. Mol. Sci.* **2016**, *17*, 34. [CrossRef]
32. Hu, B.; Jin, J.; Guo, A.Y.; Zhang, H.; Luo, J.; Gao, G. GSDS 2.0: An upgraded gene feature visualization server. *Bioinformatics* **2015**, *31*, 1296–1297. [CrossRef] [PubMed]
33. Tao, Y.; Van Peer, A.F.; Huang, Q.; Shao, Y.; Zhang, L.; Xie, B.; Jiang, Y.; Zhu, J.; Xie, B. Identification of novel and robust internal control genes from Volvariella volvacea that are suitable for RT-qPCR in filamentous fungi. *Sci. Rep.* **2016**, *6*, 29236. [CrossRef]
34. Livak, K.J.; Schmittgen, T.D. Analysis of relative gene expression data using real-time quantitative PCR and the $2^{-\Delta\Delta CT}$ method. *Methods* **2001**, *25*, 402–408. [CrossRef] [PubMed]
35. Li, X.; Guo, Y.; Zhuang, Y.; Qin, Y.; Sun, L. Nonvolatile taste components, nutritional values, bioactive compounds and antioxidant activities of three wild Chanterelle mushrooms. *Int. J. Food Sci. Technol.* **2018**, *53*, 1855–1864. [CrossRef]
36. Pelkmans, J.F.; Patil, M.B.; Gehrmann, T.; Reinders, M.J.; Wösten, H.A.; Lugones, L.G. Transcription factors of Schizophyllum commune involved in mushroom formation and modulation of vegetative growth. *Sci. Rep.* **2017**, *7*, 310. [CrossRef]
37. Byrne, M.E. A role for the ribosome in development. *Trends Plant Sci.* **2009**, *14*, 512–519. [CrossRef]
38. Norris, K.; Hopes, T.; Aspden, J.L. Ribosome heterogeneity and specialization in development. *Wiley Interdiscip. Rev. RNA* **2021**, *12*, e1644. [CrossRef]
39. Qin, Y.M.; Hu, C.Y.; Pang, Y.; Kastaniotis, A.J.; Hiltunen, J.K.; Zhu, Y.X. Saturated very-long-chain fatty acids promote cotton fiber and Arabidopsis cell elongation by activating ethylene biosynthesis. *Plant Cell* **2007**, *19*, 3692–3704. [CrossRef]
40. Kniazeva, M.; Sieber, M.; McCauley, S.; Zhang, K.; Watts, J.L.; Han, M. Suppression of the ELO-2 FA elongation activity results in alterations of the fatty acid composition and multiple physiological defects, including abnormal ultradian rhythms, in Caenorhabditis elegans. *Genetics* **2003**, *163*, 159–169. [CrossRef]
41. Piszczatowska, K.; Przybylska, D.; Sikora, E.; Mosieniak, G. Inhibition of NADPH oxidases activity by diphenyleneiodonium chloride as a mechanism of senescence induction in human cancer cells. *Antioxidants* **2020**, *9*, 1248. [CrossRef]
42. Park, Y.J.; Baek, J.H.; Lee, S.; Kim, C.; Rhee, H.; Kim, H.; Seo, J.S.; Park, H.R.; Yoon, D.E.; Nam, J.Y.; et al. Whole genome and global gene expression analyses of the model mushroom *Flammulina velutipes* reveal a high capacity for lignocellulose degradation. *PLoS ONE* **2014**, *9*, e93560. [CrossRef] [PubMed]
43. Muraguchi, H.; Umezawa, K.; Niikura, M.; Yoshida, M.; Kozaki, T.; Ishii, K.; Sakai, K.; Shimizu, M.; Nakahori, K.; Sakamoto, Y.; et al. Strand-specific RNA-seq analyses of fruiting body development in *Coprinopsis cinerea*. *PLoS ONE* **2015**, *10*, e0141586. [CrossRef] [PubMed]
44. Kamada, T. Stipe elongation in fruit bodies. In *Growth, Differentiation and Sexuality*; Wessels, J.G., Meinhardt, F., Eds.; Springer: Heidelberg, Germany, 1994; pp. 367–379.
45. Mau, J.L.; Beelman, R.B.; Ziegler, G.R. Effect of 10-oxo-trans-8-decenoic acid on growth of *Agaricus bisporus*. *Phytochemistry* **1992**, *31*, 4059–4064. [CrossRef]
46. Champavier, Y.; Pommier, M.T.; Arpin, N.; Voiland, A.; Pellon, G. 10-Oxo-trans-8-decenoic acid (ODA): Production, biological activities, and comparison with other hormone-like substances in *Agaricus bisporus*. *Enzym. Microb. Technol.* **2000**, *26*, 243–251. [CrossRef] [PubMed]
47. Mau, J.L.; Ma, J.T. Effect of 10-oxo-trans-8-decenoic acid on growth of several mushroom mycelia. *Fungi Sci.* **2001**, *16*, 1–12.

Journal of *Fungi*

MDPI

*Article*

# Multi-Omics Analysis of Low-Temperature Fruiting Highlights the Promising Cultivation Application of the Nutrients Accumulation in *Hypsizygus marmoreus*

**Ming Gong [1,2], Tianyu Huang [3], Yan Li [1], Jinxin Li [4], Lihua Tang [1], Erzheng Su [2], Gen Zou [1,*] and Dapeng Bao [1,*]**

[1] National Engineering Research Center of Edible Fungi, Key Laboratory of Edible Fungi Resources and Utilization (South), Ministry of Agriculture, Institute of Edible Fungi, Shanghai Academy of Agricultural Sciences, Shanghai 201403, China; lilysearch@163.com (M.G.); swallow2005@live.cn (Y.L.); tanglihua@saas.sh.cn (L.T.)

[2] Department of Food Science and Technology, College of Light Industry and Food Engineering, Nanjing Forestry University, Nanjing 210037, China; ezhsu@njfu.edu.cn

[3] College of Food Science and Engineering, Jiangsu Ocean University, Lianyungang 222005, China; htyhty1996@163.com

[4] Research and Development Center, Shanghai Finc Bio-Tech Inc., Shanghai 201401, China; lijinxin82@163.com

* Correspondence: zougen@sibs.ac.cn (G.Z.); baodapeng@saas.sh.cn (D.B.)

**Abstract:** *Hypsizygus marmoreus* is a representative edible mushroom with low-temperature fruiting after a long postripening (LFLP). Clarifying the mechanism of LFLP and applying a rigorous low-temperature-limited process will optimize the mushroom cultivation process. This study performed an integrative multi-omics analysis of the molecular mechanism of LFLP in combination with genetic, physiological, and cultivation confirmation. The results showed that the amino acid content was increased during LFLP, mainly arginine. pH analysis showed acidification in the postripening stage and alkalization in the substrates of the reproductive growth stage. An enzyme activity test confirmed the increased enzyme activity of arginase and citrate synthase in the postripening stage. Weighted gene coexpression network analysis of the transcriptome and metabolomics indicated that pH variation is correlated mainly with changes in citrate and arginine. Multi-omics reveals a straightforward way of providing enriched materials for amino acid biosynthesis, namely, synergistically elevating citric acid and arginine through enhanced activity of the arginine synthesis branch pathway in the citrate cycle. Our study confirmed that GCN2 mediated metabolic adaptation by enhancing protein translation, highlighting its regulatory role during LFLP. Exogenously added citric acid and arginine shortened the postripening period by 10 days and increased the fruiting body yield by 10.2~15.5%. This research sheds light on the molecular mechanism of LFLP in *H. marmoreus* and highlights the promising application of nutrient accumulation in high-efficiency cultivation.

**Keywords:** low-temperature fruiting; multi-omics; cultivation; GCN2; *Hypsizygus marmoreus*

check for updates

**Citation:** Gong, M.; Huang, T.; Li, Y.; Li, J.; Tang, L.; Su, E.; Zou, G.; Bao, D. Multi-Omics Analysis of Low-Temperature Fruiting Highlights the Promising Cultivation Application of the Nutrients Accumulation in *Hypsizygus marmoreus*. *J. Fungi* **2022**, *8*, 695. https://doi.org/10.3390/jof8070695

Academic Editor: Eriko Takano

Received: 18 May 2022
Accepted: 28 June 2022
Published: 30 June 2022



## 1. Introduction

The rapidly growing global population has also resulted in growing food demand and increased agricultural output, leading to the generation of agricultural byproducts and wastes, such as sugarcane bagasse, rice husks, cotton stalks, straw, and stover [1]. Mushrooms vegetatively grow on these stubborn residual biomasses and convert them into high-value edible and medicinal products [2–4]. In China, the economic output of edible fungi has ranked as the fifth-largest agricultural sector after grain, vegetables, fruit trees, and oilseeds, ahead of tea, sugar, and cotton [5]. Mushroom cultivation must go through vegetative and reproductive growth to form the fruiting body. For most mushrooms, the transition from vegetative to reproductive development must be induced by low-temperature treatment. Low-temperature treatment is a necessary process allowing many mushrooms to enter the reproductive stage [6,7], such as *Lentinus edodes* [8], *Agaricus bisporus* [9], *Flammulina velutipes* [10], *Pleurotus eryngii subsp. tuoliensis* (Bailinggu) [11], *Pleurotus nebrodensis* [11], *Armillaria mella* [12], and *Hypsizygus marmoreus*. In addition, the overgrown mycelia of some mushrooms need to go through a long postripening stage before low-temperature treatment. These processes consume much energy and increase production costs, limiting the cultivation of mushrooms from being low carbon and sustainable agricultural industry. It is essential to unravel the molecular mechanisms behind these transitions to guide sustainable mushroom cultivation.

*Hypsizygus marmoreus*, also known as beech mushroom, crab mushroom, sea mushroom, etc. [13,14], has a month-long postripening stage and a typical low-temperature fruiting stage after a long postripening (LFLP). Premature interruption of the postripening stage for fruiting results in a drastic reduction in production. Furthermore, the subsequent low-temperature treatment of mature mycelia is the main trigger for primordia initiation and plays a crucial role in the mushroom formation and the subsequent yields of *H. marmoreus*. Therefore, it can be used as a typical model organism to study the molecular mechanism of the postripening stage and the low-temperature fruiting of mushrooms.

The regulation of cytosolic mRNA translation is crucial for rapid adaptation to environmental stress conditions. General control nonderepressible 2 (GCN2) is a Ser-Thr kinase found in all eukaryotic organisms that phosphorylate eukaryotic translation initiation factor 2 (eIF2) under cold stress, which inhibits the initiation of protein translation and is essential for cold tolerance in Arabidopsis [15]. Recent research has shown that GCN2 is activated under various stresses, including cold treatments [16]. Therefore, it is desirable to study the regulatory role of GCN2-mediated translation in the LFLP of *H. marmoreus*.

In this study, we investigated the expression profile of the representative developmental stages in *H. marmoreus* using quantitative transcriptome, amino acid (AA), and organic acid-targeted metabolomics to reveal the molecular mechanism of LFLP in *H. marmoreus*. Arginine and citrate could act as postripening biomarkers in *H. marmoreus*. The enhanced GCN2-mediated translation pathway is essential for promoting LFLP in *H. marmoreus*. These findings have the potential to guide lower-cost, low-carbon mushroom cultivation methods.

## 2. Materials and Methods

### 2.1. Sample Preparation

The strain Finc-W-247 is the primary cultivar planted by Shanghai Fengke Science and Technology Co., Ltd., stored in China's Typical Culture Preservation Center (Wuhan University). The strain was cultured with cottonseed shell 15%, wood chips 25%, rice bran 20%, corn cob 15%, bran 19%, cornmeal 5%, lime 1%, and water content 65%. The evenly stirred medium was put into an 850 mL cultivation bottle for sterilization and inoculation after cooling. The inoculated culture bottles were transferred to the culture chamber for cultivation. The temperature of the culture chamber was generally set at 22–23 °C, the humidity was 70%, and the $CO_2$ concentration was 2500–3500 mg/kg. After 30 to 35 days of culture, mycelia were overgrown within the bottle. After approximately 40 days of culture, the mycelia reached physiological maturity (this period is called the postripening stage), and the scratching fungus operation was carried out. Then, the cultivation bottle was transferred to the mushroom room, where the temperature was 14–16 °C, the air humidity was 95–100%, and the $CO_2$ concentration was 1500–2500 mg/kg. After approximately 22 days of low-temperature cultivation, fruiting was completed from the primordium stage.

The six growth stages were sampled and stored at −80 °C until testing. They included the overgrown mycelia stage (OG), the postripening culture for 15 d (LR15), the postripening culture for 30 d (LR30), the primordium phase (PS), the droplets phase (with apparent differentiation of the cap and stalk) (DS), and the fruiting body stage (FS). The three samples in the substrates of the developmental stage of the mushroom were collected to understand the metabolic flux of substrates in fruiting stages since metabolites were secreted into the substrates. They include the substrates of the primordium phase (PSR), the substrates of the droplets phase (DSR), and the substrates of the fruiting body stage (FSR).

Finc-W-247 blocks (diameter of 6 mm) were inoculated on potato dextrose agar medium in a 90 mm diameter Petri dish and cultured at 25 °C for 13 days. Then, mycelia were collected and placed at 4 °C for 0, 2, 4, 6, and 8 h and stored at −80 °C until qPCR analysis.

All assays in this study were performed in three independent biological experiments with at least three replicates.

### 2.2. Absolute Quantitative Transcriptome of Nine Developmental Stages of H. marmoreus

The samples of nine developmental stages of *H. marmoreus* were collected and lyophilized, and total RNA was extracted using TRIzol reagent (Invitrogen, Carlsbad, CA, USA) according to the manufacturer's instructions. There were three biological replicates in this study. RNA extraction, RNA quantity and purity determination, and RNA integrity evaluation were performed according to methods described in previous research [17]. The clean reads were compared to the reference *H. marmoreus* genome [18] using Hisat2 software. PSORTb v3.03 was used for protein subcellular localization prediction [19]. The details about the transcriptome library construction and data analysis can be found in recent studies [20].

### 2.3. Amino Acid-Targeted Metabolome of the Developmental Stages of H. marmoreus

An amino acid-targeted metabolome was used to analyze the metabolites of the nine developmental stages of *H. marmoreus*. There were six biological replicates in this study. LC-MS detection was performed for each standard solution of amino acids and the treated samples. The heatmap program package in R (V3.3.2) was used for agglomerate hierarchical data clustering. The R language ropls package was used for principal component analysis (PCA). The standard for differential metabolites was $p$-value $\leq 0.05$ and VIP (variable importance for the projection) $\geq 1$. The details about constructing the Mass spectrometric (MS) detection and data analysis can be found in the supplementary material.

### 2.4. Organic Acid-Targeted Metabolome Detection

An organic acid-targeted metabolome was used to analyze the metabolites of nine developmental stages of *H. marmoreus*. There were six biological replicates in this study. Twenty-six organic acid standard substances were weighed. The single standard stock solution was prepared with methanol or water to make a mixed standard. A 30% methanol aqueous solution (containing 0.1% formic acid) was diluted to make a standard working solution, which was stored at 0 °C until LC-MS detection.

The extraction of metabolites was performed according to the methods described in previous research [21]. An ACQUITY UPLC® BEH C18 column (2.1 × 100 mm, 1.7 µm, Waters Inc., Milford, Massachusetts, USA) was used for chromatographic determination. The sample loading volume was 5 µL, the column temperature was 40 °C, and the mobile phases were A-water (containing 0.1% formic acid) and B-methanol water (containing 0.1% formic acid). Multiple response monitoring was used for scanning. Previous research explicitly referred to the chromatographic and mass spectrometric conditions for computer detection [22,23]. The details about the organic acid-targeted metabolome detection can be found in the supplementary material.

### 2.5. Gene Coexpression Network Construction

Weighted correlation network analysis (WGCNA) can find modules of highly correlated genes. Transcriptome data from the postripening stage (PRS) to the substrates of the reproductive growth stage (SRS) were used as input expression data. The organic acid content and pH sample from the PRS to SRS acted as an attribute matrix. WGCNA of the two types of data was used to analyze the critical module of pH variation from the PRS to the SRS.

Transcriptome data from the PRS, the reproductive growth stage (RGS), and SRS were used as input expression data. AA content from the PRS, RGS, and SRS was used as an input attribute matrix. WGCNA of the two types of data was used to analyze the critical module of the change in AA content in LFLP. The WGCNA R software package was used to perform various aspects of weighted correlation network analysis [24]. For details about the steps of WGCNA, refer to a previously published article [25].

### 2.6. qPCR Assays and Western Blot Analysis

Samples of nine developmental stages and the cold-treated samples of *H. marmoreus* were collected for RNA extraction. Details of the qPCR assay are described in a previous study [26]. The primers used for qPCR are listed in Supplementary Table S1. All qPCR assays were performed in three independent experiments. The expression levels of the mRNAs were normalized to Actin-1 and were calculated using the $2^{-\Delta\Delta Ct}$ method [27]. One-way ANOVA followed by the Tukey test for multiple comparisons was conducted using SigmaPlot Version 12.0 to evaluate significant differences among the groups ($p < 0.05$).

Samples from nine developmental stages of *H. marmoreus* were used for total protein extraction. The polyclonal antibody against GCN2 was purchased from Servicebio (catalog no: GB111095). Recombinant protein corresponding to mouse GCN2 was used for the immunization of a rabbit. Western blot analysis was conducted as described previously [28].

### 2.7. Determination of Enzyme Activity

Samples from nine developmental stages of *H. marmoreus* were collected to determine the enzyme activity. The arginase activity of the samples was measured using an arginase activity assay kit (Shanghai Bohu Biological Technology Co., Ltd., Shanghai, China) according to the manufacturer's protocol. In this procedure, a 1 g sample was mixed with 9 mL phosphate buffer in an ice bath and then centrifuged at 5000 r/min for 15 min. Then, 10 µL supernatant was used for enzyme activity determination. The absorbance change of the system $\Delta A$ at a wavelength of 450 nm was measured when assaying enzyme activity. $\Delta A$ is brought into the linear regression equation ($y = 0.03290\ x + 0.09103$, $R^2 = 0.9944$, where $x$ is arginase activity and y is the sample $\Delta A$ *value*) for enzyme activity calculation. The sample was diluted five times as directed by the instructions, and the calculated result multiplied by 5 is the actual sample concentration (U/L).

The citrate synthase of the samples was measured using a citrate synthase assay kit (Shanghai Bohu Biological Technology Co., Ltd., Shanghai, China) according to the manufacturer's protocol. $\Delta A$ is brought into the linear regression equation ($y = 0.06232\ x + 0.1375$, $R^2 = 0.9981$, where $x$ is citrate synthase activity and y is the sample $\Delta A$ *value*) for enzyme activity calculation. The sample was diluted five times as directed by the instructions, and the calculated result multiplied by 5 is the actual sample concentration (IU/L).

### 2.8. Cultivation Experiment with Exogenous Arginine and Citric Acid

A 10 mL solution of $H_2O$, Arg (6 mmol/L), and citric acid (200 mg/L) was injected into the OG stage. The operation of scratching the fungus was carried out after postripening the culture for 30 d. The cultivation bottle was transferred to the fruiting room after scratching the fungus, and then the weight of the fruiting body per bottle was measured (g/bottle). CK represents the normal postripening culture for 40 d without any solution addition. Each treatment was repeated with five bottles.

The lightness values (L-value) of the wood chips in the cultivation materials were determined by an SC-10 handheld color meter (Shenzhen Sunenshi Technology Co., LTD., China), and the determination time was postripening culture for 8 d, 15 d, 22 d, and 30 d. An appropriate amount of the culture material at the shoulder of the bottle was taken and laid on a 90 mm plate to measure the L-value of the wood chips in the cultivation material. Three bottles were taken from each treatment, and each bottle was measured twice.

## 3. Results

### 3.1. Enhanced Ribosome Pathway for Low-Temperature Fruiting

We obtained samples of the different developmental stages of *H. marmoreus* (Figure 1). High-throughput analysis of these samples makes it possible to obtain critical metabolic pathways, amino acids, and organic acids involved in LFLP through multi-omics analysis.

The molecular mechanism of LFLP could be further identified in combination with genetic and physiological confirmation (Figure 1). These studies will help to obtain molecular biomarkers of LFLP.

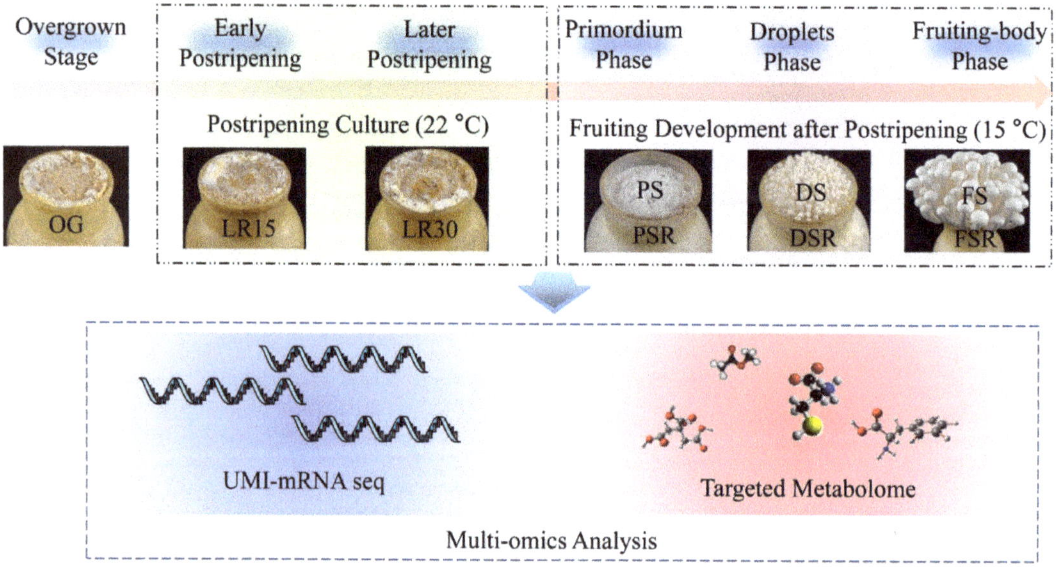


**Figure 1.** Flow chart of this study. Multi-omics analysis was used to analyze the gene and metabolite expression profiles of *H. marmoreus* at developmental stages to reveal the molecular mechanism of LFLP.

The unique molecular identifier (UMI) absolute quantitative transcriptome of *H. marmoreus* was determined through RNA extraction, purification, detection, library construction, and sequencing. PCA showed that the samples had excellent repeatability and discrimination (Supplementary Figure S1A). Heatmap analysis showed that the samples from the postripening stage (PRS) were clustered together, while the samples from the reproductive growth stage were grouped into another cluster (Supplementary Figure S1B). Venn analysis showed 234 differentially overlapped genes among the samples of the development stages (Supplementary Figure S1C).

There were five significantly enriched pathways (SEPs) in OG vs. PS using the DEGs for which the ratio of upregulated DEGs to downregulated DEGs was greater than 3 and the $p$-value < 0.05. Heatmap analysis showed that ribosomes were the most SEPs in OG vs. PS (Supplementary Figure S1D), indicating the enhanced activity of ribosomes in LFLP. There were no SEPs in OG vs. PS using the DEGs for which the ratio of downregulated DEGs to upregulated DEGs was greater than 3 and the $p$-value < 0.05. Heatmap analysis showed the upregulated ribosomal pathway in PS and DS (Supplementary Figure S2). KEGG enrichment analysis showed that the DEGs in LR30_vs_PS were enriched in the ribosome (Supplementary Figure S3). The ratio of upregulated DEGs to downregulated DEGs in LR30_vs_PS was 28.5 in the enriched ribosome pathway, representing the enhanced ribosome pathway in PS. For the comparison, KEGG enrichment analysis showed that the upregulated DEGs in OG_vs_LR30 were enriched in valine, leucine, and isoleucine degradation, methane metabolism, and glycolysis/gluconeogenesis instead of the ribosome (Supplementary Figure S4). These results indicate that enhanced activity of the ribosome pathway is essential for LFLP in *H. marmoreus*.

### 3.2. Accumulation of Dissociative Amino Acids for Low-Temperature Fruiting

The amino acid metabolism pathway showed an apparent difference in gene expression in the LFLP in *H. marmoreus* (Supplementary Figure S5A). The upregulated DEGs occurring only in the postripening stage were enriched in the cellular amino acid metabolic and biosynthetic processes (Supplementary Figure S5B). The upregulated DEGs in the reproductive growth stage (RGS) were enriched in the alpha-amino acid biosynthetic process (Supplementary Figure S5C). These results indicate that gene transcription is not restricted to the protein translation pathway in the postripening stage. Considering the enhanced activity of the ribosome pathway in the primordium phase instead of the postripening stage, protein translation is inhibited in the postripening stage. This result shifts our attention to the variation in AA content in the LFLP.

The AAs in 54 samples (6 samples per development stage or its substrates) in LFLP were detected by LC-MS/MS to obtain the metabolic flow of AAs. The hierarchical cluster of AAs showed that the substrates of the reproductive growth stage (SRS), including PSR, DSR, and FSR, were grouped and separated from the large cluster of the other developmental stages (Figure 2A). Targeted profiling of the AA metabolome showed an apparent increase in AA content in the RGS and SRS (Figure 2B). Statistical tests confirmed the significant upregulation ($p < 0.001$) of several amino acids, such as glutamine (Gln), arginine (Arg), and ornithine (Orn) (Figure 2C). These results indicated that the enriched AAs in RGS and SRS are essential for LFLP.

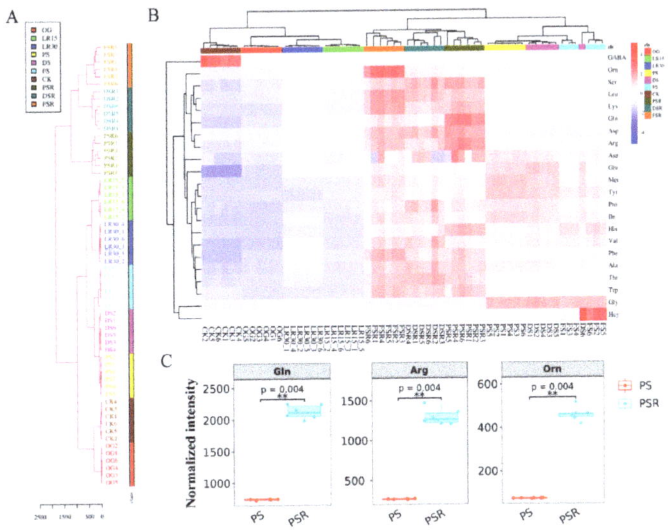


**Figure 2.** AAs targeted metabolomics of *H. marmoreus* at different developmental stages. (**A**) Hierarchical cluster of AAs. (**B**) Targeted profiling of the AA metabolome of *H. marmoreus* at different developmental stages. (**C**) Statistical tests of the top three AAs with the highest concentrations. ** represents $p < 0.001$.

Analysis of the AA content showed an apparent increase in the SRS (Supplementary Figure S6A). Gln, Arg, and lysine (Lys) exhibited the highest concentrations in the PSR, and Orn showed the highest concentrations in the FSR (Supplementary Figure S6B). PCA of the AAs in the developmental stages of *H. marmoreus* showed that Gln, Arg, Lys, Leucine (Leu), and Orn were separated from the other AAs (Supplementary Figure S7), suggesting that these AAs might play a critical role in the LFLP.

### 3.3. Formation of Arginine Pool Promotes Low-Temperature Fruiting

Correlation analysis of the significant differential AAs (SDAs) and DEGs in OG vs. PS was conducted to obtain the associated DEGs. The associated AAs showed that glutamic acid (Glu), methionine (Met), and glycine (Gly) were the top three AAs with the highest number of associated DEGs (Supplementary Table S3). The associated AAs showed that Glu, aspartic acid (Asp), and Gly were the top three AAs with the highest associated pathways (Supplementary Table S3). The EC annotations of these associated genes showed that N-methyltransferase, N-acetyltransferase, and branched-chain amino acid aminotransferase were the top three enzymes with the most significant number (Supplementary Table S4). KEGG enrichment analysis showed that the associated upregulated DEGs were enriched in arginine biosynthesis and arginine and proline metabolism (Figure 3A). The associated downregulated DEGs were enriched in alanine, aspartate, and glutamate metabolism and tryptophan metabolism. A total of 10 pathways were screened by correlative pathway analysis of the associated SDMs in OG vs. PS (Impact > 1), such as aminoacyl-tRNA biosynthesis and arginine biosynthesis (Figure 3B). The citrate cycle (TCA cycle) (map00020) showed that alanine, aspartate, and glutamate metabolism provided metabolites for downstream arginine biosynthesis. Together with the strong association of Glu and Asp with DEGs and pathways (Supplementary Tables S3 and S4), these observations indicated a critical role of Arg in LFLP. Statistics of the AA content showed the similarity between the acid AAs and the basic AAs in the PRS and RGS (Figure 3C). The content of basic AAs showed a significant increase in SRS relative to acid AAs (Figure 3C). These observations suggest that the substrates of the reproductive growth stage are the Arg pool for supplying the AAs to the reproductive growth.

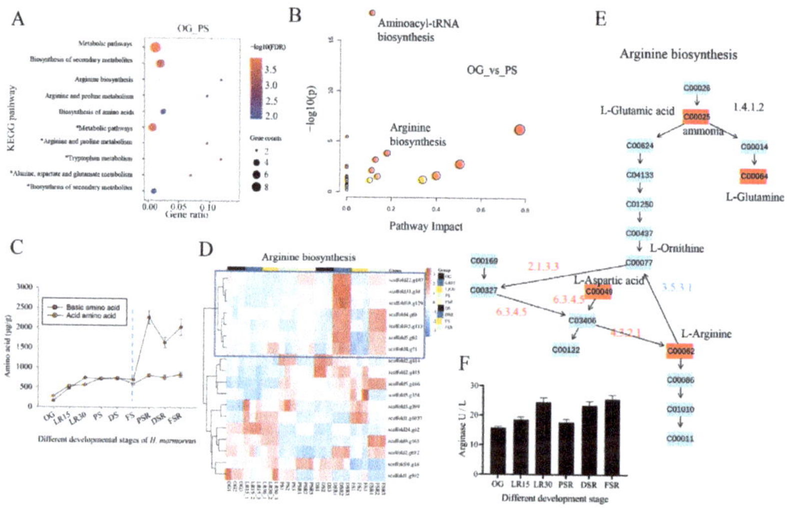


**Figure 3.** Transcriptome analysis of arginine biosynthesis in *H. marmoreus.* (**A**) KEGG enrichment analysis of the associated differentially expressed genes (DEGs) in OG vs. PS. * represents downregulated DEGs. (**B**) Correlation pathway analysis of the significantly different AAs (SDAs) in OG vs. PS (Impact > 1). (**C**) Analysis of the content of AAs in *H. marmoreus* at different developmental stages. (**D**) Heatmap analysis of gene expression in arginine biosynthesis. (**E**) KEGG pathway view (ko00350) of arginine biosynthesis. Red indicates upregulation in OG vs. PS. Blue suggests upregulation in OG vs. PS. (**F**) Determination of arginase enzyme activity in *H. marmoreus* at different developmental stages.

The absolute quantitative transcriptome was used to obtain SRS expression profiles. PCA showed that all samples were divided into three groups (Supplementary Figure S8A), including the PRS, RGS, and SRS. Heatmap analysis of the expression profile showed apparent differences in the PRS, RGS, and SRS (Supplementary Figure S8B). KEGG enrichment analysis showed decreased activity of the citrate cycle (TCA cycle) compared to the enhanced activity of the pentose phosphate pathway and pentose and glucuronate interconversions in LR30_vs_PSR (Supplementary Table S2). The enhanced activity in the latter two ways could provide large 5-carbon skeletons for the accelerated transformation from 5-carbon to 4-carbon and 7-carbon, producing enriched materials for amino acid biosynthesis.

The upregulated DEGs in PSR_vs_DSR were enriched in arginine biosynthesis (Supplementary Figure S9), indicating the enhanced activity of arginine biosynthesis in SRS and its essential role in LFLP. Homologous genes of arginine biosynthesis in *H. marmoreus* were obtained using Blastp ($1 \times 10^{-30}$). Heatmap analysis showed that the upregulated genes in arginine biosynthesis in SRS were clustered and marked by a box (Figure 3D). The KEGG pathway view (ko00350) showed that argininosuccinate synthase (scaffold5.g62) and argininosuccinate lyase (scaffold4.g71) in the box cluster were responsible for the biosynthesis of arginine (Figure 3D,E), indicating the enhanced activity of arginine biosynthesis in SRS instead of PRS. Arginase (3.5.3.1) is an enzyme that catalyzes the hydrolysis of L-arginine into L-ornithine and urea. The increased arginase in the PRS means more hydrolysis of L-arginine, leading to a lower Arg content (Figure 3D). Heatmap analysis showed the increased expression of arginase (scaffold9.g163) in the PRS (Figure 3D and Supplementary Table S5). Enzyme activity tests further confirmed the increased arginase activity in the PRS, consistent with the low concentration of Arg in the PRS (Figure 3F). Analysis of the pH values showed acidification in PRS and subsequent alkalization in SRS (Supplementary Figure S10). The apparent pH change confirms the redirection of nitrogen metabolism to the synthesis of arginine, which agrees with the essential role of the Arg pool for LFLP.

### 3.4. Accumulated Citric Acid and Arginine Promote Amino Acid Biosynthesis

The trend analysis of DEGs showed two representative clusters, Cluster 4 and Cluster 7 (Figure 4A,B). The trends in Cluster 4 showed increased gene expression in RGS and SRS but not PRS. Enrichment of DEGs in Cluster 4 showed that cytoplasmic ribosomal proteins were the most enriched terms (Figure 4C), indicating that protein translation was essential for LFLP. The trends in Cluster 7 showed increased gene expression in the PRS instead of the RGS and SRS (Figure 4B). Enrichment of DEGs in Cluster 7 showed that the organic acid metabolic process was the most enriched pathway (Figure 4D), indicating that organic acid metabolic processes play an important role in LFLP. Heatmap analysis further indicated the enhanced activity of citrate synthesis according to the upregulated CIT2, PYC2, PDB1, and ACO1 in the PRS (Supplementary Figure S11).

Considering the acidification of the postripening stage (Supplementary Figure S10), we focused on the change in organic acids using organic acid-targeted metabolome detection. Heatmap analysis of organic acids showed increased organic acids in the PSR and DSR (Supplementary Figure S12). The Z score of organic acids showed a decreased content in OG_vs_LR30, including 3-hydroxy-3-methylglutaric acid, succinic acid, and malic acid (Figure 4E). These organic acids belong mainly to the downstream products in the citrate cycle. As an upstream metabolite in the citrate cycle, citric acid showed an increase in OG_vs_LR30 (Figure 4E). The difference test box diagram confirmed the continuous apparent upregulation of citric acid in the LFLP (Figure 4F), indicating its critical role in LFLP. Expression analysis showed increased expression of citrate synthase (scaffold4.g43) in the PRS (Supplementary Figure S11). An enzyme activity test further confirmed the increasing activity of citrate synthase in PRS (Figure 4G).

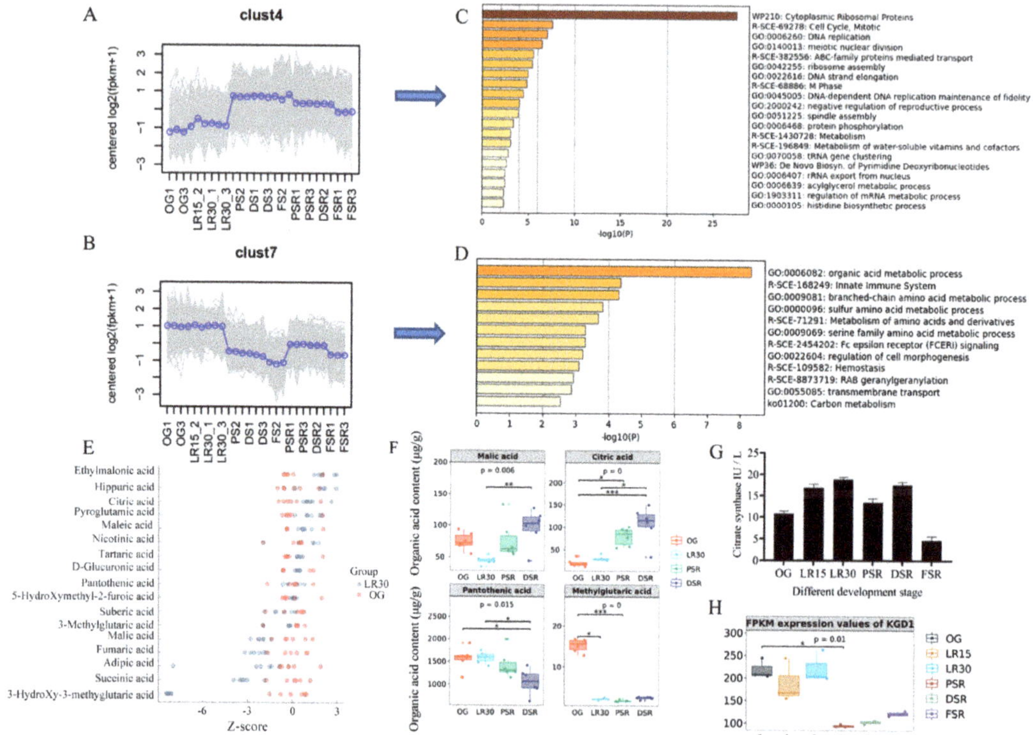


**Figure 4.** Organic acid-targeted metabolome detection of *H. marmoreus*. (**A**) Trend analysis of DEGs in Cluster 4. (**B**) Trend analysis of DEGs in Cluster 7. (**C**) Enrichment of DEGs in Cluster 4. (**D**) Enrichment of DEGs in Cluster 7. (**E**) Z score results of the content of organic acids in OG_vs_LR30. (**F**) Difference test box diagram of the four top organic acids (>10 μg/g). (**G**) Determination of citrate synthase enzyme activity in *H. marmoreus* at different developmental stages. (**H**) FPKM (fragments per kilobase million) expression values of KGD1 in *H. marmoreus* at different developmental stages. * represents $p < 0.05$; ** represents $p < 0.01$; *** represents $p < 0.001$. to the (**F**) and * represents $p < 0.05$. to the (**H**).

Since citric acid is highly acidic (pKa1 = 3.14 Ka), the upregulated citric acid could explain the acidification in the postripening stage. The difference test box diagram of organic acids showed the obvious downregulation of pantothenic acid in the PSR and DSR (Figure 4F). Based on the high concentrations of pantothenic acid, the downregulated pantothenic acid (pKa = 4.30 ± 0.10) and the increased Arg in the SRS could explain the alkalization in the SRS. The citrate cycle in KEGG showed that citrate provided the materials for the branch pathway of arginine biosynthesis through downstream 2-oxo-glutarate. The downregulated expression of the 2-oxo-glutarate metabolic enzyme (Figure 4H), alpha-ketoglutarate dehydrogenase KGD1 [EC:1.2.4.2] (scaffold6.g76), reduced the activity of the citrate cycle and shifted the branch pathway of arginine biosynthesis in SRS. The synergistically elevated citric acid and arginine reveal a straightforward way of providing enrichment of materials for amino acid biosynthesis.

### 3.5. GCN2-Mediated Translation Regulates Low-Temperature Fruiting

The large AA materials in PSR shifted our attention to translation regulation during reproductive growth. WGCNA of the AA metabolome and transcriptome from the PRS, RGS, and SRS showed 16 modules, among which pink modules correlated with most amino acid (AAs) (Supplementary Figure S13A,B). The total association score of AAs and modules showed that pink modules had the highest score (13.98). Heatmap analysis showed the prominent upregulation of the expression of the hub genes in the pink modules in the DSR and FSR (Supplementary Figure S13C,D). Enrichment analysis of the genes in the pink modules showed the most enrichment of the association of TriC/CCT with the target proteins during biosynthesis (Supplementary Figure S13E) [29], which indicated that regulation of protein biosynthesis is essential in the LFLP of *H. marmoreus*.

Considering the feature of low-temperature fruiting of *H. marmoreus*, we focused on the control of translation via the eukaryotic translation initiation factor 2 (eIF2), while four different protein kinases (GCN2, PKR, PEK, HRI) phosphorylate the $\alpha$ subunit of eIF2 in response to various environmental stresses, including cold stress [30]. The key genes' homologous genes in eIF2 translation control in *H. marmoreus* were obtained using Blastp ($1 \times 10^{-15}$). The BLASTP results showed that scaffold1.g45 was the homologous gene with the best hits among the three kinases (GCN2, PEK, and HRI), and scaffold13.g118 was the homologous gene with the best hits of PKR ($1.19 \times 10^{-22}$). The Blastp e-values of GCN2, PEK, and HRI to scaffold1.g45 were $5.02 \times 10^{-111}$, $3.96 \times 10^{-21}$, and $4.46 \times 10^{-32}$, respectively, indicating that the conserved GCN2 (scaffold1.g45) was an ancient type of kinase involved in response to different environmental stresses. Heatmap analysis showed the upregulated expression of GCN2 and translation control genes, such as eIF2, eIF2B, and EFT1, in the reproductive growth PS, DS, and DSR (Figure 5A). qPCR confirmed the obviously upregulated expression of GCN2 and eIF2 ($p < 0.05$) in PS compared to OG (Figure 5B). Notably, GCN2 showed obviously ($p < 0.05$) upregulated expression in PS, DSR, and DS compared to OG, indicating its primary role in LFLP (Figure 5B). A western blot experiment with GCN2 showed approximately 160 kD bands enriched in DSR (Figure 5C). The molecular weight of GCN2 is 162.77 kDa, and it falls in this range. qPCR experiments further confirmed that cold stress (4 °C) stimulated the obviously upregulated expression of GCN2 (scaffold1.g45) in *H. marmoreus* mycelia (Supplementary Figure S14). Analysis of genes interacting with GCN2 showed many interacting genes, including GCN20 and TOR2 (Figure 5D). Heatmap analysis showed the upregulated expression of GCN2, GCN20, and TOR2 in PS and DS, and they consisted of a cluster (Figure 5E). These results indicated that low temperature stimulated the GCN2-mediated translation pathway to promote LFLPs.

### 3.6. Cultivation Optimization with Citric Acid and Arginine Addition

Weighted correlation network analysis (WGCNA) of the organic acid-targeted metabolome and transcriptome from the PRS and SRS showed six modules: blue, brown, green, gray, red, turquoise, and yellow (Figure 6A). The cluster dendrogram showed that the turquoise and blue modules were the two largest modules (Figure 6B). The genes in the turquoise modules were correlated mainly with organic acids, including citric acid and malic acid, while the genes in the blue modules were primarily associated with the pH values (Figure 6B). The genes in the turquoise modules were enriched in the ribosome, arginine biosynthesis, pyruvate metabolism, and biosynthesis of amino acids (Figure 6C). The genes in the blue modules were enriched in protein processing in the endoplasmic reticulum, citrate cycle (TCA cycle), phagosome, and biosynthesis of amino acids (Figure 6D). These observations further indicated that the pH variation in the PRS and SRS was correlated mainly with the changes in arginine and citrate.

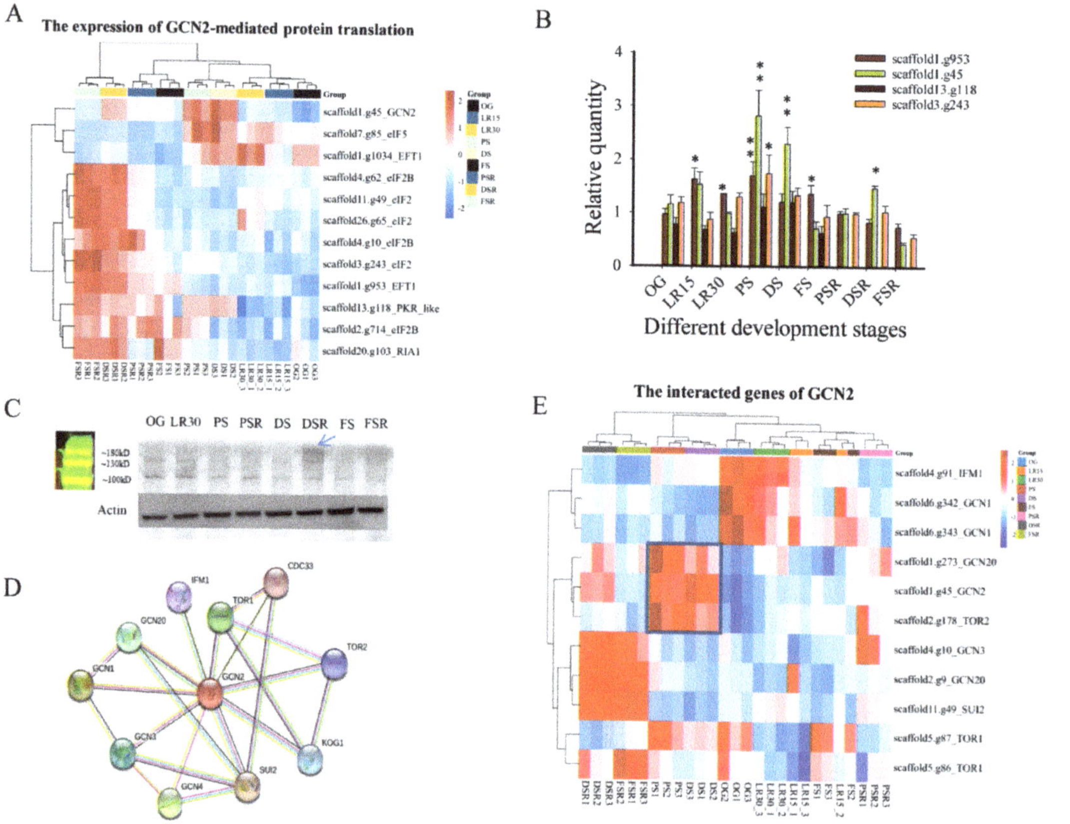


**Figure 5.** GCN2-mediated translation regulation in low-temperature fruiting after long postripening (LFLP) in *H. marmoreus*. (**A**) Heatmap analysis of key genes in GCN2-mediated translation control in *H. marmoreus* at different developmental stages. (**B**) qPCR analysis of GCN2 and eIF2 in LFLP of *H. marmoreus*. * indicates $p < 0.05$; ** indicates $p < 0.001$. (**C**) Western blot experiment of GCN2 in LFLP. Arrow: the enriched GCN2 with expected MW 160 kD. (**D**) Analysis of GCN2-interacting genes. (**E**) Heatmap analysis of the interacting genes of GCN2.

The cultivation results showed that the L-value of cultivation substrates supplemented with additive citric acid and arginine on postripening culture for 8 d, 15 d, and 24 d was higher than those supplemented with $H_2O$ (Figure 6E). Fruiting experiments confirmed that the fruiting body yield of the citric acid treatment (postripening culture for 30 d) was significantly higher than that of CK (postripening culture for 40 d) ($p = 0.0021$) (Figure 6F). The average fruiting body yield per bottle under the citric acid treatment and Arg treatment was 15.468% and 10.176% higher than that under CK. The average fruiting body yield per bottle under the citric acid and Arg treatments was 12.27% and 7.12% higher than the control, respectively. These results confirmed that citric acid and Arg addition could shorten the postripening culture by 10 d and increase the production by at least 10% compared to CK. These observations confirmed that the accumulation of arginine and citrate could promote the fruiting efficiency of *H. marmoreus*.

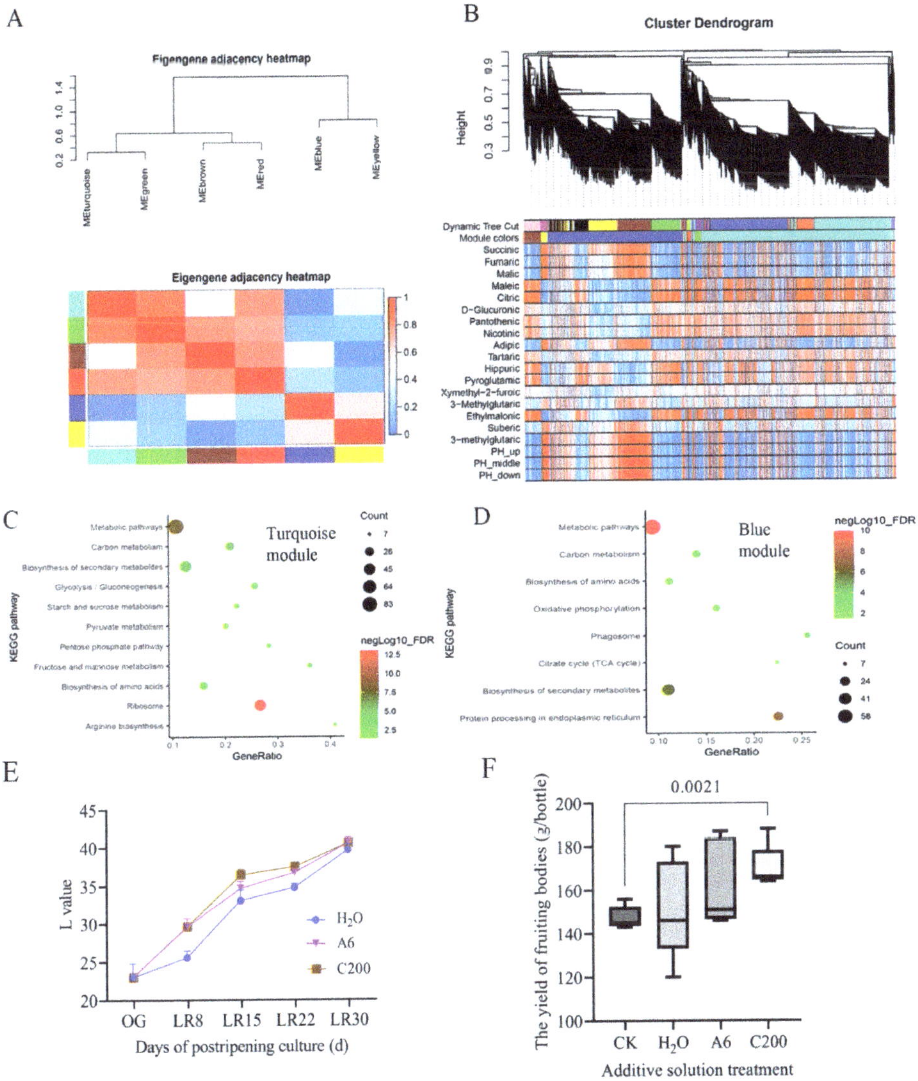


**Figure 6.** Cultivation experiment with arginine and citric acid addition. (**A**) Clustering dendrograms of module eigengenes and heatmap of eigengene adjacency. Each row and column corresponds to one eigengene within the heatmap. Red indicates high adjacency (positive correlation) within the heatmap, and turquoise indicates low adjacency (negative correlation), as shown by the color legend. (**B**) Hierarchical cluster dendrogram showing coexpressed modules and module trait heatmap. Each leaf on the tree represents a gene. Each colored row indicates a color-coded module that contains a group of highly interconnected genes. (**C**) The KEGG enrichment of genes in turquoise modules. (**D**) The KEGG enrichment of genes in blue modules. (**E**) Change in the L-value of wood chips in the postripening culture under different treatments; $H_2O$ represents 10 mL $H_2O$ solution. A6 represents 10 mL arginine solution with 6 mmol/L; C200 represents 10 mL citrate solution with 200 mg/L. (**F**) Statistics of fruiting body yield per bottle in the postripening culture under different treatments. CK represents the traditional postripening culture without any solution addition for 40 days. The illustrations of other treatments can be found in (**E**).

## 4. Discussion

Vernalization accelerates plant development from the vegetative to the reproductive stage, leading to considerable changes in the metabolite profiles of plants [31]. This study also confirmed that the vernalization-like low-temperature treatment process in mushrooms has similar metabolic changes. Some of these metabolites can be used as biomarkers in mushroom factory culture to optimize the cultivation process. Compared to RGS, the abundant AAs in SRS indicated that SRS develops an AAs pool. Arg (one of the major biomarkers) is the dominant component of the AA pool in SRS (Figure 2B), indicating its essential role in LFLP. The arginine pool plays an essential role in loblolly pine seedlings without prior conversion [32], suggesting that the newly synthesized Arg in SRS is transported rapidly and efficiently for fruiting development without prior conversion. The increased Arg content in SRS indicates that Arg is synthesized mainly in SRS. The pH value of the upper substrates in the cultivation bottle was lower than that in the lower part in LR15 (Supplementary Figure S10), compared to the opposite trend of pH value in DSR. Our results showed the opposite trend of pH change in the spatial distribution inner the cultivation bottle in LFLP (Supplementary Figure S10), suggesting an upwards metabolic flux of arginine for supplying the fruiting.

Arginase (scaffold9.g163) showed high gene expression in the postripening stage (Figure 4B,D), which supports the increased enzyme activity of arginase in the postripening stage (Figure 4E). The absence of arginase during the initiation of fruit growth indicated a redirection of nitrogen metabolism to the synthesis of arginine [33]. The lower arginase activity in the PSR stage (Figure 4E) explained the increased content of Arg in SRS, suggesting a redirection of nitrogen metabolism in the PRS to arginine synthesis in SRS. The addition of Arg in the OG stage promoted fruiting efficiency (Figure 6F), indicating that enhanced nitrogen metabolism activity is essential for postripening. This result agrees with the developmental velocity influenced by the external addition of arginine to the ascomycete *Sordaria macrospora* Auersw [34].

Another biomarker found in this study is citrate, which has a high concentration in the PSR. In contrast, citrate synthase enzyme activity decreased in this stage. Previous research reported that a mitochondrial methyltransferase inhibited citrate synthase activity through metabolite-sensitive lysine methylation [35]. Multi-omics analysis showed that methyltransferase was the top associated enzyme with the most significant number in OG vs. PS (Figure 3B). Subcellular localization prediction indicated that two methyltransferases (scaffold27.g50 and scaffold2.g228) were located in mitochondria, and they showed increased expression in the postripening stage, suggesting that methylation also plays an inhibitory role against citrate synthase activity in the postripening stage. The methylation level might be reduced due to the decreased expression of methyltransferases in the low-temperature treatment, which is consistent with a previous study that showed that cold treatment selectively decreased the methylation level in the root of maize [36]. The regular citrate synthase activity may account for the enriched citrate in the PSR. The increased citrate is positively correlated with the high AA content in the LFLP of *H. marmoreus*, which is consistent with the finding that increased citrate may induce AA synthesis in citrus fruit [37]. A high concentration of citrate might inhibit the activity of the citrate cycle and shift it to the branch AA synthesis pathways.

A close link between the upregulation of low-temperature-associated proteins and vernalization fulfillment in wheat [38] suggests a role of cold stress-responsive genes in the vernalization-like process in *H. marmoreus*. GCN2 is a serine/threonine-protein kinase that regulates translation in response to stressors such as amino acid and purine deprivation, cold shock, wounding, cadmium exposure, and UV-C exposure [39]. In yeast and animals, phosphorylation of the α-subunit of eIF2 is the most thoroughly characterized event regulating global translation under stress. GCN2 is activated under various stresses, including cold treatments [16]. GCN2 and eIF2 were upregulated under low-temperature fruiting (Figure 5B), indicating that low temperature stimulated the GCN2-mediated eIF2 pathway to promote global protein translation (Figure 7). Activated GCN2 phosphorylates the α-subunit of eIF2, drastically inhibiting protein synthesis in most species [39]. However, GCN2-mediated protein translation in *H. marmoreus* promotes the activation of protein synthesis, and this constitutes a significant exception; it also occurs in wheat under cold shock [40]. These results suggest that the stress response mechanism via GCN2-mediated eIF2α phosphorylation is not identical in all eukaryotes [40].

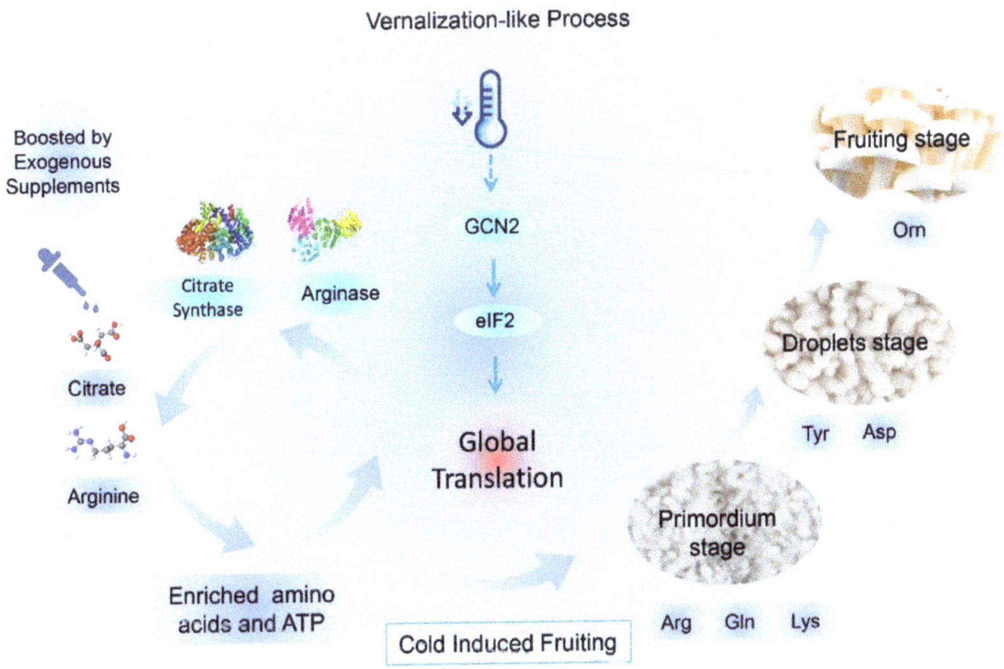


**Figure 7.** A potential model of GCN2-mediated translation regulation for low-temperature fruiting after long postripening (LFLP) in *H. marmoreus*. Low temperature stimulated the GCN2-mediated eIF2 translation pathway with the enriched amino acids and ATP in LFLP, which are essential for promoting global translation. The synergistically increased citrate and Arg may promote amino acid synthesis and ATP production for global translation in *H. marmoreus*.

Protein translation is controlled by versatile ATP-driven machines [41], which means that enriched ATP is essential for global protein translation. A high concentration of citrate means the accumulation of ATP due to the enhanced activity of the citrate cycle during the postripening process of mycelia. Reduced activity of the citrate cycle means a shift to the branch AA synthesis pathways. Our observations reveal a straightforward way of providing enrichment materials for amino acid biosynthesis, namely, synergistically elevated citric acid and arginine levels. The downregulated expression of the enzyme KGD1 (Figure 4H) indicates the reduced activity of the downstream pathway of the citrate cycle. The accumulation of citric acids could provide enriched materials to the arginine synthesis branch pathway, promoting arginine accumulation. The accumulation of arginine is necessary for protein translation in the LTLP of *H. marmoreus*, which is consistent with previous studies showing that sufficient arginine concentrations are required for normal protein accumulation rates [42]. Our model suggested that the synergistically rising citrate and arginine induced amino acid synthesis and ATP production for global translation (Figure 7). The enhanced activity of the GCN2-mediated eIF2 translation pathway promoted global protein translation in the LFLP of *H. marmoreus* with enriched AAs and ATP (Figure 7).

The analysis of this process contributes to a low-cost mushroom cultivation process and provides a breeding direction for mushrooms suitable for sustainable cultivation.

In addition to the pivotal cold induction, other factors are also involved in forming fruiting bodies. $H_2O$ addition positively affected the yield of the fruiting body compared to CK (Figure 6F). The decreased $CO_2$ concentration during the fruiting of *H. marmoreus* may agree with the increased respiration activity during primordia formation in the development of fruiting bodies [43]. Future research needs to integrate these factors to reveal this complex process's mechanism further.

## 5. Conclusions

Our findings reveal that protein translation is essential for the LFLP in *H. marmoreus* with the synergistic accumulation of citric acid and arginine. The highly accumulated concentration of citrate might inhibit the activity of the citrate cycle and shift it to the branch arginine synthesis pathways. Low temperature stimulated GCN2-mediated global protein translation control in LFLPs with enriched ATP and AAs. The SRS acts as the AAs pool for supplying the AAs for the LFLP in *H. marmoreus*. Citric acid and arginine additions could increase fruiting efficiency in *H. marmoreus* and reduce the cost of large-scale cultivation. The increasing activity of citrate synthase and arginase could be used as screening criteria for superior breeding varieties. More importantly, our pilot study of LFLP in *H. marmoreus* contributes to a better understanding of the physiological and genetic mechanisms involved in fungal growth and development.

**Supplementary Materials:** The following supporting information can be downloaded at: https://www.mdpi.com/article/10.3390/jof8070695/s1, Figure S1: Unique molecular identifier (UMI) absolute quantitative transcriptome analysis of developmental stages of *H. marmoreus*; Figure S2: Heatmap analysis of gene expression in the ribosomal pathways; Figure S3: KEGG enrichment analysis of the DEGs in LR30_vs_PS; Figure S4: KEGG enrichment analysis of the DEGs in OG vs. LR30; Figure S5: Expression profile of the amino acid metabolism pathway in low-temperature fruiting after long postripening (LFLP) in *H. marmoreus*; Figure S6: Targeted profiling of the amino acid metabolome of different developmental stages of *H. marmoreus*; Figure S7: PCA of the AAs in the developmental stages of *H. marmoreus*; Figure S8: UMI absolute quantitative transcriptome analysis of the developmental stages of *H. marmoreus*, including the substrates of the reproductive growth stage (SRG); Figure S9: KEGG enrichment analysis of the DEGs in PSR vs. DSR; Figure S10: pH value of substrates in the developmental stages of *H. marmoreus*. The pH values of the substrates from the upper, middle, and lower parts of the cultivation bottle were measured; Figure S11: Heatmap analysis of the expression of citrate cycle in the developmental stages of *H. marmoreus*; Figure S12: Heatmap analysis of organic acids in the postripening stage (PRS) and SRG; Figure S13: WGCNA of the AA-targeted metabolome and transcriptome. (A) Hierarchical cluster dendrogram showing coexpressed modules and a module trait heatmap. Each leaf on the tree represents a gene. Each coloured row indicates a colour-coded module that contains a group of highly interconnected genes. (B) Heatmap showing the correlation between the modules and AAs. Each row corresponds to a module, whereas each column corresponds to an AA. The correlation coefficient between a given module and an AA is indicated by the colour of the cell at the row-column intersection. Blue and red indicate positive and negative correlations, respectively. (C) The hub genes in the pink modules. (D) Heatmap of the hub genes in the pink module. (E) Enrichment analysis of the genes in the pink modules; Figure S14: qPCR analysis of GCN2 and eIF2 in the 4 °C cold stress experiment of *H. marmoreus* mycelia. Table S1: Primers used for qPCR; Table S2: KEGG enrichment analysis of the DEGs in LR30 vs. PSR; Table S3: Correlation analysis of differentially expressed genes (DEGs) and significantly different AAs (SDAs) in OG vs. PS. The associated AAs obtained by correlation analysis of DEGs and SDAs; Table S4: The associated enzymes obtained by correlation analysis of DEGs and SDAs; Table S5 KEGG annotation of arginine biosynthesis.

**Author Contributions:** Conceptualization, Investigation, Writing—Original Draft, Writing—Review and Editing, Funding Acquisition, M.G.; Methodology, Investigation, T.H.; Methodology, Y.L.; Methodology, J.L.; Resources, E.S.; Methodology, L.T.; Investigation, Writing—Original Draft, Writing—Review and Editing, Funding Acquisition, Validation, G.Z.; Conceptualization, Resources, Supervision, Validation, D.B. All authors have read and agreed to the published version of the manuscript.

**Funding:** This study was sponsored by the National Natural Science Foundation of China (32072642) and the Science and Technology Innovation Plan of the Shanghai Science and Technology Commission (20310741900). The funders had no roles in study design, data collection and analysis, decision to publish, or preparation of the manuscript.

**Institutional Review Board Statement:** Not applicable.

**Informed Consent Statement:** Not applicable.

**Data Availability Statement:** The data supporting the findings of this work are available within the paper and its supplementary files. The data sets generated and analyzed during this study are available from the corresponding author upon request.

**Conflicts of Interest:** The authors declare that they have no known competing financial interests or personal relationships that could have appeared to influence the work reported in this paper. The authors (Jinxin Li) declare no conflict of interest with Shanghai Finc Bio-Tech Inc.

## References

1. Zou, G.; Li, B.; Wang, Y.; Yin, X.; Gong, M.; Shang, J.; Wei, Y.; Li, X.; Bao, D. Efficient conversion of spent mushroom substrate into a high value-added anticancer drug pentostatin with engineered Cordyceps militaris. *Green Chem.* **2021**, *23*, 10030–10038. [CrossRef]
2. Zou, G.; Bao, D.; Wang, Y.; Zhou, S.; Xiao, M.; Yang, Z.; Wang, Y.; Zhou, Z. Alleviating product inhibition of Trichoderma reesei cellulase complex with a product-activated mushroom endoglucanase. *Bioresour. Technol.* **2021**, *319*, 124119. [CrossRef] [PubMed]
3. Amirullah, N.A.; Zainal Abidin, N.; Abdullah, N. The potential applications of mushrooms against some facets of atherosclerosis: A review. *Food Res. Int.* **2018**, *105*, 517–536. [CrossRef] [PubMed]
4. Zhou, J.; Chen, M.; Wu, S.; Liao, X.; Wang, J.; Wu, Q.; Zhuang, M.; Ding, Y. A review on mushroom-derived bioactive peptides: Preparation and biological activities. *Food Res. Int.* **2020**, *134*, 109230. [CrossRef] [PubMed]
5. Zhang, Y.; Geng, W.; Shen, Y.; Wang, Y.; Dai, Y.-C. Edible mushroom cultivation for food security and rural development in China: Bio-innovation, technological dissemination and marketing. *Sustainability* **2014**, *6*, 2961–2973. [CrossRef]
6. Stamets, P. *Growing Gourmet and Medicinal Mushrooms*, 3rd ed.; Ten Speed Press: California, CN, USA, 2000.
7. Pang, J.; Sun, G.; Jing, Y.; Meng, H.; Wang, H.; Liu, Y.; Yang, Z. Cluster Analysis Based on the Growth Temperature of 35 Edible Fungi. *Edible Fungi China* **2015**, *34*, 34–37.
8. Nakazawa, T.; Miyazaki, Y.; Kaneko, S.; Shishido, K. Stimulative effects of light and a temperature downshift on transcriptional expressions of developmentally regulated genes in the initial stages of fruiting-body formation of the basidiomycetous mushroom Lentinula edodes. *FEMS Microbiol. Lett.* **2008**, *289*, 67–71. [CrossRef]
9. Morin, E.; Kohler, A.; Baker, A.R.; Foulongne-Oriol, M.; Lombard, V.; Nagy, L.G.; Ohm, R.A.; Patyshakuliyeva, A.; Brun, A.; Aerts, A.L.; et al. Genome sequence of the button mushroom Agaricus bisporus reveals mechanisms governing adaptation to a humic-rich ecological niche. *Proc. Natl. Acad. Sci. USA* **2012**, *109*, 17501–17506. [CrossRef]
10. Sakamoto, Y.; Ando, A.; Tamai, Y.; Miura, K.; Yajima, T. Protein expressions during fruit body induction of Flammulina velutipes under reduced temperature. *Mycol. Res.* **2002**, *106*, 222–227. [CrossRef]
11. Fu, Y.P.; Liang, Y.; Dai, Y.T.; Yang, C.T.; Duan, M.Z.; Zhang, Z.; Hu, S.N.; Zhang, Z.W.; Li, Y. De Novo Sequencing and Transcriptome Analysis of Pleurotus eryngii subsp. tuoliensis (Bailinggu) Mycelia in Response to Cold Stimulation. *Molecules* **2016**, *21*, 560. [CrossRef]
12. Ford, K.L.; Baumgartner, K.; Henricot, B.; Bailey, A.M.; Foster, G.D. A reliable in vitro fruiting system for Armillaria mellea for evaluation of Agrobacterium tumefaciens transformation vectors. *Fungal Biol.* **2015**, *119*, 859–869. [CrossRef] [PubMed]
13. Liu, Y.; Chen, H.; Feng, Z.; Chen, M.; Wang, H.; Long, Y.; Xie, M.; Zhang, J. Analysis on correlation between LBL evaluation and agronomic characters of Hypsizygus marmoreus. *Acta Agric. Shanghai* **2011**, *2*, 247–257.
14. Park, Y.J.; Jung, E.S.; Singh, D.; Lee, D.E.; Kim, S.; Lee, Y.W.; Kim, J.-G.; Lee, C.H. Spatial (cap & stipe) metabolomic variations affect functional components between brown and white beech mushrooms. *Food Res. Int.* **2017**, *102*, 544–552. [PubMed]
15. Wang, L.; Li, H.; Zhao, C.; Li, S.; Kong, L.; Wu, W.; Kong, W.; Liu, Y.; Wei, Y.; Zhu, J.K.; et al. The inhibition of protein translation mediated by AtGCN1 is essential for cold tolerance in Arabidopsis thaliana. *Plant Cell Environ.* **2017**, *40*, 56–68. [CrossRef] [PubMed]
16. Lokdarshi, A.; Morgan, P.W.; Franks, M.; Emert, Z.; Emanuel, C.; von Arnim, A.G. Light-Dependent Activation of the GCN2 Kinase Under Cold and Salt Stress Is Mediated by the Photosynthetic Status of the Chloroplast. *Front. Plant Sci.* **2020**, *11*, 431. [CrossRef] [PubMed]

17. Gong, M.; Wang, Y.; Zhang, J.; Zhao, Y.; Wan, J.; Shang, J.; Yang, R.; Wu, Y.; Li, Y.; Tan, Q.; et al. Chilling Stress Triggers VvAgo1-Mediated miRNA-Like RNA Biogenesis in Volvariella volvacea. *Front. Microbiol.* **2020**, *11*, 523593. [CrossRef]
18. Ming, G.; Yan, L.; Dapeng, B.; Junjun, S.; Ruiheng, Y.; Chenli, Z.; Jianing, W.; Yingying, W. Genome-wide association study on mycelial growth rate of Hypsizygus marmoreus cultivation strains. *Acta Edulis. Fungi.* **2021**, *28*, 8.
19. Yu, N.Y.; Wagner, J.R.; Laird, M.R.; Melli, G.; Rey, S.; Lo, R.; Dao, P.; Sahinalp, S.C.; Ester, M.; Foster, L.J.; et al. PSORTb 3.0: Improved protein subcellular localization prediction with refined localization subcategories and predictive capabilities for all prokaryotes. *Bioinformatics* **2010**, *26*, 1608–1615. [CrossRef]
20. Gong, M.; Wang, Y.; Su, E.; Zhang, J.; Tang, L.; Li, Z.; Zhang, L.; Zou, G.; Wan, J.; Bao, D. The promising application of a β-glucosidase inhibitor in the postharvest management of Volvariella volvacea. *Postharvest Biol. Technol.* **2022**, *185*, 111784. [CrossRef]
21. Selli, S.; Kelebek, H.; Kesen, S.; Sonmezdag, A.S. GC-MS olfactometric and LC-DAD-ESI-MS/MS characterization of key odorants and phenolic compounds in black dry-salted olives. *J. Sci. Food Agric.* **2018**, *98*, 4104–4111. [CrossRef]
22. Fiori, J.; Amadesi, E.; Fanelli, F.; Tropeano, C.V.; Rugolo, M.; Gotti, R. Cellular and mitochondrial determination of low molecular mass organic acids by LC-MS/MS. *J. Pharm. Biomed. Anal.* **2018**, *150*, 33–38. [CrossRef] [PubMed]
23. Pawlak, M.; Klupczynska, A.; Kokot, Z.J.; Matysiak, J. Extending Metabolomic Studies of Apis mellifera Venom: LC-MS-Based Targeted Analysis of Organic Acids. *Toxins* **2019**, *12*, 14. [CrossRef] [PubMed]
24. Langfelder, P.; Horvath, S. WGCNA: An R package for weighted correlation network analysis. *BMC Bioinform.* **2008**, *9*, 559. [CrossRef] [PubMed]
25. Liang, Y.; Wang, S.; Zhao, C.; Ma, X.; Zhao, Y.; Shao, J.; Li, Y.; Li, H.; Song, H.; Ma, H.; et al. Transcriptional regulation of bark freezing tolerance in apple (Malus domestica Borkh.). *Hortic. Res.* **2020**, *7*, 205. [CrossRef] [PubMed]
26. Gong, M.; Wang, H.; Chen, M.; Bao, D.; Zhu, Q.; Tan, Q. A newly discovered ubiquitin-conjugating enzyme E2 correlated with the cryogenic autolysis of Volvariella volvacea. *Gene* **2016**, *583*, 58–63. [CrossRef] [PubMed]
27. Livak, K.J.; Schmittgen, T.D. Analysis of relative gene expression data using real-time quantitative PCR and the 2(-Delta Delta C(T)) Method. *Methods* **2001**, *25*, 402–408. [CrossRef]
28. Lu, D.; Pava-Ripoll, M.; Li, Z.; Wang, C. Insecticidal evaluation of Beauveria bassiana engineered to express a scorpion neurotoxin and a cuticle degrading protease. *Appl. Microbiol. Biotechnol.* **2008**, *81*, 515–522. [CrossRef]
29. Yam, A.Y.; Xia, Y.; Lin, H.T.; Burlingame, A.; Gerstein, M.; Frydman, J. Defining the TRiC/CCT interactome links chaperonin function to stabilization of newly made proteins with complex topologies. *Nat. Struct. Mol. Biol.* **2008**, *15*, 1255–1262. [CrossRef]
30. Flores-Méndez, M.A.; Martínez-Lozada, Z.; Monroy, H.C.; Hernández-Kelly, L.C.; Barrera, I.; Ortega, A. Glutamate-dependent translational control in cultured Bergmann glia cells: eIF2α phosphorylation. *Neurochem. Res.* **2013**, *38*, 1324–1332. [CrossRef]
31. Nugroho, A.B.D.; Lee, S.W.; Pervitasari, A.N.; Moon, H.; Choi, D.; Kim, J.; Kim, D.H. Transcriptomic and metabolic analyses revealed the modulatory effect of vernalization on glucosinolate metabolism in radish (Raphanus sativus L.). *Sci. Rep.* **2021**, *11*, 24023. [CrossRef]
32. Todd, C.D.; Cooke, J.E.; Mullen, R.T.; Gifford, D.J. Regulation of loblolly pine (Pinus taeda L.) arginase in developing seedling tissue during germination and post-germinative growth. *Plant Mol. Biol* **2001**, *45*, 555–565. [CrossRef] [PubMed]
33. Alabadi, D.; Aguero, M.S.; Perez-Amador, M.A.; Carbonell, J. Arginase, Arginine Decarboxylase, Ornithine Decarboxylase, and Polyamines in Tomato Ovaries (Changes in Unpollinated Ovaries and Parthenocarpic Fruits Induced by Auxin or Gibberellin). *Plant Physiol.* **1996**, *112*, 1237–1244. [CrossRef] [PubMed]
34. Molowitz, R.; Bahn, M.; Hock, B. The control of fruiting body formation in the ascomycete Sordaria macrospora Auersw. by arginine and biotin: A two-factor analysis. *Planta* **1976**, *128*, 143–148. [CrossRef] [PubMed]
35. Małecki, J.; Jakobsson, M.E.; Ho, A.Y.Y.; Moen, A.; Rustan, A.C.; Falnes, P. Uncovering human METTL12 as a mitochondrial methyltransferase that modulates citrate synthase activity through metabolite-sensitive lysine methylation. *J. Biol. Chem.* **2017**, *292*, 17950–17962. [CrossRef] [PubMed]
36. Steward, N.; Kusano, T.; Sano, H. Expression of ZmMET1, a gene encoding a DNA methyltransferase from maize, is associated not only with DNA replication in actively proliferating cells, but also with altered DNA methylation status in cold-stressed quiescent cells. *Nucleic Acids Res.* **2000**, *28*, 3250–3259. [CrossRef] [PubMed]
37. Degu, A.; Hatew, B.; Nunes-Nesi, A.; Shlizerman, L.; Zur, N.; Katz, E.; Fernie, A.R.; Blumwald, E.; Sadka, A. Inhibition of aconitase in citrus fruit callus results in a metabolic shift towards amino acid biosynthesis. *Planta* **2011**, *234*, 501–513. [CrossRef]
38. Sarhadi, E.; Mahfoozi, S.; Hosseini, S.A.; Salekdeh, G.H. Cold acclimation proteome analysis reveals close link between the up-regulation of low-temperature associated proteins and vernalization fulfillment. *J. Proteome Res.* **2010**, *9*, 5658–5667. [CrossRef] [PubMed]
39. Llabata, P.; Richter, J.; Faus, I.; Słomińska-Durdasiak, K.; Zeh, L.H.; Gadea, J.; Hauser, M.T. Involvement of the eIF2α Kinase GCN2 in UV-B Responses. *Front. Plant Sci.* **2019**, *10*, 1492. [CrossRef] [PubMed]
40. Zhigailov, A.V.; Alexandrova, A.M.; Nizkorodova, A.S.; Stanbekova, G.E.; Kryldakov, R.V.; Karpova, O.V.; Polimbetova, N.S.; Halford, N.G.; Iskakov, B.K. Evidence That Phosphorylation of the α-Subunit of eIF2 Does Not Essentially Inhibit mRNA Translation in Wheat Germ Cell-Free System. *Front. Plant Sci.* **2020**, *11*, 936. [CrossRef]
41. Gerovac, M.; Tampé, R. Control of mRNA Translation by Versatile ATP-Driven Machines. *Trends Biochem. Sci.* **2019**, *44*, 167–180. [CrossRef] [PubMed]

42. Flint, H.J.; Dible, S.; Kacser, H. Depression of enzyme synthesis in response to arginine limitation in Neurospora crassa. *J. Gen. Microbiol.* **1985**, *131*, 2891–2900.
43. Sakamoto, Y. Influences of environmental factors on fruiting body induction, development and maturation in mushroom-forming fungi. *Fungal Biol. Rev.* **2018**, *32*, 236–248. [CrossRef]

Journal of *Fungi*

MDPI

*Article*

# Characteristics of the Genome, Transcriptome and Ganoderic Acid of the Medicinal Fungus *Ganoderma lingzhi*

**Qiang Wu [1,2,3,†], Huan Liu [1,2,3,†], Yixin Shi [1,2,3], Wanting Li [1,2,3], Jia Huang [1,2,3], Feifei Xue [1,2,3], Yongnan Liu [1,2,3,*] and Gaoqiang Liu [1,2,3,*]**

1 Hunan Provincial Key Laboratory of Forestry Biotechnology, Central South University of Forestry & Technology, Changsha 410004, China
2 International Cooperation Base of Science and Technology Innovation on Forest Resource Biotechnology of Hunan Province, Central South University of Forestry & Technology, Changsha 410004, China
3 Microbial Variety Creation Center, Yuelushan Laboratory of Seed Industry, Changsha 410004, China
* Correspondence: ynliu@csuft.edu.cn (Y.L.); gaoliuedu@csuft.edu.cn (G.L.); Tel./Fax: +86-731-8562-3490 (Y.L.)
† These authors contributed equally to this work.

**Abstract:** *Ganoderma* (Ganodermaceae) is a genus of edible and medicinal mushrooms that create a diverse set of bioactive compounds. *Ganoderma lingzhi* has been famous in China for more than 2000 years for its medicinal properties. However, the genome information of *G. lingzhi* has not been characterized. Here, we characterized its 49.15-Mb genome, encoding 13,125 predicted genes which were sequenced by the Illumina and PacBio platform. A wide spectrum of carbohydrate-active enzymes, with a total number of 519 CAZymes were identified in *G. lingzhi*. Then, the genes involved in sexual recognition and ganoderic acid (GA, key bioactive metabolite) biosynthesis were characterized. In addition, we identified and deduced the possible structures of 20 main GA constituents by UPLC-ESI-MS/MS, including a new special ganochlearic acid A. Furthermore, 3996 novel transcripts were discovered, and 9276 genes were predicted to have the possibility of alternative splicing from RNA-Seq data. The alternative splicing genes were enriched for functional categories involved in protein processing, endocytosis, and metabolic activities by KEGG. These genomic, transcriptomic, and GA constituents' resources would enrich the toolbox for biological, genetic, and secondary metabolic pathways studies in *G. lingzhi*.

**Keywords:** *Ganoderma lingzhi*; whole genome characteristics; ganoderic acid; RNA sequencing; alternative splicing

check for updates

**Citation:** Wu, Q.; Liu, H.; Shi, Y.; Li, W.; Huang, J.; Xue, F.; Liu, Y.; Liu, G. Characteristics of the Genome, Transcriptome and Ganoderic Acid of the Medicinal Fungus *Ganoderma lingzhi*. *J. Fungi* **2022**, *8*, 1257. https://doi.org/10.3390/jof8121257

Academic Editor: Gen Zou

Received: 28 September 2022
Accepted: 25 November 2022
Published: 28 November 2022



## 1. Introduction

*Ganoderma* (Ganodermaceae) is a genus of edible and medicinal mushrooms that are widely distributed in the tropic and subtropics of Asia, Africa, and America. The medicinal values of this mushroom were documented thousands of years ago [1]. The *G. lingzhi* species is a mushroom that has been renowned in China, which has been incorrectly considered to be another species, *G. lucidum*, which is distributed in Europe, for many years. The most striking characteristics which differentiate *G. lingzhi* from *G. lucidum* are the presence of melanoid bands in the context, a yellow pore surface and thick dissepiments at maturity by morphological studies, rDNA nuc-ITS sequences, and additional gene fragments analysis [2]. The attractive characteristic of *Ganoderma* is its immunomodulatory and anti-tumor activities, which are mainly attributed to the major active components, ganoderic acids (GAs), a type of triterpenoids [3]. GAs are effective as supplemental therapies and improve health when combined with other medications to treat hepatitis, fatigue syndrome, and prostate cancer [4]. Up to now, there are four whole genome sequencing projects for *G. lucidum* (Accession no. PRJDA61381, *G. lucidum* strain BCRC 37177, whole genome shotgun sequences available; PRJNA77007 *G. lucidum* strain Xiangnong No. 1, high-quality genome assembly available; PRJNA71455 *G. lucidum* strain G.260125-1, high-quality genome assembly available; *G. lucidum* strain Ling-Jian No. 2, high-quality genome assembly available) have been published and registered in the National Center of Biotechnology Information (NCBI) [5–8]. Recently, with the combination of Illumina and PacBio sequencing strategies, we de novo sequenced and assembled the genome of *G. lingzhi* [9]. However, the whole genome information of *G. lingzhi* has not been well characterized yet.

In this work, the genes involved in lignocellulose degradation and sexual recognition, and ganoderic acid biosynthesis of the *G. lingzhi* genome were characterized and analyzed. Then, the key bioactive metabolite, ganoderic acid, was identified by mass spectrometry. Furthermore, novel transcripts were discovered and differential alternative splicings were detected from RNA-Seq data. Collectively, these foundational genomic, transcriptomic, and ganoderic acid bioactive compound resources would enrich the toolbox for further biological, genetic, and secondary metabolic pharmacological studies in *G. lingzhi.*

## 2. Materials and Methods

### 2.1. Fungal Strains and Culture Conditions

The *Ganoderma lingzhi* strain SCIM 1006 (No. CGMCC 18819) kindly provided by Prof. Yu-Cheng Dai was selected for characteristics of the whole genome information, identification of ganoderic acids (GAs), and RNA-Seq [2]. The *G. lingzhi* strain was cultured at 27 °C with shaking at 160 rpm in the dark on an artificial medium (glucose, 44.0; corn flour, 0.5; peptone, 6.5; $KH_2PO_4$, 0.75; $MgSO_4 \cdot 7H_2O$, 0.45; vitamin B1, 0.01; in g/L) [10].

### 2.2. Genome Sequencing, Assembly, and Annotation

The genome of the G. lingzhi strain was de novo sequenced by high-throughput Illumina HiSeqX-Ten and PacBio RSII long-read sequencing platforms (PacBio P6-C4) at Shanghai OE Biotech. Co., Ltd. in our previous studies [9]. These two sequencing platforms were widely used in the genome sequencing analysis of many species. Briefly, FALCON (version 0.7.0) was used for the de novo assembly of contigs [11]. Further, the PacBio-corrected contigs with accurate Illumina short reads and generating the genome assembly was performed with Pilon (v 1.23) [12]. Finally, BUSCO 3.1.0 was used with comparison to the lineage dataset fungi_odb9 to evaluate the genome assembly [13]. Homologous comparison and de novo prediction were used to annotate the G. lingzhi genome sequences according to the previous description [14].

### 2.3. Identification of CAZymes and Mating Type (MAT) Genes

The BLASTP search (covered fraction ratio $\geq 0.2$; e-value $\leq 1 \times 10^{-6}$; maximum hit number = 500) of dbCAN HMMs 6.0 and Hidden Markov Model (HMM) search (default cutoff threshold) were used for the analysis of the Carbohydrate-active enzymes (CAZymes) family [15]. The final CAZyme annotation was obtained from the common results of these two methods.

By using TBLASTN and MAT-A genes from *G. lucidum,* the MAT-A genes in *G. lingzhi* were identified including homeodomain type 1 and homeodomain type 2 mating-type genes (HD1 and HD2 genes) [8]. The same method is used to identify the mitochondrial intermediate peptidase gene (mip) in *G. lingzhi.* The pheromone receptor genes (MAT-B) in *G. lingzhi* were identified by the Swissprot annotation with the keyword "pheromone receptor".

### 2.4. Ganoderic Acid Extraction and Mass Spectrometry Analysis

Fermentation mycelium samples of *G. lingzhi* were freeze-dried by a vacuum freeze dryer (Scientz-100F). The sample was ground into powder and then dissolved in a 70% methanol solution. Vortex extraction was performed 6 times (30 s every 30 min), and then overnight extraction was performed at 4 °C. After centrifugation (12,000 rpm, 10 min), the supernatant is used for subsequent UPLC–MS/MS analysis.

A UPLC–ESI–MS/MS system (UPLC, Shim-pack UFLC SHIMADZU CBM A system) was used for the analysis of GAs. The UPLC C18 column (1.8 μm, 2.1 mm × 100 mm) was used to sample diversion in water (0.1% formic acid)-acetonitrile (0.1% acetic acid) solution by gradient at 40 °C, 0.4 mL/min flow rate.

Mass spectrometry was performed using an AB4500 Q TRAP UPLC/MS/MS System, equipped with an ESI Turbo Ion-Spray interface. The ESI source, turbo spray at 550 °C; 5500 V and −4500 V at positive and negative ion mode respectively. Gas I, II, and curtain gas were set at 50, 60, and 25.0 psi, respectively. Triple quadrupole scans were acquired as Multiple Reaction Monitoring (MRM) experiments (nitrogen as the medium), and MRM transitions were monitored.

### 2.5. RNA-Seq

*G. lingzhi* fresh fermented mycelium (six biological repeats) was used to perform RNA extraction by TRIzol Reagent (Life Technologies, Carlsbad, CA, USA) according to the manufacturer's instructions. The cDNA library was constructed following the instruction manual of the NEBNext Ultra RNA Library Prep Kit (NEB, E7530) and NEBNext Multiplex Oligos (NEB, E7500).

The library was sequenced using the Illumina HiSeq platform. Transcriptome analysis was performed using *G. Lingzhi* reference genome-based read mapping. Raw data (raw reads) of fastq format were first processed through in-house perl scripts. Clean data (clean reads) were obtained by using Tophat2 software to remove the low-quality reads from the raw data, such as unknown nucleotides >5%, reads containing ploy-N or a Q20 <20% (error rates < 1%) [16]. The aligned records were further examined to remove potential duplicate molecules.

Discovery of novel transcripts and genes was achieved by StringTie on the base of the reference genome [17]. The transcripts without annotations compared with original genome annotations are defined as novel transcripts. The transcripts containing only one exon or short transcripts (less than 150 bp) were excluded. Novel genes were annotated by DIAMOND [18] against databases including NR [19], Swiss-Prot [20], COG [21], KOG [22], and KEGG [23].

StringTie was applied to assemble the mapped reads generated by Hisat2 [24]. Asprofile [25] was employed to predict alternative splicing events in each sample and sort them into 12 types.

## 3. Results and Discussion

### 3.1. The Genome Characteristics of G. lingzhi

Using a whole-genome shotgun sequencing strategy, the genome of *G. lingzhi* was sequenced [9]. Genome sequences of 49.15 Mb were assembled into 30 scaffolds, with an average length of 1.65 Mb. An amount of 13,125 protein-coding genes were predicted, with an average mRNA and CDS length of 1.95 kb and 1.45 kb, respectively. On average, each predicted gene contained 6.04 exons, with an average length of 240.48 bp (Table S1).

Further, we identified 7,275,816 bp repeat sequences, accounting for 14.80% of the genome (Table 1). Of them, long terminal repeats (LTRs), DNA transposons, and simple repeats were the main categories, accounting for 5.71%, 2.06%, and 0.64% of the genome, respectively. Short/long interspersed nuclear elements (SINEs/LINEs) and satellite DNA comprised less than 0.10% of the genome. In addition, 178 noncoding RNAs containing 141 tRNAs, 22 snRNAs, 13 rRNAs, and 2 sRNAs with a total length of 32,535 bp were identified (Table 1).

**Table 1.** Characteristics of the repeat sequences and non-coding RNA of *G. lingzhi*.

| Repeat Sequences | | | |
|---|---|---|---|
| **Item** | **Number** | **Length (bp)** | **Coverage** |
| SINE | 6 | 488 | 0.00% |
| LINE | 121 | 22,417 | 0.05% |
| LTR | 1849 | 2,806,397 | 5.71% |

**Table 1.** *Cont.*

| Repeat Sequences | | | |
|---|---|---|---|
| **Item** | **Number** | **Length (bp)** | **Coverage** |
| DNA transposon | 538 | 1,013,038 | 2.06% |
| Satellite DNA | 26 | 2435 | 0.00% |
| Simple repeat | 7431 | 313,771 | 0.64% |
| Low complexity | 1008 | 51,127 | 0.10% |
| Other | 42 | 10,370 | 0.02% |
| Unknown | 6091 | 3,055,773 | 6.22% |
| Total | 17,112 | 7,275,816 | 14.80% |
| **Non-Coding RNA** | | | |
| **Class** | **Number** | **Total Length (bp)** | **Mean Length (bp)** |
| rRNA | 13 | 17,274 | 1328 |
| sRNA | 2 | 709 | 354 |
| snRNA | 22 | 2869 | 130 |
| tRNA | 141 | 11,683 | 82 |
| Total | 178 | 32,535 | 182 |

### 3.2. Gene Annotation

Eight public databases were used to annotate the function of predicted genes, including UniProt, Gene Ontology (GO), Kyoto Encyclopedia of Genes and Genomes (KEGG), Pfam, RefSeq, NCBI non-redundant, NCBI clusters of orthologous groups of proteins (COG), and Pathway. Overall, 12,802 genes (accounting for 97.54% of all genes) were annotated to at least one function. A total of 6464 genes (49.25%), 9166 genes (69.84%), 4634 genes (35.31%), 12,762 genes (97.23%), 6406 genes (48.81%), 4734 genes (36.07%), 2898 genes (22.08%), and 1274 genes (9.71%) were annotated with UniProt, Pfam, Refseq, NCBI nr, GO, KEGG, Pathway, and COG, respectively. The annotation results were combined and shown in Supplementary Materials S1.

### 3.3. GO and KEGG Analysis

The GO enrichment analysis showed that transmembrane transport was the most representative term, followed by intracellular protein transduction and cell division in biological processes (Figure 1A, left). In cell components, the nucleus was the most representative term, followed by cytosol and the integral component of the membrane (Figure 1A, centre). In molecular function, ATP binding was the most representative term, followed by metal ion binding and RNA binding (Figure 1A, right). The KEGG analysis revealed that all the predicted genes were enriched in five classes with 31 pathways in total (Figure 1B; organismal systems with 9 pathways, metabolism with 12 pathways, genetic information processing with 4 pathways, environmental information processing with 2 pathways, and cellular processes with 4 pathways). Similar to the KEGG annotation, metabolic and biosynthesis categories in COG were highly enriched (Figure 1C), such as carbohydrate transport and metabolism, lipid transport and metabolism, and secondary metabolic biosynthesis (Figure 1C Group G, I, and Q).

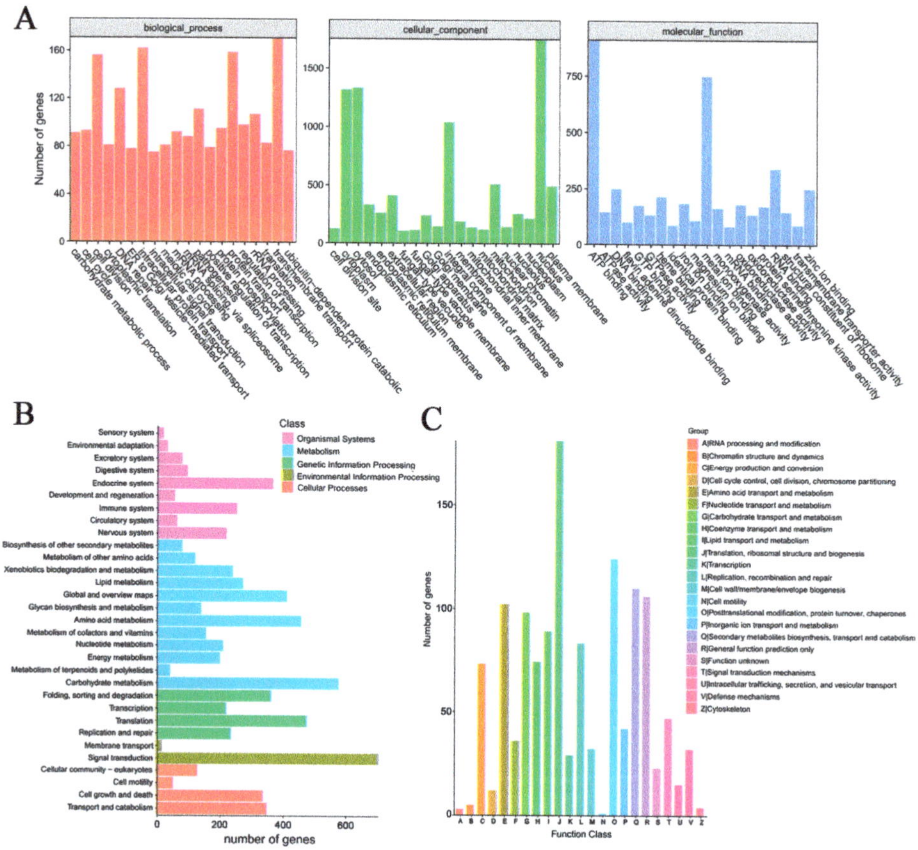


**Figure 1.** Function annotation of predicted genes in *G. lingzhi*. (**A**) The GO function annotation of predicted genes in *G. lingzhi*. Biological processes, cell components, and molecule functions are three categories of functional analysis. Counts for each category represent the total associated terms in the database with the query gene list. Terms with a $p$-value < 0.05 are statistically significant. The 20 most significantly enriched terms were listed, and the number of involved genes in a term is shown on the left y-axis. (**B**) KEGG analysis of predicted genes in *G. lingzhi*. Enriched KEGG pathways are clustered into five categories of cellular processes, environmental information processing, genetic information, processing, metabolism, and organismal systems. The number of involved genes in a term is shown on the left x-axis. (**C**) Distribution of genes in different COG function classifications of *G. lingzhi*. The number of involved genes in a term is shown on the left y-axis.

### 3.4. Identification of CAZymes

In nature, *G. lingzhi* often grows on rotten or living wood by degrading plant cell walls with carbohydrate-active enzymes (CAZymes). A total of 510 *G. lingzhi* genes could be assigned to CAZyme families (Figure 2, Supplementary Materials S2) as defined in the CAZy database by a Hidden Markov Model (HMM) search [15]. In detail, 254 glycoside hydrolases (GHs), 79 glycosyl transferases (GTs), 9 polysaccharide lyases (PLs), 117 auxiliary activities (AAs), 47 carbohydrate esterases (CEs), and 4 carbohydrate-binding modules (CBMs) were searched. In the wide spectrum of GH families, many genes for digesting cellulose (such as GH5 with 20 genes), hemicellulose (such as GH3, GH10, and GH43 with 13, 10, and 15 genes, respectively), and chitin (such as GH18 with 21 genes) were found in the *G. lingzhi* genome (Figure 2).

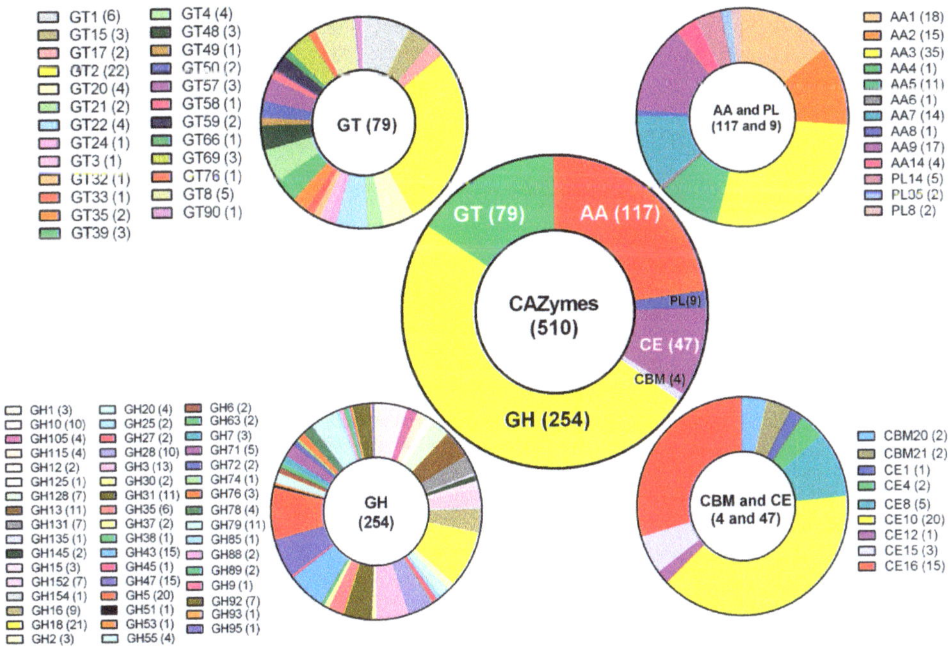


**Figure 2.** The number of carbohydrate-active enzyme genes (CAZymes) in *G. lingzhi*. GH: glycoside hydrolase; GT: glycosyl transferase; PL: polysaccharide lyase; CE: carbohydrate esterase; CBM: carbohydrate-binding module; AA: auxiliary activity. The number of involved genes in a term is shown in brackets.

The CAZyme profile in *G. lingzhi* was also compared to those of 14 other fungi containing two *G. lucidum* species (Table S2). The global number of GTs, GHs, PLs, CEs, and CBMs of *G. lingzhi* was comparable with those of two *G. lucidum* species, Xiangnong No. 1 [6] and 260125-1 [8], and other white-rot fungi, such as *Lentinula edodes* [26] and *P. ostreatus* [27], but larger than those of *Hericium erinaceus* [14] and brown-rot fungi *Serpula lacrymans* [28]. In addition, the number of CBMs of *G. lingzhi* was significantly less than that of *G. lucidum* and the other species except for *H. erinaceus* (Table S2), indicating different CAZyme activity properties between *G. lingzhi* and *G. lucidum* in plant cell wall degradation.

### 3.5. Mating Gene Loci in G. lingzhi

Mating is the crucial step in the fruiting and sexual reproduction of mushroom-forming fungi and is governed by mating genes that are located at distinct loci called mating type (MAT) loci [29]. *Ganoderma* spp. are tetrapolar species with two distinct mating type loci (MAT-A and MAT-B) [30]. The MAT-A locus contains genes encoding homeodomain transcription factors, including homeodomain type 1 and type 2 mating-type genes (HD1 and HD2 genes). As shown in Figure 3A, these two genes are adjacent to one another in opposite orientations on scaffold 1 and adjacent to mitochondrial intermediate peptidase (mip) and beta flanking gene (*β-fg*), as observed in other basidiomycete species [31,32]. The MAT-B locus genes encode pheromones and pheromone receptors. Eleven potential pheromone receptor genes were identified and clustered on scaffold 9, which is homologous to the STE3 gene (Figure 3B). However, no pheromone precursor genes were found in the proximity ~20 kb flanking region of these pheromone receptors, which is similar to reports in the medicinal mushroom *Hericium erinaceus* [14].

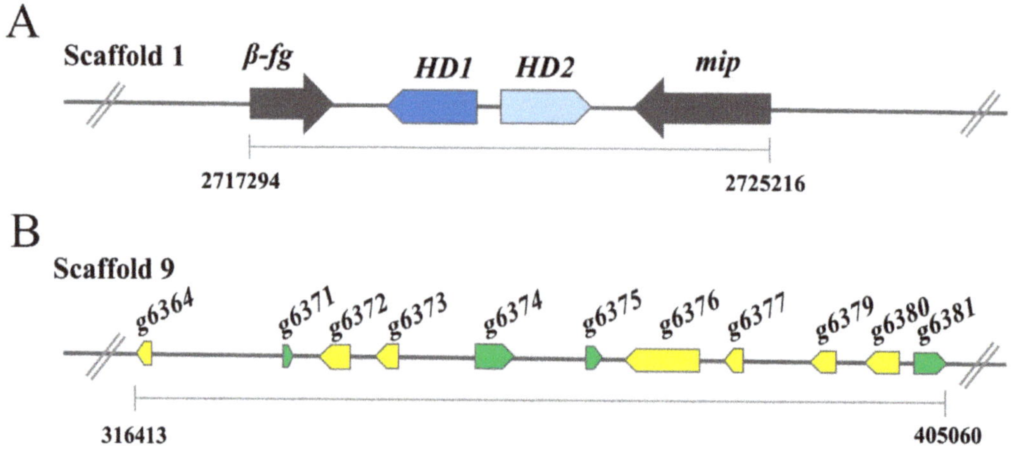


**Figure 3.** Analysis of the *G. lingzhi* MAT-A and MAT-B gene locus. (**A**) The MAT-A includes homeodomain type 1 and type 2 mating-type genes (HD1 and HD2 genes). *mip*: mitochondrial intermediate peptidase, *β-fg*: beta flanking gene. (**B**) The MAT-B locus genes encode eleven potential pheromone receptor genes.

### *3.6. The Pathway of Ganoderic Acid Biosynthesis*

At present, the synthetic pathway of GAs, which is an important medicinal component found in *Ganoderma* spp., is still not well elucidated. GAs are synthesized via the mevalonate (MVA) pathway, which is conserved in all eukaryotes. A total of 13 genes encoding 11 enzymes existed in the terpenoid backbone biosynthesis (map00900) pathway (Table S3). The genome of *G. lingzhi* contained two farnesyl diphosphate synthases and two squalene monooxygenases, which is different from that of *G. lucidum* which contained two acetyl-CoA acetyltransferase genes and two farnesyl diphosphate synthase genes [8]. The catalytic synthesis of lanosterol from acetyl coenzyme A by the MVA pathway was well-studied. The following catalytic synthesis of GAs from lanosterol was performed by the cytochrome P450 superfamily (CYPs), which includes various complex and unclear oxidation, reduction, and acylation modification reactions. As shown in Supplementary Materials S3, 145 CYP coding genes have been annotated, which requires further research to clarify their functions in GA biosynthesis.

### *3.7. Identification of GAs by Mass Spectrometry*

The GAs were further identified by ultrahigh-performance liquid chromatography-electrospray ionization-tandem mass spectrometry (UPLC-ESI-MS/MS, fragmentation ions of MS spectra are shown in Supplementary Materials S4). We tentatively identified and deduced the possible structure of 20 main GA constituents, including 8 GAs of C20, 5 GAs of C34, 4 GAs of C32, and 1 GAs of C24, C33, and C36 each (Figure 4 and Table 2). Seventeen ganoderic acids were identified in *G. lingzhi* including well-known ganoderic acids A, T, Me, and C2. Some less-known ganoderic acids, AP2 and GS-3, were also identified in *G. lingzhi* which were previously found in *Ganoderma applanatum* [33] and *Ganoderma sinense* [34], respectively. In addition, we identified ganodermanontriol, ganoderenic acid C, and a new special GA, ganochlearic acid A ($C_{24}H_{34}O_5$), which is a rearranged hexanorlanostane triterpenoid featuring a γ-lactone ring (A ring) and a five-membered carbon ring (B ring) (Figure 4 compound **20**). The A ring is connected with a B ring through a single bond (C-5−C-10) unambiguously deduced in *Ganoderma cochlear* [35].

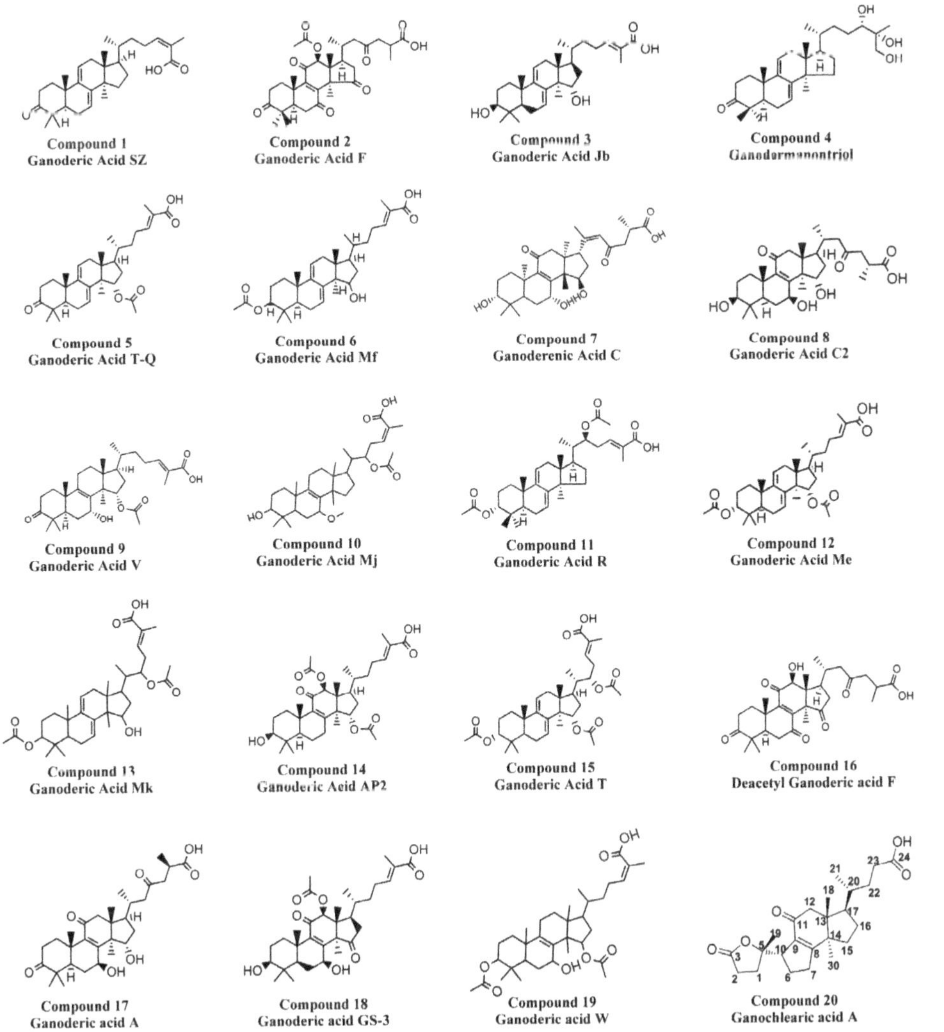


**Figure 4.** Chemical structures of the identified ganoderic acids.

**Table 2.** Identification of possible ganoderic acids from *G. lingzhi*.

| NO. | Compounds | Molecular Wght (Da) | Q1, *m/z* (Da) | Fragmentation, *m/z* (Da) |
|---|---|---|---|---|
| 1 | Ganoderic Acid SZ ($C_{30}H_{44}O_3$) | 452.3 | 453.3 ($[M+H]^+$) | 435.3, 185.1, 187.1, 239.2, 173.1, 201.2, 225.2 |
| 2 | Ganoderic Acid F ($C_{30}H_{46}O_3$) | 454.3 | 455.3 ($[M+H]^+$) | 437.3, 229.2, 299.3, 123.1, 201.2, 135.1 |
| 3 | Ganoderic Acid Jb ($C_{30}H_{46}O_4$) | 470.3 | 471.3 ($[M+H]^+$) | 435.3, 453.3, 201.2, 187.1, 471.3, 159.1, 175.1 |
| 4 | Ganodermanontriol ($C_{30}H_{48}O_4$) | 472.4 | 473.4 ($[M+H]^+$) | 329.2, 455.4, 415.3, 243.2, 261.2, 437.3, 189.1 |
| 5 | Ganoderic Acid T-Q ($C_{32}H_{46}O_5$) | 510.3 | 511.3 ($[M+H]^+$) | 433.3, 493.3, 311.2, 451.3, 293.2, 399.3, 337.2 |

**Table 2.** *Cont.*

| NO. | Compounds | Molecular Wght (Da) | Q1, m/z (Da) | Fragmentation, m/z (Da) |
|---|---|---|---|---|
| 6 | Ganoderic Acid Mf ($C_{32}H_{48}O_5$) | 512.4 | 513.4 ($[M+H]^+$) | 435.3, 495.3, 201.2, 295.2, 203.2, 453.3, 133.1 |
| 7 | Ganoderenic Acid C ($C_{30}H_{44}O_7$) | 516.3 | 517.3 ($[M+H]^+$) | 371.3, 499.3, 399.3, 463.3, 481.3, 353.2, 381.2 |
| 8 | Ganoderic Acid C2 ($C_{30}H_{46}O_7$) | 518.3 | 519.3 ($[M+H]^+$) | 355.3, 483.3, 501.3, 465.3, 447.3, 373.3 |
| 9 | Ganoderic Acid V ($C_{32}H_{48}O_6$) | 528.3 | 529.3 ($[M+H]^+$) | 469.3, 243.2, 423.3, 329.2, 451.3, 369.3, 355.3 |
| 10 | Ganoderic Acid Mj ($C_{33}H_{52}O_6$) | 544.4 | 545.2 ($[M+H]^+$) | 527.3, 449.3, 467.3, 431.3, 353.2, 421.3, 327.2 |
| 11 | Ganoderic Acid R ($C_{34}H_{50}O_6$) | 554.4 | 555.4 ($[M+H]^+$) | 435.3, 495.3, 201.2, 187.1, 145.1, 239.2, 341.2 |
| 12 | Ganoderic Acid Me ($C_{34}H_{50}O_6$) | 554.4 | 555.4 ($[M+H]^+$) | 435.3, 495.3, 295.2, 201.2, 203.2, 187.1, 189.2 |
| 13 | Ganoderic Acid Mk ($C_{34}H_{50}O_7$) | 570.4 | 571.4 ($[M+H]^+$) | 433.3, 451.3, 493.3, 201.2, 511.3, 293.2, 339.2 |
| 14 | Ganoderic Acid AP2 ($C_{34}H_{50}O_8$) | 586.4 | 587.4 ($[M+H]^+$) | 587.4, 569.4, 491.3, 509.3, 431.3, 395.3, 409.3, 463.3 |
| 15 | Ganoderic Acid T ($C_{36}H_{52}O_8$) | 612.4 | 613.4 ($[M+H]^+$) | 433.3, 493.3, 553.4, 201.2, 293.2, 451.3, 227.2 |
| 16 | Deacetyl Ganoderic acid F ($C_{30}H_{40}O_8$) | 528.3 | 527.3 ($[M-H]^-$) | 509.6, 479.4, 465.5, 435.6, 365.8, 315.2, 299.4 |
| 17 | Ganoderic acid A ($C_{30}H_{44}O_7$) | 516.3 | 515.3 ($[M-H]^-$) | 497.5, 453.6, 435.5, 355.3, 337.3, 299.5, 285.5 |
| 18 | Ganoderic acid GS-3 ($C_{32}H_{46}O_8$) | 558.3 | 557.3 ($[M-H]^-$) | 539.6, 497.4, 453.5, 435.6, 303.4, 287.4 |
| 19 | Ganoderic acid W ($C_{34}H_{52}O_7$) | 572.4 | 571.4 ($[M-H]^-$) | 553.6, 523.8, 509.3, 479.3, 465.4, 419.6, 345.5, 303.5, 285.4 |
| 20 | Ganochlearic acid A ($C_{24}H_{34}O_5$) | 402.5 | 401.2 ($[M-H]^-$) | 357.4, 329.4, 313.3, 287.4 |

It has been reported that there are more than 150 kinds of GAs isolated from *Ganoderma*, which are mainly from the fruiting body and spore [36,37]. The GA in the fermentation mycelia of *Ganoderma* is still unclear. Only 19 kinds of GAs were found in the mycelia of *Ganoderma* [4]. In this work, we first identified and deduced the possible structures of 20 main GA constituents by UPLC-ESI-MS/MS from *G. lingzhi mycelia*, including a new special ganochlearic acid A.

### *3.8. RNA-Seq Analysis*

Further, we analyzed transcripts and alternative splicing via RNA sequencing (RNA-Seq). For deep RNA-Seq, we used the Illumina paired-end sequencing strategy to assemble six repeat *G. lingzhi* RNA libraries. The clean reads and mapping ratio against the reference genome of each repeat ranged from 21.11 to 27.41 million and 92.74% to 92.91%, respectively (Figure 5A,B). By comparing with the original annotation of the genome, a total of 3996 new transcripts were found in the RNA-Seq. The annotation results of new transcripts are shown in Supplementary Materials S5, and the novel transcript sequence information has been deposited at SRR17081370.

Then, ASprofile was employed to predict alternative splicing (AS) events in each sample. More than 30,000 AS events were predicted in each sample of six replicates and sorting them into 12 types (details shown in Supplementary Materials S6). As shown in Figure 6A, alternative 5′ first exon (transcription start site, TSS) splicing and alternative 3′ last exon (transcription terminal site, TTS) splicing were the most frequent AS events. Multi-skipped exon (MSKIP), approximate multi-intron retention (XMIR), and approximate multi-exon skipped exon (XMSKIP) were the AS events with less frequency. Venn analysis showed that 9276 genes were predicted to have AS events in six replicates (Figure 6B). The 9276 genes were enriched for functional categories involved in protein processing, endocytosis, and metabolic activities by KEGG. In particular, pyrimidine metabolism ($p$-value = $3.62 \times 10^{-2}$), arginine and proline metabolism ($p$-value = $6.58 \times 10^{-2}$), citrate cycle ($p$-value = $7.065 \times 10^{-2}$), mismatch repair ($p$-value = $1.25 \times 10^{-1}$) were enriched (Figure 6C and Supplementary Materials S7). In addition, 21 and 3 AS event genes were enriched in steroid biosynthesis ($p$-value = $9.90 \times 10^{-1}$) and sesquiterpenoid and triterpenoid biosynthesis ($p$-value = $7.54 \times 10^{-1}$) pathways, respectively (Supplementary Materials S7), which provided a reference for further study on the regulation mechanism of AS involved in GAs biosynthesis.

**A**

**RNA-Seq data statistics**

| Samples | Clean reads | Clean bases | GC Content | %≥Q30 | Mapping ratio |
|---|---|---|---|---|---|
| Repeat 1 | 22,015,584 | 6,581,536,918 | 59.43% | 94.61% | 92.74% |
| Repeat 2 | 21,109,033 | 6,312,542,820 | 59.42% | 94.62% | 92.79% |
| Repeat 3 | 22,242,838 | 6,653,646,214 | 59.50% | 94.58% | 92.76% |
| Repeat 4 | 27,405,168 | 8,200,635,556 | 59.71% | 94.44% | 92.78% |
| Repeat 5 | 21,853,351 | 6,536,384,384 | 59.68% | 94.67% | 92.91% |
| Repeat 6 | 24,642,710 | 7,375,210,188 | 59.59% | 94.46% | 92.77% |

**B**

**Figure 5.** Summary of transcriptome sequencing data. (**A**) The statistics of RNA sequencing data. ≥Q30%: Percentage of bases with a Q-score no less than Q30. (**B**) Distribution of mapped reads on reference genome: position and depth. *X*-axis: Position on the chromosome; *Y*-axis: Log2 of coverage depth (coverage depth was defined as reads counted within a chromosome window of 10 kb in length); Blue represents + strand and green represents − strand.

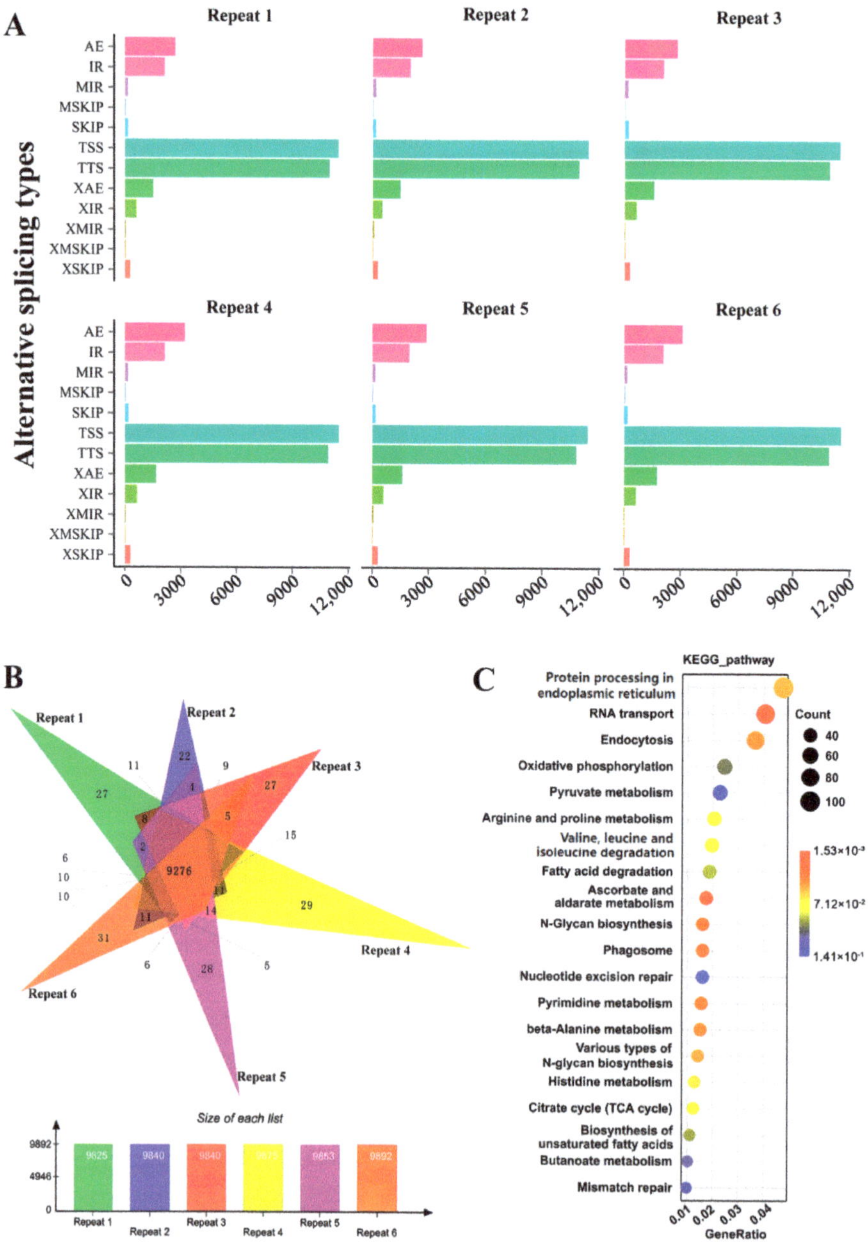


**Figure 6.** Alternative splicing analysis. (**A**) Statistics of alternative splicing events. *X*-axis: Number of transcripts in specific alternative splicing type; *Y*-axis: 12 alternative splicing types. TSS: alternative 5′ first exon (transcription start site, TSS) splicing; TTS: alternative 3′ last exon (transcription terminal site, TTS) splicing; SKIP: single exon skipping; XSKIP: approximate SKIP; MSKIP: multi-SKIP; XMSKIP: approximate MSKIP; IR: single intron retention; XIR: approximate IR; MIR: multi-IR; XMIR: approximate MIR; AE: alternative exon ends; XAE: approximate AE. (**B**) Venn analysis the alternative splicing genes in six repeat RNA-Seq. (**C**) KEGG analysis of the alternative splicing genes.

## 4. Conclusions

In summary, we characterized the *G. lingzhi* genome with a length of 49.15-Mb, consisting of 30 scaffolds, encoding 13,125 predicted genes, which were sequenced with the combination of the Illumina HiSeq X-Ten and PacBio RSII strategy. We identified 519 CAZymes and analyzed the difference in CAZymes genes of *G. lingzhi* with those of other fungi. In addition, the mating gene loci and ganoderic acid biosynthesis pathway were characterized. Then, 20 ganoderic acids were first identified from the mycelia of *G. lingzhi* by mass spectrometry, including a new special ganochlearic acid A. Furthermore, 3996 novel transcripts were discovered and 9276 genes were predicted to have the possibility of alternative splicing from RNA-Seq data. Collectively, this study develops foundational genomic, transcriptomic, and ganoderic acid bioactive compound resources that can be used in further breeding, pharmacological research, and molecular genetic analyses in *G. lingzhi*.

**Supplementary Materials:** The following supporting information can be downloaded at: https://www.mdpi.com/article/10.3390/jof8121257/s1. Supplementary Materials S1: Gene annotation of *G. lingzhi*; Supplementary Materials S2: CAZy; Supplementary Materials S3: p450; Supplementary Materials S4: Fragmentation ions of MS spectra; Supplementary Materials S5: New gene annotation; Supplementary Materials S6: Alternative splicing; Supplementary Materials S7: AS genes KEGG enricht; Table S1: General characteristics of the *G. lingzhi* genome; Table S2: Comparison of CAZyme gene numbers of *G. lingzhi* with those of other fungi; Table S3: Genes involved in terpenoid backbone in *G. lingzhi*.

**Author Contributions:** Conceptualization, Y.L.; methodology, Q.W. and H.L.; software, F.X. and J.H.; Y.S. and W.L. collected the samples; writing—original draft preparation, Q.W. and H.L.; writing—review and editing, Y.L. and G.L.; supervision, Y.L. and G.L.; project administration, Y.L. and G.L.; funding acquisition, Y.L. and G.L. All authors have read and agreed to the published version of the manuscript.

**Funding:** This work was supported by grants from the National Natural Science Foundation of China (31900027 to Y.L., 31772374 and 32071673 to G.L.), the China Postdoctoral Science Foundation (2020M682601 to Y.L.), the Science and Technology Innovation Program of Hunan Province (2020RC2059 to Y.L. and 2021RC4063 to G.L.), the Natural Science Foundation of Hunan Province (2020JJ5972 to Y.L. and 2021JJ31151 to G.L.), and the Scientific Research Fund of Hunan Provincial Education Department, China (NO. 18B167 to Y.L.).

**Institutional Review Board Statement:** This study did not involve humans or animals.

**Informed Consent Statement:** Not applicable.

**Data Availability Statement:** The assembly genome in this paper is associated with NCBI BioProject: PRJNA738334. This Whole Genome Shotgun project has been deposited at DDBJ/ENA/GenBank under the accession JAPJYM000000000. The version described in this paper is version JAPJYM010000000. Raw sequencing data for PacBio RSII reads and Illumina HiSeq X-Ten reads of the genome have been deposited in the NCBI Sequence Read Archive under accession no. SRR22226938 and SRR22226939, respectively. Six repeat RNA raw sequencing data have been deposited in NCBI's Sequence Read Archive under accession no. SRR22226940-45. The genome sequencing information and transcripts sequencing information can be downloaded from SRR14933280 and SRR17081370.

**Conflicts of Interest:** The authors declare no conflict of interest.

## References

1. Kladar, N.V.; Gavarić, N.S.; Božin, B.N. Ganoderma: Insights into anticancer effects. *Eur. J. Cancer Prev.* **2016**, *25*, 462–471. [CrossRef] [PubMed]
2. Cao, Y.; Wu, S.H.; Dai, Y.C. Species clarification of the prize medicinal *Ganoderma* mushroom "Lingzhi". *Fungal Divers.* **2012**, *56*, 49–62. [CrossRef]
3. Ren, A.; Shi, L.; Zhu, J.; Yu, H.S.; Jiang, A.L.; Zheng, H.H.; Zhao, M.W. Shedding light on the mechanisms underlying the environmental regulation of secondary metabolite ganoderic acid in *Ganoderma lucidum* using physiological and genetic methods. *Fungal Genet. Biol.* **2019**, *128*, 43–48. [CrossRef] [PubMed]

4. Liang, C.Y.; Tian, D.N.; Liu, Y.Z.; Li, H.; Zhu, J.L.; Li, M.; Xin, M.H.; Xia, J. Review of the molecular mechanisms of *Ganoderma lucidum* triterpenoids: Ganoderic acids A, C2, D, F, DM, X and Y. *Eur. J. Med. Chem.* **2019**, *174*, 130–141. [CrossRef] [PubMed]
5. Huang, Y.H.; Wu, H.Y.; Wu, K.M.; Liu, T.T.; Liou, R.F.; Tsai, S.F.; Shiao, M.S.; Ho, L.T.; Tzean, S.S.; Yang, U.C. Generation and analysis of the expressed sequence tags from the mycelium of *Ganoderma lucidum*. *PLoS ONE* **2013**, *8*, e61127. [CrossRef]
6. Liu, D.B.; Gong, J.; Dai, W.K.; Kang, X.C.; Huang, Z.; Zhang, H.M.; Liu, W.; Liu, L.; Ma, J.P.; Xia, Z.L.; et al. The Genome of *Ganderma lucidum* provide insights into triterpense biosynthesis and wood degradation. *PLoS ONE* **2012**, *7*, e36146. [CrossRef]
7. Tian, Y.Z.; Wang, Z.F.; Liu, Y.D.; Zhang, G.Z.; Li, G. The whole-genome sequencing and analysis of a *Ganoderma lucidum* strain provide insights into the genetic basis of its high triterpene content. *Genomics* **2021**, *113*, 840–849. [CrossRef]
8. Chen, S.L.; Xu, J.; Liu, C.; Zhu, Y.J.; Nelson, D.R.; Zhou, S.G.; Li, C.F.; Wang, L.Z.; Guo, X.; Sun, Y.Z.; et al. Genome sequence of the model medicinal mushroom *Ganoderma lucidum*. *Nat Commun.* **2012**, *3*, 913. [CrossRef]
9. Liu, Y.-N.; Wu, F.-Y.; Tian, R.-Y.; Shi, Y.-X.; Xu, Z.-Q.; Liu, J.-Y.; Huang, J.; Xue, F.-F.; Liu, G.-Q. The Regulatory and Transcriptional Landscape Associated with Triterpenoid and Lipid Metabolisms by the bHLH-Zip Transcription Factor SREBP in the Medicinal Fungus *Ganoderma lingzhi*. Available online: https://doi.org/10.21203/rs.3.rs-1244597/v1 (accessed on 5 August 2022). [CrossRef]
10. Liu, Y.N.; Tong, T.; Zhang, R.R.; Liu, L.M.; Shi, M.L.; Ma, Y.C.; Liu, G.Q. Interdependent nitric oxide and hydrogen peroxide independently regulate the coix seed oil-induced triterpene acid accumulation in *Ganoderma lingzhi*. *Mycologia* **2019**, *111*, 529–540. [CrossRef]
11. Chin, C.S.; Peluso, P.; Sedlazeck, F.J.; Nattestad, M.; Concepcion, G.T.; Clum, A.; Dunn, C.; O'Malley, R.; Figueroa-Balderas, R.; Morales-Cruz, A.; et al. Phased diploid genome assembly with single-molecule real-time sequencing. *Nat. Methods* **2016**, *13*, 1050–1054. [CrossRef] [PubMed]
12. Walker, B.J.; Abeel, T.; Shea, T.; Priest, M.; Earl, A.M. Pilon: An integrated tool for comprehensive microbial variant detection and genome assembly improvement. *PLoS ONE* **2014**, *9*, e112963. [CrossRef] [PubMed]
13. Simao, F.A.; Waterhouse, R.M.; Ioannidis, P.; Kriventseva, E.V.; Zdobnov, E.M. BUSCO: Assessing genome assembly and annotation completeness with single-copy orthologs. *Bioinformatics* **2015**, *31*, 3210–3212. [CrossRef] [PubMed]
14. Gong, W.B.; Wang, Y.H.; Xie, C.L.; Zhou, Y.J.; Zhu, Z.H.; Peng, Y.D. Whole genome sequence of an edible and medicinal mushroom, *Hericium erinaceus* (Basidiomycota, Fungi). *Genomics* **2020**, *112*, 2393–2399. [CrossRef]
15. Han, Z.; Tanner, Y.; Le, H.; Sarah, E.; Peizhi, W.; Zhenglu, Y.; Busk, P.K.; Ying, X.; Yanbin, Y. dbCAN2: A meta server for automated carbohydrate-active enzyme annotation. *Nucleic Acids Res.* **2018**, W95–W101. [CrossRef]
16. Kim, D.; Pertea, G.; Trapnell, C.; Pimentel, H.; Kelley, R. TopHat2: Accurate alignment of transcriptomes in the presence of insertions, deletions and gene fusions. *Genome Biol.* **2013**, *14*, R36. [CrossRef]
17. Pertea, M.; Pertea, G.M.; Antonescu, C.M.; Chang, T.C.; Mendell, J.T.; Salzberg, S.L. StringTie enables improved reconstruction of a transcriptome from RNA-Seq reads. *Nat. Biotechnol.* **2015**, *33*, 290. [CrossRef]
18. Buchfink, B.; Xie, C.; Huson, D.H. Fast and sensitive protein alignment using DIAMOND. *Nat. Methods* **2015**, *12*, 59–60. [CrossRef]
19. Deng, Y.; Jianqi, L.I.; Songfeng, W.U.; Zhu, Y.; Chen, Y.; Fuchu, H.E. Integrated nr database in protein annotation system and its localization. *Comput. Eng.* **2006**, *32*, 71–72.
20. Bateman, A.; Martin, M.J.; Orchard, S.; Magrane, M.; Agivetova, R.; Ahmad, S.; Alpi, E.; Bowler-Barnett, E.H.; Britto, R.; Bursteinas, B.; et al. UniProt: The universal protein knowledgebase in 2021. *Nucleic Acids Res* **2021**, *49*, D480–D489. [CrossRef]
21. Consortium, T. Gene ontology: Tool for the unification of biology. *Nat. Genet.* **2000**, *25*, 25–29. [CrossRef]
22. Tatusov, R.L.; Fedorova, N.D.; Jackson, J.D.; Jacobs, A.R.; Kiryutin, B.; Koonin, E.V.; Krylov, D.M.; Mazumder, R.; Mekhedov, S.L.; Nikolskaya, A.N.; et al. The COG database: An updated version includes eukaryotes. *BMC Bioinform.* **2003**, *4*, 41. [CrossRef] [PubMed]
23. Minoru, K.; Susumu, G.; Shuichi, K.; Yasushi, O.; Masahiro, H. The KEGG resource for deciphering the genome. *Nucleic Acids Res* **2004**, *32*, D277–D280. [CrossRef]
24. Kim, D.; Landmead, B.; Salzberg, S.L. HISAT: A fast spliced aligner with low memory requirements. *Nat. Methods* **2015**, *12*, 357–360. [CrossRef]
25. Cingolani, P. A program for annotating and predicting the effects of single nucleotide polymorphisms, SnpEff: SNPs in the genome of *Drosophila melanogaster* strain w1118; iso-2; iso-3. *Fly* **2012**, *6*, 80–92. [CrossRef]
26. Chen, L.F.; Gong, Y.H.; Cai, Y.L.; Liu, W.; Zhou, Y.; Xiao, Y.; Xu, Z.Y.; Liu, Y.; Lei, X.Y.; Wang, G.Z.; et al. Genome sequence of the edible cultivated mushroom *Lentinula edodes* (Shiitake) reveals insights into lignocellulose degradation. *PLoS ONE* **2016**, *11*, e0160336. [CrossRef]
27. Qu, J.B.; Huang, C.Y.; Zhang, J.X. Genome-wide functional analysis of SSR for an edible mushroom *Pleurotus Ostreatus*. *Gene* **2016**, *575*, 524–530. [CrossRef]
28. Eastwood, D.C.; Floudas, D.; Binder, M.; Majcherczyk, A.; Schneider, P.; Aerts, A.; Asiegbu, F.O.; Baker, S.E.; Barry, K.; Bendiksby, M.; et al. The plant cell wall-decomposing machinery underlies the functional diversity of forest fungi. *Science* **2011**, *333*, 762–765. [CrossRef]
29. Brown, A.J.; Casselton, L.A. Mating in mushrooms: Increasing the chances but prolonging the affair. *Trends Genet.* **2001**, *17*, 393–400. [CrossRef]
30. Gilbertson, A. Cultural studies and genetics of sexuality of *Ganoderma lucidum* and *G. tsugae* in relation to the taxonomy of the *G. lucidum* complex. *Mycologia* **1986**, *78*, 694–705. [CrossRef]

31. Casselton, L.A.; Olesnicky, N.S. Molecular genetics of mating recognition in basidiomycete fungi. *Microbiol. Mol. Biol. Rev.* **1998**, *62*, 55–70. [CrossRef] [PubMed]
32. James, T.Y.; Kues, U.; Rehner, S.A.; Vilgalys, R. Evolution of the gene encoding mitochondrial intermediate peptidase and its cosegregation with the A mating-type locus of mushroom fungi. *Fungal Genet. Biol.* **2004**, *41*, 381–390. [CrossRef] [PubMed]
33. Wang, F.; Liu, J.K. Highly oxygenated lanostane triterpenoids from the fungus *Ganoderma applanatum*. *Chem. Pharm. Bull.* **2008**, *56*, 1035–1037. [CrossRef]
34. Liu, J.Q.; Wang, C.F.; Li, Y.; Luo, H.R.; Qiu, M.H. Isolation and bioactivity evaluation of terpenoids from the medicinal fungus *Ganoderma sinense*. *Planta Med.* **2012**, *78*, 368–376. [CrossRef] [PubMed]
35. Peng, X.R.; Wang, X.; Zhou, L.; Hou, B.; Zuo, Z.L.; Qiu, M.H. Ganocochlearic acid A, a rearranged hexanorlanostane triterpenoid, and cytotoxic triterpenoids from the fruiting bodies of *Ganoderma cochlear*. *RSC Adv.* **2015**, *5*, 95212–95222. [CrossRef]
36. Zhao, X.R.; Zhang, B.J.; Deng, S.; Zhang, H.L.; Huang, S.S.; Huo, X.K.; Wang, C.; Liu, F.; Ma, X.C. Isolation and identification of oxygenated lanostane-type triterpenoids from the fungus *Ganoderma lucidum*. *Phytochem. Lett.* **2016**, *16*, 87–91. [CrossRef]
37. Peng, X.R.; Liu, J.Q.; Han, Z.H.; Yuan, X.X.; Luo, H.R.; Qiu, M.H. Protective effects of triterpenoids from *Ganoderma resinaceum* on $H_2O_2$-induced toxicity in HepG2 cells. *Food Chem.* **2013**, *141*, 920–926. [CrossRef]

Journal of *Fungi*

MDPI

*Article*

# The Effect of Mitochondria on *Ganoderma lucidum* Growth and Bioactive Components Based on Transcriptomics

**Liyun Ye [1], Xiaofang He [1], Congbao Su [1], Haiying Feng [1], Guoliang Meng [1], Bingzhi Chen [2,3] and Xiaoping Wu [1,*]**

[1] College of Life Sciences, Fujian Agriculture and Forestry University, Fuzhou 350002, China
[2] College of Food Sciences, Fujian Agriculture and Forestry University, Fuzhou 350002, China
[3] Key Laboratory of Subtropical Characteristic Fruits, Vegetables and Edible Fungi Processing (Co-Construction by Ministry and Province), Ministry of Agriculture and Rural Affairs, Fuzhou 350002, China
* Correspondence: fjwxp@126.com

**Abstract:** Mitochondria are the power source of living cells and implicated in the oxidative metabolism. However, the effect of mitochondria on breeding is usually ignored in conventional research. In this study, the effect of mitochondria on *Ganoderma lucidum* morphology, yield, and main primary bioactive components was analyzed via structuring and comparing isonuclear alloplasmic strains. The crucial biological pathways were then explored based on the transcriptome. The results showed that isonuclear alloplasmic exhibited difference in mycelial growth rate in potato dextrose agar medium (PDA), basidiospore yield, and polysaccharide and triterpenoid content. Otherwise, mitochondria did not change colony and fruit body morphology, mushroom yield, or mycelial growth rate in solid-state fermentation cultivation material. The transcriptome data of two significant isonuclear alloplasmic strains S1 and S5 revealed that the involvement of differentially expressed genes (DEGs) was mainly in pentose and glucuronate interconversions, starch and sucrose metabolism, and steroid biosynthesis. The result was further confirmed by the other isonuclear alloplasmic strains. The above results further proved that mitochondria could affect the active components of *G. lucidum*. Our results provide information which will contribute to understanding of mitochondria and will be helpful for breeding improved varieties.

**Keywords:** isonuclear alloplasmic; yield; polysaccharide; triterpenoid; transcriptome



**Citation:** Ye, L.; He, X.; Su, C.; Feng, H.; Meng, G.; Chen, B.; Wu, X. The Effect of Mitochondria on *Ganoderma lucidum* Growth and Bioactive Components Based on Transcriptomics. *J. Fungi* **2022**, *8*, 1182. https://doi.org/10.3390/jof8111182

Academic Editors: Mingwen Zhao, Gen Zou and Jing Zhu

Received: 27 September 2022
Accepted: 30 October 2022
Published: 9 November 2022



## 1. Introduction

Mitochondria are organelles of eukaryotic cells that provide the platform for efficient energy metabolism and participate in the Krebs cycle to form ATP. Moreover, many studies showed that mitochondria are related to apoptosis, senescence, virulence, and drug resistance [1–3]. Elucidating the structure and function of the genome is important to understand species. Fungi mitochondria are semi-autonomous organelles and carry their own mitochondrial genomes (mtDNAs), most of which are closed circular superhelical DNA molecules. Different fungi mtDNAs usually present variability in size due to intergenic spacers, duplications, proliferation of repeats, and insertions of plasmid components or other elements [4–6]. With the continuous research and the development of new techniques, many edible fungi mtDNAs have been analyzed and annotated, such as *Agaricus bisporus*, *Agrocybe aegerita*, *Grifola frondose*, and *Sparassis crispa* [7–10]. In term of *G. lucidum*, Li et al. [11] assembled and analyzed the mtDNA of *G. lucidum* and clarified that the protein-coding genes were expressed higher in mycelia or primordial stages compared with those in the fruiting bodies. Li et al. [12] compared features, evolution, and phylogeny of five *Ganoderma* species mtDNAs. These results laid a foundation for the further study of mitochondrial function of *G. lucidum*.

Although most of the genetic information is stored in the nucleus, mitochondria also contain their own mtDNAs and mechanisms of replication and transcription. Mitochondria will not only directly affect the changes of organisms, but interact with nuclear genes to play a regulatory role. Presently, several studies clarified that mitochondrial and nuclear interactions, nuclear mitochondrial compatibility, and co-adaptation played an important role in fungal evolution and adaptation [1,13]. Moon et al. [14] found that the change of mitochondrial genetic background in mice would lead to significant changes in metabolites, obesity, and nuclear gene expression. Latorre-Pellicer et al. [15] study indicated that mtDNA influences mitochondrial proteostasis and reactive oxygen species (ROS) generation, insulin signaling, obesity, and ageing parameters, resulting in profound differences in health longevity.

*G. lucidum* is a medicinal fungus belonging to the family Polyporaceae in Basidiomycete. Modern pharmacologic studies have revealed that *G. lucidum* contains a variety of abundant pharmacological and biological active substances, such as polysaccharides, nucleosides, triterpenoids, peptides, sterols, protein, and alkaloids [16–18]. Therein, triterpenoids and polysaccharides are considered as the primary bioactive components that contribute to its medicinal properties. *G. lucidum* attracted various research interests due to its auxiliary function on anti-inflammatory, antitumor (both in vitro and in vivo), hepatoprotection, and some other activities [19–23]. Therefore, improvement of bioactive component production efficiency in *G. lucidum* has become a research focus in recent years.

Isonuclear alloplasmic, first named from isonuclear-alloplasmic male sterile lines in plant, indicated strains with the same nucleus and different mitochondrial genomes. The objective of the study reported here was to explore the effect of mitochondria on *G. lucidum* morphology, yield, and main primary bioactive components via comparing isonuclear alloplasmic strains. More than that, the mechanism of difference among test strains was analyzed based on the transcriptome. This information will be helpful for the further understanding of mitochondria function and provide a reference for the screening of excellent strains.

## 2. Materials and Methods

### 2.1. Strains

*G. lucidum* monokaryon strain M119-M71, M119-M54, MWBL, MTL, MSL4, and MSL10 were obtained from dikaryon strains 119, WBL, TL, SL4, and SL10 via protoplast method. All strains are currently preserved at the Mycological Research Center of Fujian Agriculture and Forestry University, Fujian, China. The mtDNAs of the *G. lingzhi* strains were submitted to GenBank and the accession numbers are shown as follows: M119-M71/M119-54: MT843216; MWBL: MT765268; MTL: MT843209; MSL4: MT843211; and MSL10: MT843212.

### 2.2. Isonuclear Alloplasmic Strains Obtention

As shown in Figure 1, monokaryon strain M119-M71 and MSL10 were crossed on the plate. On the near MSL10 side, a dikaryon strain with M71-MSL10 nuclei and MSL10 mitochondria was obtained [24]. Next, protoplast method was used to produce a monokaryon strain with M119 or MSL10 nucleus and MSL10 mitochondria. The newly obtained monokaryon strain was backcrossed with M119-M71 and the type of nucleus was preliminary judged by clamp connection. Picked-up strains without clamp connection as the target monokaryon strain, for which nucleus was M119-M71-type and mitochondria was MSL10-type, were crossed with monokaryon strain M119-M54 to obtain dikaryon strain S1 and S5. S1 and S5 contained the same type nucleus but different mitochondria, M71 and MSL10, separately. Similarly, monokaryon strain M119-M71 was crossed with monokaryon strain MTL, MWBL, and MSL4 to obtain the target monokaryon strain with M119-M71-type nucleus and different mitochondria. Dikaryon strains S2 (MTL-type mitochondria), S3 (MWBL-type mitochondria), and S4 (MSL4-type mitochondria) were obtained by crossing target monokaryon strain with strain M119-M54. Strain S1, S2, S3, S4, and S5 were tested as isonuclear alloplasmic strains, which needed to be verified.

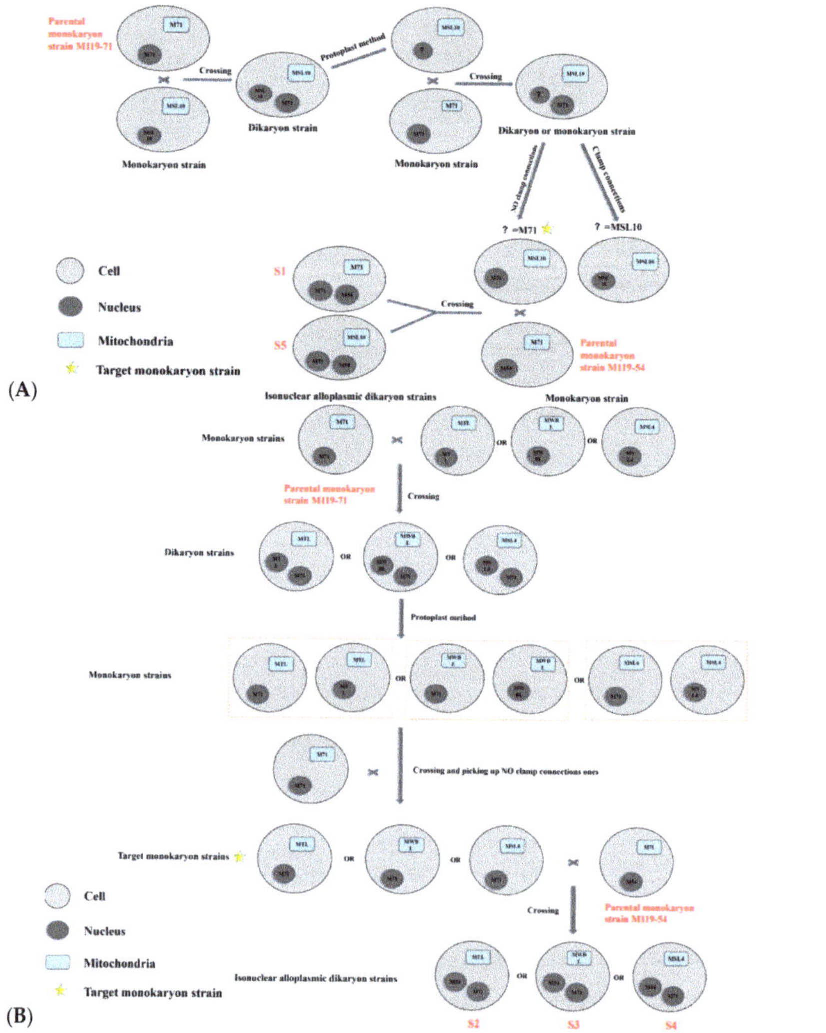


**Figure 1.** The process of constructing isonuclear alloplasmic. (**A**) The constructed strains S1 and S5. (**B**) The constructed strains S2, S3, and S4.

### 2.3. Isonuclear Alloplasmic Strains' Nuclei DNA and mtDNA Proof

The whole genome of strains S1–S5 were sequenced using a HiSeq 2500 Illumina sequencer at Novogene Bioinformatics Institute (Shanghai, China). Obtained clean data were assembled into contigs by software SPAdes-3.7.1 with a Kmer 127. Relative mitochondrial proteins NCBI website was downloaded and mitochondria-related contigs were picked out from the assemblies by program Blastx. Mitochondria-related contigs were linked and then assembled into an intact circular molecule using the pair-end relationship of reads [24]. The mtDNAs of strains S1–S5 were compared using the Clustal W program (http://www.genome.jp/tools/clustalw/, accessed on: 16 October 2021). Non-identical regions in the mtDNAs sequences were identified and four pairs of primers were designed to amplify the mtDNAs of *G. lucidum* strains S1–S5 using online tool Primer-BLAST (https://www.ncbi.nlm.nih.gov/tools/primer-blast/, accessed on: 16 October

2021) (Table S1). Similarly, primers of different nuclear specific fragments were designed among those *G. lucidum* strains (Table S1).

*2.4. Biological Characteristics Observation*

Mycelia were grown routinely at 25 °C on potato dextrose agar medium (PDA). Mycelial growth rate was determined by the ratio of mycelial extension distance to time. The cultivation material used for solid-state fermentation was composed of 39% sawdust, 39% cottonseed shell, 17% bran, 3% corn flour, 1% gypsum, and 1% light calcium carbonate, with 65% moisture content. The mixed substrate (1000 g) was transferred into a polypropylene bag (18 cm in width × 36 cm in length). Isonuclear alloplasmic strains' fruit body and basidiospore yield were determined.

*2.5. Total Polysaccharide and Triterpenoid Detection*

The dried mycelia were ground into a fine powder. Total polysaccharide and triterpenoid extraction and detection was performed according to existing reports [25,26]. Polysaccharide was extracted by water and detected by the phenol-sulfuric acid method. The extraction method of triterpenoid was alcohol extraction and the detection method was vanillin–glacial acetic acid.

*2.6. Total RNA Isolation and Transcriptome Analysis*

Total RNA was extracted using a Total RNA Extraction Kit (Omega Bio-Tek, Guangzhou, China) according to the manufacturer's instructions, and then the concentration and quality of RNA were detected. High-quality RNA was sent to Novogene (Beijing, China) for RNA-seq libraries preparation and sequenced using an Illumina HiSeq. Low-quality sequences and contamination were removed using FASTX toolkit (http://hannonlab.cshl.edu/fastx_toolkit, accessed on: 10 February 2022) to obtain clean reads, followed by mapping using HISAT2 [27]. The gene expression levels for each sample were expressed according to the fragments per kilobase per million (FPKM). $|\log_2(\text{Fold Change})| > 1$ and q value < 0.05 were used as criteria of significant difference to identify differentially expressed genes (DEGs) [28]. For DEGs, GOseq software was used for gene ontology (GO) enrichment analysis based on Wallenius noncentral hypergeometric distribution [29], and Kyoto Encyclopedia of Genes and Genomes (KEGG) enrichment analysis was performed online according to KEGG database (https://www.kegg.jp/, accessed on: 18 May 2022).

*2.7. Statistical Analysis*

All data are expressed as the mean ± standard deviation of three independent experiments. Data were subjected to analysis of variance using the Student's *t*-test, and the mean values indicating statistical significance were compared by Duncan's multiple-range test using SPSS 20.0 (SPSS Inc., Chicago, IL, USA). $p < 0.05$ denoted statistical significance.

## 3. Results

*3.1. Isonuclear Alloplasmic Proof*

According to different fragments between nuclear genomes, five pairs of primers were designed and the amplificated results showed that all strains only present the same bright band by the primer Nc-119, which was designed to distinguish M119 nuclear genome (Figure 2A). This means that all tested isonuclear alloplasmic strains contained the same style nuclear genomes M119. The amplification results of the mtDNAs of the constructed strains are shown in Figure 2B. Strain S1 amplified four different bands as expected. It indicated that S1 mtDNA was M119-style. Strain S2 only had a band by primer designed for MTL style, explaining that S2 mtDNA was MTL-style. Similarly, it was confirmed that strains S3, S4, and S5 contained MWBL, MSL4, and MSL10 mtDNA style, separately. This further concluded that strains S1–S5 were isonuclear alloplasmic strains containing the same style nucleus but different mitochondrion.

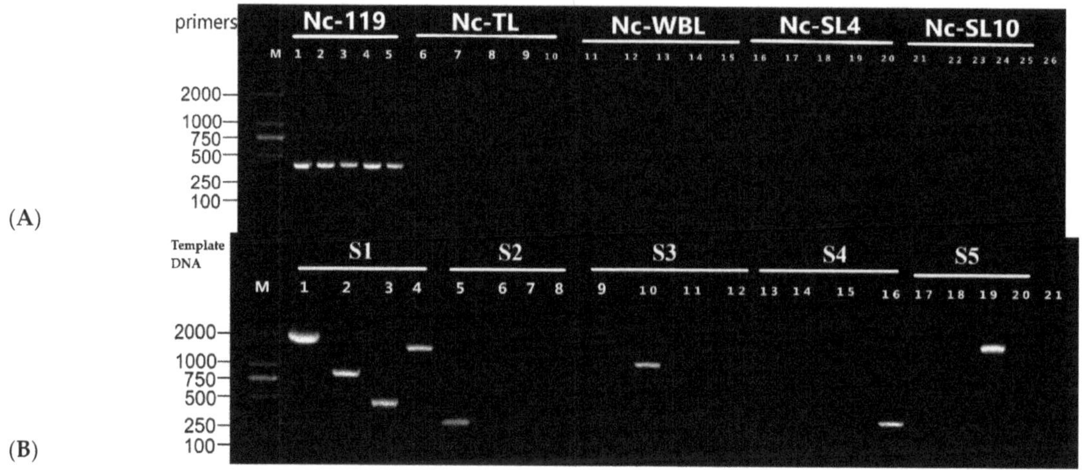


**Figure 2.** The PCR amplification of isonuclear alloplasmic strain S1–S5 nuclei and mitochondrion DNA. (**A**) The PCR amplification of strain S1–S5 nuclear genome. M indicates the 2000 bp DNA marker; the primers of lanes 1–5, 6–10, 11–15, 16–20, and 21–25 were Nc-119, Nc-TL, Nc-WBL, Nc-SL4, and Nc-SL10, separately; the template DNA of lane 1, 6, 11, 16, and 21 was S1; the template DNA of lane 2, 7, 12, 17, and 22 was S2; the template DNA of lane 3, 8, 13, 18, and 23 was S3; the template DNA of lane 4, 9, 14, 19, and 24 was S4; the template DNA of lane 5, 10, 15, 20, and 25 was S5; lane 26 is negative control with double-distilled water instead of DNA. (**B**) The PCR amplification of strain S1–S5 mtDNA. M indicates the 2000 bp DNA marker; the template DNA of lane 1–4, 5–8, 9–12, 13–16, and 17–20 were S1–S5, separately; the primers of 1, 5, 9, 13, and 17 were primer 6; the primers of 2, 6, 10, 14, and 18 were primer 7; the primers of 3, 7, 11, 15, and 19 were primer 8; the primers of 4, 8, 12, 16, and 20 were primer 9; lane 21 is negative control with double-distilled water instead of DNA.

### *3.2. Mycelia and Fruit Body Morphology Observation*

The colony morphology and growth rate of five isonuclear alloplasmic strains were studied on PDA culture medium. The morphology of colonies was similar, with dense hyphae in the center and regular edges (Figure 3A). Conversely, the tested strains showed different mycelial growth rate. The fastest growth rate strain S5 was 4.04 mm/day, which was 1.81 times that of the slowest strain S2. Furthermore, the growth rate among the five strains was significantly different at the level of 0.05 (Figure 3A,B). Interestingly, their growth in solid-state fermentation cultivation material is similar, about $3.095 \pm 0.125$ mm/day (Figure 3B). The fruiting bodies of the five strains were difficult to distinguish. They all had umbrella shape and light reddish-brown pileus with a few rings (Figure 3A). It is concluded that different mitochondria affected the mycelial growth rate in PDA but did not affect the mycelial growth rate in solid-state fermentation cultivation material, colony, and fruit body morphology.

### *3.3. Comparative Analysis of Yield and Polysaccharides and Triterpenoid Content*

The fruit body yield, basidiospore yield, and polysaccharides and triterpenoid content of isonuclear alloplasmic strains S1–S5 were measured and compared. As shown in Table 1, there was no statistical difference ($p > 0.05$) in mushroom yield among five tested strains. However, isonuclear alloplasmic strains showed significant difference in basidiospore yield ($p < 0.05$). The basidiospore yield of strain S1 was about 2.54 times higher than that in strain S5 ($5.55 \pm 0.99$ g/bag). In addition, the content of polysaccharides was different among these tested strains. Polysaccharide content of strain S1 was highest, which was 20.17% higher than that in strain S3 ($25.93 \pm 0.85$ mg/g). In contrast, the content of triterpenoid of strain S1 was significantly lower than that in the other four strains ($p < 0.05$). Hence, one can see that different mitochondria did not affect mushroom yield, but they affect the basidiospore yield and polysaccharide and triterpenoid content.

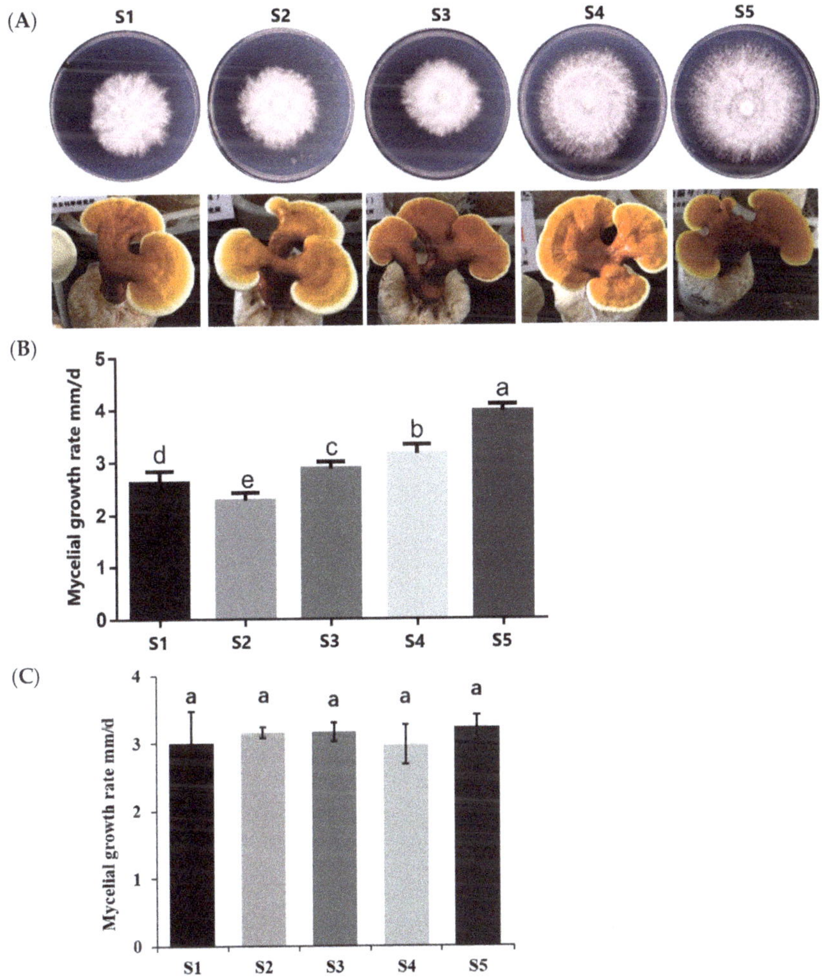


**Figure 3.** Biological characteristics of five isonuclear alloplasmic strains. (**A**) Colony and fruit body morphology. (**B**) Mycelial growth rate in PDA. Different lowercase a, b, c, and d indicate the significant difference among strains at the significance level of 0.05 ($p < 0.05$). (**C**) Mycelial growth rate in solid-state fermentation. The same lowercase a indicates no significant difference among strains at the significance level of 0.05 ($p > 0.05$).

### 3.4. Assessment and Analysis of Transcriptome

Compared to others strains, strains S1 and S5 exhibited significant difference in mycelial growth rate in PDA, basidiospore yield, and polysaccharide and triterpenoid content. In order to further understand the effects of different mitochondria on the growth, the complete transcriptome of *G. lucidum* isonuclear alloplasmic strain S1 and S5 with the same nuclear DNA and different mtDNAs was studied to explore the main mechanism of differences forming. Pearson's correlation analysis between samples was performed and the result showed that $R^2 > 0.941$, indicating that the biological replicates were reliable and transcriptome assembly was robust and available (Figure 4A, Table S2). The transcriptome analysis revealed a total of 1067 DEGs between strain S1 and S5. Compared to strain S1, 348 genes were upregulated and 719 genes were downregulated in strain S5 (Figure 4B).

**Table 1.** Mushroom yield, basidiospore yield, and polysaccharide and triterpenoid content in isonuclear alloplasmic strains S1–S5.

| Strains | Mushroom Yield g/bag | Basidiospores Yield g/bag | Polysaccharide Content (mg/g) | Triterpenoid Content (mg/g) |
|---|---|---|---|---|
| S1 | 70.67 ± 3.42 [a] | 19.67 ± 1.34 [a] | 31.16 ± 0.97 [a] | 19.34 ± 0.38 [d] |
| S2 | 70.67 ± 5.41 [a] | 15.94 ± 1.86 [b] | 29.19 ± 1.57 [b] | 20.04 ± 0.19 [bc] |
| S3 | 70.83 ± 3.75 [a] | 16.89 ± 1.24 [ab] | 25.93 ± 0.85 [d] | 20.87 ± 0.50 [b] |
| S4 | 73.17 ± 4.91 [a] | 15.33 ± 2.17 [b] | 26.58 ± 1.05 [cd] | 22.51 ± 0.76 [a] |
| S5 | 66.67 ± 4.13 [a] | 5.55 ± 0.99 [c] | 28.00 ± 0.43 [bc] | 19.94 ± 0.26 [c] |

For polysaccharide and triterpenoid content: values are the mean ± standard deviation of three independent samples. For all other parameters: values are the mean ± standard deviation of 30 independent bag samples. Mean values in each column followed by lowercase indicate the difference among strains at the significance level of 0.05, and the same letters mean not significantly different ($p > 0.05$).

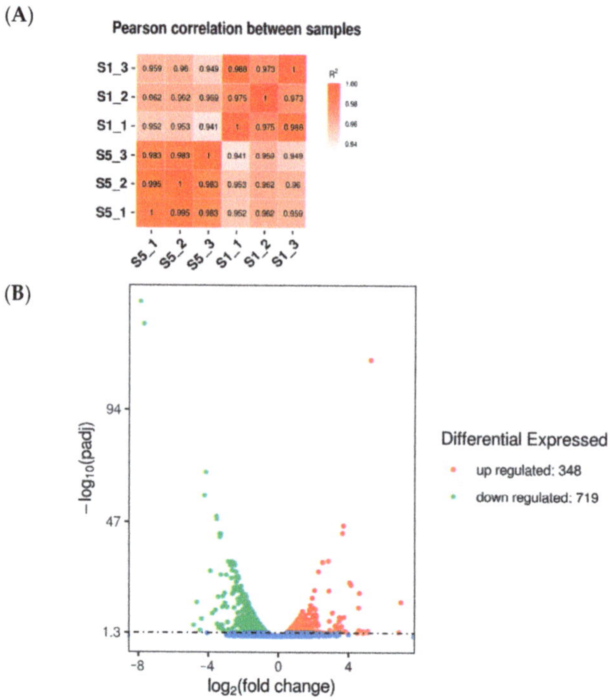


**Figure 4.** Correlation and DEGs analysis between strain S1 and S5. (**A**) Pearson correlation analysis between samples of strain S1 and S5. "_" indicated three biological duplications in each group. (**B**) DEGs between strain S1 and S5.

### *3.5. Enrichment Analysis of DEGs*

GO and KEGG enrichment analyses of DEGs were performed to explore the main difference formation mechanism between isonuclear alloplasmic strains. The result of GO enrichment analysis is shown in Figure 5A. In the biological process category, DEGs were mainly enriched in the oxidation–reduction process, carbohydrate metabolic process, and transmembrane transport. In the molecular function category, DEGs were mainly enriched in catalytic activity, hydrolase activity, and oxidoreductase activity. Gene enrichment analysis of DEGs based on the KEGG database revealed that these genes were mainly involved in pentose and glucuronate interconversions, starch and sucrose metabolism, and steroid biosynthesis, which were closely related to the active components of *G. lucidum* (Figure 5B).

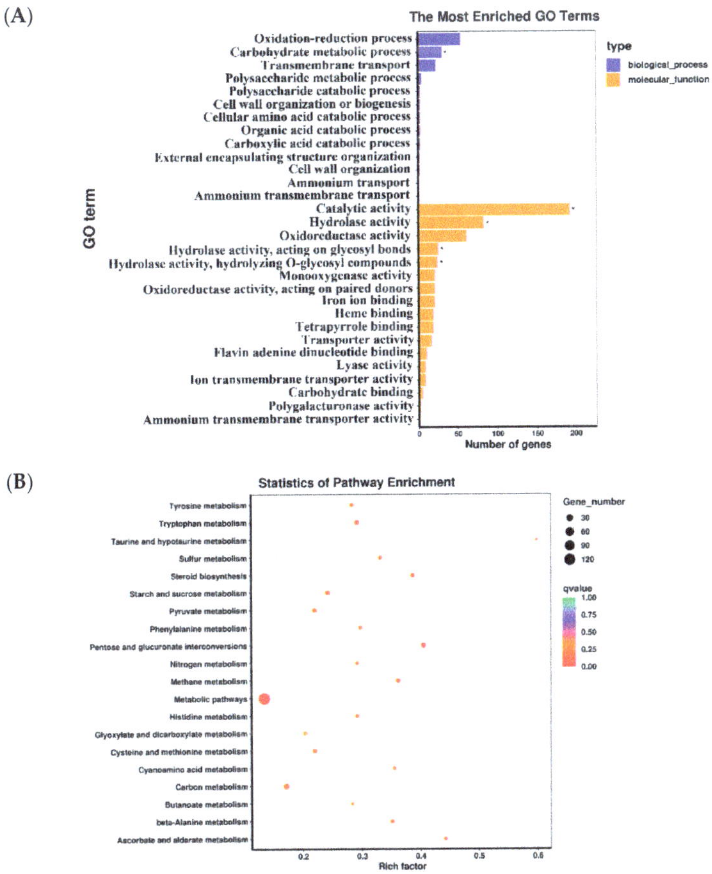


**Figure 5.** Enrichment analysis of DEGs. (**A**) GO enrichment of DEGs. (**B**) Bubble plot of KEGG enrichment of DEGs.

### *3.6. Analysis of Bioactive-Component-Related Biological Pathways*

From the above results, it was concluded that mitochondria affected the content of polysaccharide and triterpenoid, which may relate to starch and sucrose metabolism, steroid biosynthesis, and pentose and glucuronate interconversions, according to the enrichment analysis of DEGs of strain S1 and S5 (Figure 5B). In these pathways, the expression of most genes differed not only in strain S1 and S5, but also differed in the other three isonuclear alloplasmic strains. In the pathway of pentose and glucuronate interconversions, most genes showed high expression in strain S5 and low expression in strain S1 and S3 (Figure 6A,D). In the starch and sucrose metabolism pathway, gene expression in strain S1 was significantly higher than that in strain S4, except for 1,4-alpha-glucan branching enzyme gene (*g6017*) (Figure 6B,D). In the pathway of steroid biosynthesis, except for sterol 14-alpha-demethylase gene (*g7346*, Figure 6D) and 3-keto steroid reductase gene (*g9390*, Figure 6D), genes showed a low expression level in strain S1 and high expression level in strain S4 (Figure 6C,D). These results confirmed the important role of the pathway of pentose and glucuronate interconversions, starch and sucrose metabolism, and steroid biosynthesis in the content of polysaccharide and triterpenoid.

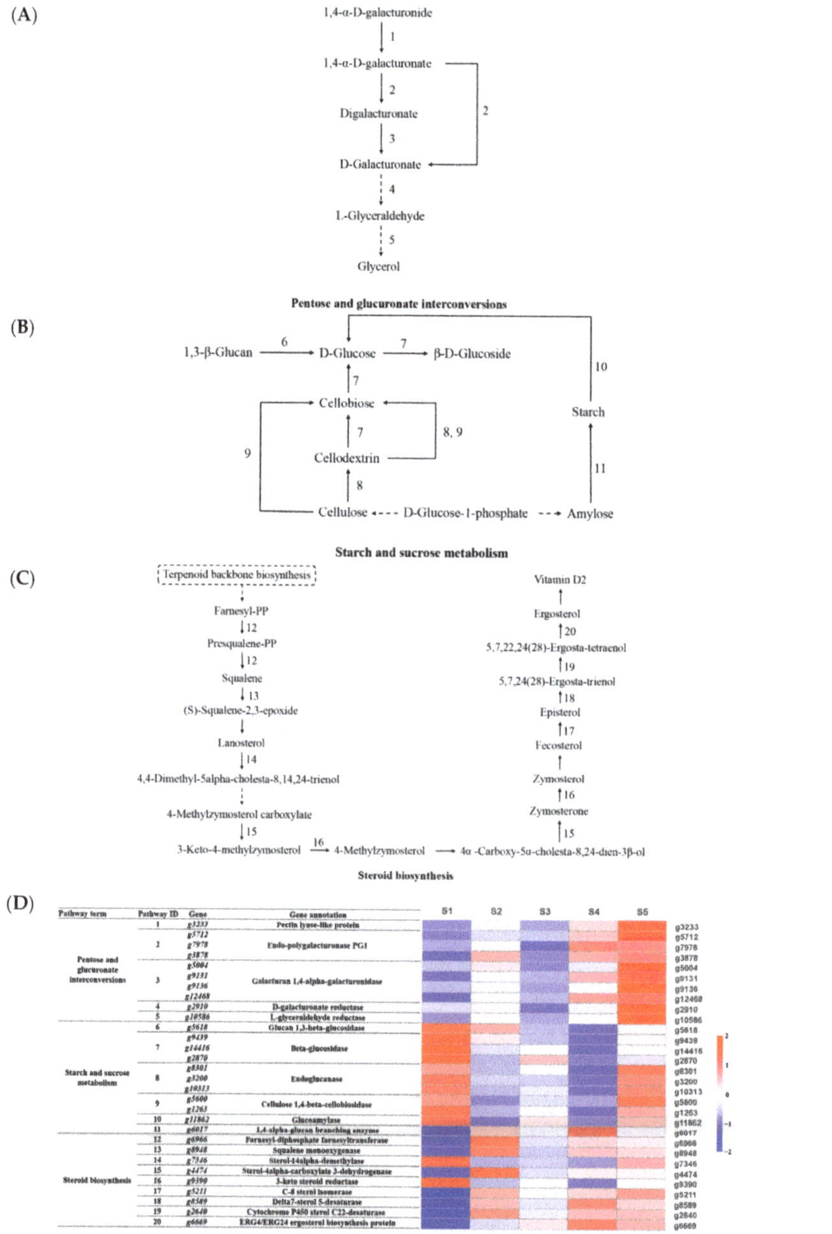


**Figure 6.** Polysaccharide- and triterpenoid-related pathways of isonuclear alloplasmic based on transcriptome analysis. Dashed lines indicate indirect connections or unknown reactions. The number along the arrows indicate pathway IDs in (**C**). (**A**) Pentose and glucuronate interconversions pathway. (**B**) Starch and sucrose metabolism pathway. (**C**) Steroid biosynthesis pathway. (**D**) Changes in expression of genes involved in pathways. Red to blue color indicated the highest to lowest relative expression level.

## 4. Discussion

In this study, five isonuclear alloplasmic strains S1–S5 of *G. lucidum* were built and proved in order to explore the effect of mitochondria on *G. lucidum* morphology, yield, and main primary bioactive components. The results showed that different mitochondria lead to different mycelial growth rate in PDA, basidiospore yield, and the content of polysaccharide and triterpenoid.

Polysaccharides and triterpenoid, as the main active components, directly reflect the quality of *G. lucidum*. For a long time, many workers have tried many different approaches to improve the production of polysaccharides and triterpenoid. Wei et al. [30] obtained the highest intracellular polysaccharide production by a sucrose fed-batch strategy. Tang et al. [31] reported that added concentration of 5 mM $Cu^{2+}$ on the culture of day 4 led to higher production of polysaccharides. The work of Sun et al. [32] demonstrated that promoting sporulation effectively improves GA production in *Ganoderma* species. Ma et al. [33] screened mutated high production of polysaccharide strains from plasma mutagenesis. It is worth noting that the effect of mitochondria on active components is usually ignored in conventional research. However, the result of this paper showed that different mitochondria led to different content of polysaccharides and triterpenoid in *G. lucidum*. To our knowledge, no published studies discuss the relation of mitochondria and active components. Shi et al. [34] found that the change in ROS would affect the production of *Ganoderma* acid. ROS were produced during energy metabolism in mitochondria [35,36]. Here, we suspect that the different result of isonuclear alloplasmic may be due to the difference in respiratory transmission chain, which then affects ROS, resulting in the difference in triterpenoid content. The mechanism of mitochondria's effect was further investigated, analyzed based on transcriptome. In this paper, strains S1 and S5 were representative isonuclear alloplasmic because of the significant difference among mycelial growth rate in PDA, basidiospore yield, and polysaccharide and triterpenoid content. Their complete transcriptomes were analyzed. DEGs between strains S1 and S5 were enriched in biological process and molecular function category according to GO analysis results and involved the pathway of pentose and glucuronate interconversions, starch and sucrose metabolism, and steroid biosynthesis according to KEGG pathway analysis.

Most polysaccharides of *G. lingzhi* are miscellaneous polysaccharides, usually containing glucose and a small amount of arabinose, galactose, mannose, xylose, and other monosaccharides [37]. In the present study, DEGs of *G. lingzhi* isonuclear alloplasmic mainly involved the conversion of 1,4-α-D-galacturonide to glycerol in the pathway of pentose and glucuronate interconversions, which may affect the content of arabinose and galactose in polysaccharides. Remarkably, enriched starch and sucrose pathway metabolism included 1,3-β-Glucan and D-Glucose, which were considered to be the main chain of *G. lingzhi* polysaccharide [38]. In *G. lingzhi*, triterpenoids are synthesized via the mevalonate pathway (MVA). Farnesyl-PP to squalene and, finally, to lanosterol is an important part of MVA [26]. Enriched steroid biosynthesis pathway not only involved these metabolites, but also the downstream metabolic process to ergosterol. It well explains why mitochondria could affect the content of triterpenoids. Furthermore, the expression level of DEGs of the three pathways behaved differently in the other isonuclear alloplasmic strains, which proved the accuracy of the result. The three pathways were related to the synthesis and metabolism of the secondary metabolites in *G. lucidum*, polysaccharides and triterpenoid. This was roughly consistent with the previous results of content and reflected the influence of mitochondria on active components from the side. Although it is hard to explain how the mechanism of mitochondria affect the mycelial growth rate and basidiospore yield based on present transcriptome results, we believe that this work will contribute to a better understanding of mitochondria function and pave the way for screening of excellent strains and improving active components.

## 5. Conclusions

The present study illustrated that mitochondria may affect mycelial growth rate, basidiospore yield, and the content of polysaccharide and triterpenoid via structuring and comparing isonuclear alloplasmic strains. Moreover, based on the analysis of the transcriptome, pentose and glucuronate interconversions, starch and sucrose metabolism, and steroid biosynthesis play an important role in the variations in polysaccharide and triterpenoid content in isonuclear alloplasmic strains. This study will be a useful foundation for understanding of mitochondria and will be helpful for breeding improved varieties.

**Supplementary Materials:** The following supporting information can be downloaded at: https://www.mdpi.com/article/10.3390/jof8111182/s1, Figure S1: Pearson correlation analysis between samples among strain S1–S5; Table S1: Specific primers for nonconserved regions among nucleus DNA and mtDNA of *G. lucidum* strains; Table S2: Summary of RNA-Seq data.

**Author Contributions:** Data curation, L.Y., C.S. and H.F.; Formal analysis, L.Y.; Funding acquisition, B.C. and X.W.; Investigation, X.H.; Methodology, C.S.; Project administration, X.W.; Resources, X.H.; Software, G.M.; Visualization, H.F. and G.M.; Writing—original draft, L.Y.; Writing—review and editing, B.C. and X.W. All authors have read and agreed to the published version of the manuscript.

**Funding:** This research was funded by grants from the National Key Research and Development Program of China (2019YFC1710501) and Outstanding Young Scientific Research Talent Project of Fujian Agriculture and Forestry University (xjq202113).

**Institutional Review Board Statement:** Ethics approval was not required for this research.

**Informed Consent Statement:** Not applicable.

**Data Availability Statement:** All experimental data in this study will be made available upon reasonable request from readers.

**Conflicts of Interest:** The authors declare that they have no competing interest.

## References

1. Giordano, L.; Sillo, F.; Garbelotto, M.; Gonthier, P. Mitonuclear interactions may contribute to fitness of fungal hybrids. *Sci. Rep.* **2018**, *8*, 1706. [CrossRef] [PubMed]
2. Medina, R.; Franco, M.E.E.; Bartel, L.C.; Alcántara, V.M.; Saparrat, M.C.N.; Balatti, P.A. Fungal mitogenomes: Relevant features to planning plant disease management. *Front. Microbiol.* **2020**, *11*, 978. [CrossRef] [PubMed]
3. Kouvelis, V.N.; Hausner, G. Editorial: Mitochondrial genomes and mitochondrion related gene insights to fungal evolution. *Front. Microbiol.* **2022**, *13*, 897981. [CrossRef] [PubMed]
4. Al-Reedy, R.M.; Malireddy, R.; Dillman, C.B.; Kennell, J.C. Comparative analysis of *Fusarium* mitochondrial genomes reveals a highly variable region that encodes an exceptionally large open reading frame. *Fungal Genet. Biol.* **2012**, *49*, 2–14. [CrossRef] [PubMed]
5. Freel, K.C.; Friedrich, A.; Schacherer, J. Mitochondrial genome evolution in yeasts: An all-encompassing view. *FEMS Yeast Res.* **2015**, *15*, fov023. [CrossRef] [PubMed]
6. Liu, W.; Cai, Y.; Zhang, Q.; Shu, F.; Chen, L.; Ma, X.; Bian, Y.B. Subchromosome-scale nuclear and complete mitochondrial genome characteristics of *Morchella crassipes*. *Int. J. Mol. Sci.* **2020**, *21*, 483. [CrossRef]
7. O'Connor, E.; McGowan, J.; McCarthy, C.; Amini, A.; Grogan, H.; Fitzpatrick, D.A. Whole genome sequence of the commercially relevant mushroom strain *Agaricus bisporus* var. bisporus ARP23. *G3 Genes Genomes Genet.* **2019**, *9*, 3057–3066.
8. Liu, X.R.; Wu, X.P.; Tan, H.; Xie, B.G.; Deng, Y.J. Large inverted repeats identified by intra-specific comparison of mitochondrial genomes provide insights into the evolution of *Agrocybe aegerita*. *Comput. Struct. Biotechnol. J.* **2020**, *18*, 2424–2437. [CrossRef]
9. Bashir, K.M.I.; Rheu, K.M.; Kim, M.S.; Cho, M.G. The complete mitochondrial genome of an edible mushroom, *Sparassis crispa*. *Mitochondrial DNA Part B Resour.* **2020**, *5*, 862–863. [CrossRef]
10. Song, Y.; Wan, J.N.; Shang, J.J.; Feng, Z.; Jin, Y.; Li, H.; Guo, T.; Wu, Y.Y.; Bao, D.P.; Zhang, M.; et al. The complete mitochondrial genome of the edible mushroom *Grifola frondosa*. *Mitochondrial DNA Part B Resour.* **2022**, *7*, 286–288. [CrossRef]
11. Li, J.Q.; Zhang, J.H.; Chen, H.M.; Chen, X.D.; Lan, J.; Liu, C. Complete mitochondrial genome of the medicinal mushroom *Ganoderma lucidum*. *PLoS ONE* **2013**, *8*, e72038. [CrossRef] [PubMed]
12. Li, Q.; Xiang, D.B.; Wan, Y.; Wu, Q.; Wu, X.Y.; Ma, C.R.; Song, Y.; Zhao, G.; Huang, W.L. The complete mitochondrial genomes of five important medicinal *Ganoderma* species: Features, evolution, and phylogeny. *Int. J. Biol. Macromol.* **2019**, *139*, 397–408. [CrossRef] [PubMed]
13. Steensels, J.; Gallone, B.; Verstrepen, K.J. Interspecific hybridization as a driver of fungal evolution and adaptation. *Nat. Rev. Microbiol.* **2021**, *19*, 485–500. [CrossRef] [PubMed]

14. Moon, S.K.; Thompson, L.J.; Madamanchi, N.; Ballinger, S.; Papaconstantinou, J.; Horaist, C.; Runge, M.S.; Patterson, C. Aging, oxidative responses, and proliferative capacity in cultured mouse aortic smooth muscle cells. *Am. J. Physiol. Heart Circ. Physiol.* **2001**, *280*, H2779–H2788. [CrossRef]
15. Latorre-Pellicer, A.; Moreno-Loshuertos, R.; Lechuga-Vieco, A.V.; Sánchez-Cabo, F.; Torroja, C.; Acín-Pérez, R.; Calvo, E.; Aix, E.; González-Guerra, A.; Logan, A.; et al. Mitochondrial and nuclear DNA matching shapes metabolism and healthy ageing. *Nature* **2016**, *535*, 561–565. [CrossRef]
16. Sun, X.M.; Wang, H.H.; Han, X.F.; Chen, S.W.; Zhu, S.; Dai, J. Fingerprint analysis of polysaccharides from different *Ganoderma* by HPLC combined with chemometrics methods. *Carbohydr. Polym.* **2014**, *114*, 432–439. [CrossRef]
17. Qu, Z.W.; Zhou, S.Y.; Guan, S.X.; Gao, R.; Duan, Z.W.; Zhang, X.; Sun, W.Y.; Fan, W.L.; Chen, S.S.; Chen, L.J.; et al. Recombinant expression and bioactivity comparison of four typical fungal immunomodulatory proteins from three main *Ganoderma* Species. *BMC Biotechnol.* **2018**, *18*, 80. [CrossRef]
18. Liu, Y.C.; Tang, X.C.; Hu, H.P.; Chen, D.L.; Xie, Y.Z.; Liang, X.W.; Li, X.M.; Xiao, C.; Huang, L.H.; Wu, Q.P. Genetic diversity and main functional composition of *Lingzhi* strains from main producing areas in China. *AMB Express.* **2021**, *11*, 119. [CrossRef]
19. Xu, J.W.; Zhao, W.; Zhong, J.J. Biotechnological production and application of ganoderic acids. *Appl. Microbiol. Biotechnol.* **2010**, *87*, 457–466. [CrossRef]
20. Gong, X.; Ji, M.Y.; Xu, J.P.; Zhang, C.H.; Li, M.H. Hypoglycemic effects of bioactive ingredients from medicine food homology and medicinal health food species used in China. *Crit. Rev. Food Sci. Nutr.* **2020**, *60*, 2303–2326. [CrossRef]
21. He, X.R.; Fang, J.C.; Guo, Q.; Wang, M.; Li, Y.S.; Meng, Y.B.; Huang, L.H. Advances in antiviral polysaccharides derived from edible and medicinal plants and mushrooms. *Carbohydr. Polym.* **2020**, *229*, 115548. [CrossRef] [PubMed]
22. Hu, Y.; Wang, S.X.; Wu, F.Y.; Wu, K.J.; Shi, R.P.; Qin, L.H.; Lu, C.F.; Wang, S.Q.; Wang, F.F.; Zhou, S.B. Effects and mechanism of *Ganoderma lucidum* polysaccharides in the treatment of diabetic nephropathy in streptozotocin-induced diabetic rats. *BioMed Res. Int.* **2022**, *2022*, 4314415. [CrossRef] [PubMed]
23. Chen, S.D.; Guan, X.Y.; Yong, T.Q.; Gao, X.; Xiao, C.; Xie, Y.Z.; Chen, D.L.; Hu, H.P.; Wu, Q.P. Structural characterization and hepatoprotective activity of an acidic polysaccharide from *Ganoderma lucidum*. *Food Chem. X* **2022**, *13*, 100204. [CrossRef]
24. Ye, L.Y.; Deng, Y.J.; Mukhtar, I.; Meng, G.L.; Song, Y.J.; Cheng, B.; Hao, J.B.; Wu, X.P. Mitochondrial genome and diverse inheritance patterns in *Pleurotus pulmonarius*. *J. Microbiol.* **2020**, *58*, 142–152. [CrossRef] [PubMed]
25. Chen, B.Z.; Ke, B.R.; Ye, L.Y.; Jin, S.S.; Jie, F.; Zhao, L.L.; Wu, X.P. Isolation and varietal characterization of *Ganoderma resinaceum* from areas of *Ganoderma lucidum* production in China. *Sci. Hortic.* **2017**, *224*, 109–114. [CrossRef]
26. Ye, L.Y.; Liu, S.R.; Xie, F.; Zhao, L.L.; Wu, X.P. Enhanced production of polysaccharides and triterpenoids in *Ganoderma lucidum* fruit bodies on induction with signal transduction during the fruiting stage. *PLoS ONE* **2018**, *13*, e0196287. [CrossRef] [PubMed]
27. Pertea, M.; Kim, D.; Pertea, G.M.; Leek, J.T.; Salzberg, S.L. Transcript-level expression analysis of RNA-seq experiments with HISAT, StringTie and Ballgown. *Nat. Protoc.* **2016**, *11*, 1650–1667. [CrossRef] [PubMed]
28. Hao, J.B.; Ye, L.Y.; Meng, G.L.; Song, Y.J.; Fu, J.S.; Wu, X.P. The protective effect and crucial biological pathways analysis of *Trametes lactinea* mycelium polysaccharides on acute alcoholic liver injury in mice based on transcriptomics and metabolomics. *Food Sci. Hum. Well.* **2021**, *10*, 480–489. [CrossRef]
29. Young, M.D.; Wakefield, M.J.; Smyth, G.K.; Oshlack, A. Gene ontology analysis for RNA-seq: Accounting for selection bias. *Genome Biol.* **2010**, *11*, R14. [CrossRef]
30. Wei, Z.H.; Liu, L.L.; Guo, X.F.; Li, Y.J.; Hou, B.C.; Fan, Q.L.; Wang, K.X.; Luo, Y.D.; Zhong, J.J. Sucrose fed-batch strategy enhanced biomass, polysaccharide, and ganoderic acids production in fermentation of *Ganoderma lucidum* 5.26. *Bioprocess Biosyst. Eng.* **2016**, *39*, 37–44. [CrossRef]
31. Tang, Y.J.; Zhu, L.W. Improvement of ganoderic acid and *Ganoderma* polysaccharide biosynthesis by *Ganoderma lucidum* fermentation under the inducement of $Cu^{2+}$. *Biotechnol. Prog.* **2010**, *26*, 417–423. [PubMed]
32. Sun, B.; You, H.; Xu, J.W. Enhancement of ganoderic acid production by promoting sporulation in a liquid static culture of *Ganoderma* species. *J. Biotechnol.* **2021**, *328*, 72–77. [CrossRef] [PubMed]
33. Ma, Y.H.; Zhang, Q.Q.; Zhang, Q.F.; He, H.Q.; Chen, Z.; Zhao, Y.; Wei, D.; Kong, M.G.; Huang, Q. Improved production of polysaccharides in *Ganoderma lingzhi* mycelia by plasma mutagenesis and rapid screening of mutated strains through infrared spectroscopy. *PLoS ONE* **2018**, *13*, e0204266. [CrossRef] [PubMed]
34. Shi, D.K.; Zhu, J.; Sun, Z.H.; Zhang, G.; Liu, R.; Zhang, T.J.; Wang, S.L.; Ren, A.; Zhao, M.W. Alternative oxidase impacts ganoderic acid biosynthesis by regulating intracellular ROS levels in *Ganoderma lucidum*. *Microbiology* **2017**, *163*, 1466–1476. [CrossRef] [PubMed]
35. Mezhnina, V.; Ebeigbe, O.P.; Poe, A.; Kondratov, R.V. Circadian control of mitochondria in reactive oxygen species homeostasis. *Antioxid. Redox Signal.* **2022**, *37*, 647–663. [CrossRef]
36. Suski, J.; Lebiedzinska, M.; Bonora, M.; Pinton, P.; Duszynski, J.; Wieckowski, M.R. Relation between mitochondrial membrane potential and ROS formation. *Methods Mol. Biol.* **2018**, *1782*, 357–381.
37. Miyazaki, T.; Nishijima, M. Studies on fungal polysaccharides. XXVII. Structural examination of a water-soluble, antitumor polysaccharide of *Ganoderma lucidum*. *Chem. Pharm. Bull.* **1981**, *29*, 3611–3616. [CrossRef]
38. Liu, G.; Zhang, J.; Kan, Q.; Song, M.; Hou, T.; An, S.; Lin, H.; Chen, H.; Hu, L.; Xiao, J.; et al. Extraction, structural characterization, and immunomodulatory activity of a high molecular weight polysaccharide from *Ganoderma lucidum*. *Front. Nutr.* **2022**, *9*, 846080. [CrossRef]

Journal of *Fungi*

MDPI

*Article*

# Whole-Genome Sequencing and Transcriptome Analysis of *Ganoderma lucidum* Strain Yw-1-5 Provides New Insights into the Enhanced Effect of Tween80 on Exopolysaccharide Production

**Tuheng Wu [1,2,†], Manjun Cai [3,†], Huiping Hu [3], Chunwei Jiao [1], Zhi Zhang [1,3], Yuanchao Liu [3], Jian Chen [1,2], Chun Xiao [3], Xiangmin Li [3], Xiong Gao [3], Shaodan Chen [3], Qingping Wu [2,3,*] and Yizhen Xie [1,3,*]**

1 Guangdong Yuewei Edible Fungi Technology Co., Guangzhou 510000, China
2 School of Bioscience and Bioengineering, South China University of Technology, Guangzhou 510000, China
3 Guangdong Provincial Key Laboratory of Microbial Safety and Health, State Key Laboratory of Applied Microbiology Southern China, Institute of Microbiology, Guangdong Academy of Sciences, Guangzhou 510000, China
* Correspondence: wuqp203@163.com (Q.W.); xieyizhen@126.com (Y.X.)
† These authors contributed equally to this work.

check for updates

**Citation:** Wu, T.; Cai, M.; Hu, H.; Jiao, C.; Zhang, Z.; Liu, Y.; Chen, J.; Xiao, C.; Li, X.; Gao, X.; et al. Whole-Genome Sequencing and Transcriptome Analysis of *Ganoderma lucidum* Strain Yw-1-5 Provides New Insights into the Enhanced Effect of Tween80 on Exopolysaccharide Production. *J. Fungi* **2022**, *8*, 1081. https://doi.org/10.3390/jof8101081

Academic Editors: Mingwen Zhao, Gen Zou and Jing Zhu

Received: 24 August 2022
Accepted: 1 October 2022
Published: 14 October 2022

**Publisher's Note:** MDPI stays neutral with regard to jurisdictional claims in published maps and institutional affiliations.



**Abstract:** *Ganoderma lucidum* is an important medicinal mushroom widely cultured in Asian countries. Exopolysaccharides are bioactive compounds of *G. lucidum* with health benefits. Limited exopolysaccharide content hinders its extraction from *G. lucidum*. The addition of Tween80 had an enhanced effect on *G. lucidum* exopolysaccharide production in submerged fermentation. However, the mechanism of this effect remains unclear. In this study, we report on a high-quality assembly of *G. lucidum* strain yw-1-5 to lay the foundation for further transcriptome analysis. The genome sequence was 58.16 Mb and consisted of 58 scaffolds with an N50 of 4.78 Mb. A total of 13,957 protein-coding genes were annotated and Hi-C data mapped to 12 pseudo-chromosomes. Genes encoding glycosyltransferases and glycoside hydrolases were also obtained. Furthermore, RNA-seq was performed in a Tween80-treated group and control group for revealing the enhanced effect of Tween80 on exopolysaccharide production. In total, 655 genes were identified as differentially expressed, including 341 up-regulated and 314 down-regulated. Further analysis of differentially expressed genes showed that groups of MAPK, amino sugar and nucleotide sugar metabolism, autophagy, ubiquitin-mediated proteolysis, peroxisome, starch and sucrose metabolism, TCA cycle, glycolysis/gluconeogenesis KEGG pathway, glycosyltransferases and glycoside hydrolases played important roles in the enhanced effect of Tween80 on exopolysaccharide production. This work provides a valuable resource for facilitating our understanding of the synthesis of polysaccharides and accelerating the breeding of new strains with a high content of exopolysaccharides.

**Keywords:** *Ganoderma lucidum*; genome sequencing; polysaccharides; Tween80; transcriptome; functional gene analysis

## 1. Introduction

*Ganoderma lucidum*, commonly known as Lingzhi, is an extremely valuable fungi with many medical properties that has been cultured for a thousand years in Asia. Due to its antioxidant, anti-tumor, anti-inflammatory, antiviral, immune-system strengthening, cognitive-impairment reducing and anti-diabetic bioactivity, polysaccharides from *G. lucidum* have been applied in functional food and pharmaceutical industries [1–4]. Importantly, *G. lucidum* polysaccharide RF3 had been reported to inhibit the infection of SARS-CoV-2 in animal models [5].

Submerged fermentation of mycelium is an effective approach to obtain *G. lucidum* exopolysaccharides, due to its shorter culture period and seasonal independence [6]. In particular, a polysaccharide isolated from submerged cultured *G. lucidum*, namely GLEP-2, presented enhanced effects on the lymphocyte proliferation [7]. However, according to our previous research, the total content of exopolysaccharides from submerged cultured *G. lucidum* normally ranged from 0.5 g/L to 1 g/L, indicating that content of *G. lucidum* exopolysaccharides cannot meet industrial requirements. Therefore, the improvement of exopolysaccharide production of *G. lucidum* is an important trend. Although many physical approaches such as controlling light [8] can improve exopolysaccharide production, the addition of chemical agents is a more appropriate strategy, due to their low cost and easy separation [9]. Recently, the non-ionic surfactant Tween80, namely polyoxyethylene glycol sorbitan monooleate, has been applied to increase the exopolysaccharide production of mushrooms (e.g., *Lentinus edodes* [10], *Schizophyllum commune* [11]) in submerged fermentation. Our previous work showed that the addition of Tween80 increased the production of *G. lucidum* strain yw-1 exopolysaccharide in submerged fermentation [12] Although Tween80 increased the expression level of key genes, which were involved in the synthesis of exopolysaccharides [12], the mechanism of the enhanced effect of Tween80 on exopolysaccharide production in *G. lucidum* remains unclear.

Studies from microbial exopolysaccharides suggest that the biosynthesis of exopolysaccharides in *G. lucidum* might include precursor production, backbone assembly, side-chain modification, polymerization and export [13]. Current studies of the biosynthesis of polysaccharides in *G. lucidum* have mainly focused on polysaccharide-precursor production. Results of RNAi and over-expression experiments in *G. lucidum* showed that genes encoding phosphoglucomutase (PGM), glucose phosphate isomerase (PGI) and UDP-glucose pyrophosphorylase (UGP) were involved in the biosynthesis of polysaccharides [14,15]. Besides this, glycoside-hydrolase-family proteins and glycosyltransferase-family proteins participate in processes of backbone assembly, side-chain modification and polymerization [16,17].

Since Sanger sequencing technology was created in 1977, sequencing technologies and assembly methods have been developed rapidly. The first report of the *G. lucidum* genome occurred in 2012, using a whole-genome shotgun-sequencing strategy [18]. In 2021, a more complete *G. lucidum* strain Lingjian-2 genome was assembled using the PacBio Sequel system and provided new insights into the genetic mechanism behind its high triterpene content [19]. However, a high-quality genome assembly is needed for the discovery of genetic variation and the breeding of new strains with a high content of bioactive products. Interestingly, a chromosome-level genome of *Lentinula edodes* was assembled using long-read PacBio, Illumina short-read sequencing, and the high-throughput chromatin conformation capture (Hi-C) technique for revealing genetic architecture [20]. Therefore, a third-generation sequencing technique, Illumina short-read sequencing, along with the Hi-C technique, is an effective sequencing strategy for sequencing the *G. lucidum* genome to achieve high-quality genome assembly.

Over the years, a transcriptome based on a next-generation sequencing technique (RNA-seq) has been widely applied to discover the genes for the biosynthesis of polysaccharides. For instance, transcriptomic analysis of three cultivars of *Poria cocos* showed that *malZ*, *galA*, *sord*, *gnl* and *bglX* are key genes related to the polysaccharide biosynthetic pathway [21]. Genes encoding PGM, PGI, UGP, glycoside-hydrolase-family proteins and glycosyltransferase-family proteins were identified by comparative transcriptome analysis of six *Hericium erinaceus* strains as participating in the biosynthesis of polysaccharides [22]. De novo genome sequencing of *Lentinula edodes* and comparative transcriptomic analysis during its developmental stages showed that β-1,3-glucan synthase regulatory genes and β-1,6-glucan synthesis-associated protein (SKN1) genes were involved in polysaccharide synthesis [23]. Therefore, from a technological perspective, the RNA-seq technique can be used in the identification of genes associated with the biosynthesis of exopolysaccharides in *G. lucidum*.

In this study, we first assembled the genome of *G. lucidum* strain yw-1-5 by Oxford Nanopore (ONT), Hi-C and next-generation sequencing (Illumina) technologies, providing genomic information for revealing the biosynthesis of exopolysaccharides in *G. lucidum*. Subsequently the gene-expression changes of *G. lucidum* mycelium with and without the addition of Tween80 were investigated by RNA-seq. This study will provide valuable insights into the mechanism of exopolysaccharide synthesis and Tween80-enhanced *G. lucidum* exopolysaccharide production.

## 2. Materials and Methods

### 2.1. Strain and Mycelium Nucleic Acid Preparation

The *G. lucidum* monokaryon strain yw-1-5 was derived from the dikaryotic strain yw-1 by protoplasting. The strain yw-1-5 was preserved in Guangdong Microbial Culture Collection Center (GDMCC), Institute of Microbiology, Guangdong Academy of Science. The vegetative mycelium of strain yw-1-5, cultured on potato dextrose agar (PDA) medium (200 g potato, 20 g glucose, 20 g agar, 1.5 g $KH_2PO_4$, 1.5 g $MgSO_4$, 10 g peptone, 100 mL water) with cellophane for 7 days at 28 °C in darkness, was harvested. The mycelium were then frozen in liquid nitrogen and stored at −80 °C.

Genomic DNA was extracted by QIAGENVR Genomic DNA extraction kit. The concentration and purity of extracted DNA was evaluated by NanoDropTM One UV-Vis spectrophotometer (Thermo Fisher Scientific, Waltham, MA, USA) and then QubitVR 3.0 Fluorometer (Invitrogen, Waltham, MA, USA) was applied to quantify DNA accuracy. Total RNA was extracted from vegetative mycelium of strain yw-1-5. The concentration and purity of total RNA were evaluated by Nanodrop2000. Agilent2100 was then used to determine the RNA integrity number (RIN) value.

### 2.2. De Novo Sequencing and Assembly

Genomic DNA of strain yw-1-5 was used to construct Oxford Nanopore Technologies library with a 20-kb library size according to Oxford Nanopore Technologies standard protocol and the Illumina library with 350-bp library size according to Illumina standard protocol. Sequencing of strain yw-1-5 was performed using Nanopore and Illumina sequel platforms at the Biomarker Technologies Corporation (Beijing, China). The filtered reads were then assembled using NECAT software. The assembled genome was corrected with Illumina data using Pilon software [24]. The high-throughput chromosome conformation capture technique (Hi-C) was used to construct a more accurate genome assembly for *G. lucidum* strain yw-1-5. The vegetative mycelium of strain yw-1-5 was used to construct a Hi-C library and sequenced at Illumina Hiseq platform at Biomarker Technologies Corporation (Beijing, China) according to a previously reported method [25]. Importantly, the construction of the Hi-C libraries are described below. Abstractly, vegetative mycelium cells were cross-linked with formaldehyde and cross-linked cells were subsequently lysed. Chromatin DNA was digested and labeled with biotin-14-dCTP, and then ligated by T4 DNA ligase. The ligated DNA was extracted, purified and sheared into fragments of 300–500 bp. The Blunt-end repair, addition of an A-tail, adaptor addition and purification through biotin-streptavidin-mediated pull-down, PCR amplification were performed, resulting in the construction of the Hi-C libraries.

Finally, these assembly results were integrated into the final assembly. For evaluating the completeness of genome assembly, Illumina data were mapped to assembly sequence by Burrows–Wheeler Aligner (bwa) software [26] and the evaluation with Benchmarking Universal Single-Copy Ortholog (BUSCO) analysis (fungi_odb9 database and BUSCO v2.0 software) [27] was performed.

### 2.3. Genomic Component Analysis and Genome Annotation

Genome component analysis included the prediction of protein-coding genes, noncoding RNAs, pseudogenes and repetitive sequences. For prediction of repetitive sequences, LTR_FINDER v1.05 [28], MITE-Hunter [29], RepeatScout v1.0.5 [30], and PILER-DF v2.4 software [31] were used to construct a repetitive sequence database of the genome of *G. lucidum*. It was then merged with the Repbase database as the final repetitive sequence database. Finally, RepeatMasker v4.0.6 [32] was applied to predict repeat sequences from the *G. lucidum* strain yw-1-5 genome. For the prediction of noncoding RNAs (ncRNAs), transfer RNA (tRNA) genes were predicted with tRNAscan-SE v1.3.1 [33] and Ribosome RNA (rRNA), microRNA (miRNA), and small nucleolar RNA (snoRNA) sequences were predicted with Infernal v1.1 [34] based on the Rfam database. The secondary metabolism gene cluster was predicted by antiSMASH v6.0.0 software [35].

For prediction of protein-coding genes, we used *ab initio* prediction, homologous protein prediction and transcriptome data-prediction methods. We then integrated these three prediction results into the final result. We used Genscan [36], Augustus v2.4 [37], GlimmerHMM v3.0.4 [38], GeneID v1.4 [39], SNAP (2006-07-28) software [40] for de novo prediction. GeMoMa v1.3.1 [41] was used for homologous protein prediction. RNA sequencing was performed using the Illumina HiSeq platform, and standard bioinformatic analysis was performed at the Biomarker Technologies Corporation (Beijing, China). In particular, Hisat2 v2.0.4 [42] and Stringtie v1.2.3 software [43] were used to perform assembly based on transcripts. TransDecoder v2.0 and PASA v2.0.2 software [44] were applied for unigene sequence prediction. Finally we use EVM v1.1.1 [45] to integrate the prediction results and PASA v2.0.2 software was used to modify. For functional annotation of protein-coding genes, predicted protein-coding genes were blasted (e-value: $1 \times 10^{-5}$) against Nr, Swiss-Prot, TrEMBL, KEGG, KOG databases. The GO annotation of protein-coding genes was achieved using Blast2go software [46]. Hmmer software [47] was used for Pfam annotation of protein-coding genes. Furthermore, protein-coding genes were annotated by blast against the Carbohydrate-Active enZYmes Database (CAZy).

### 2.4. *Tween80 Treatment and Transcriptome Analysis*

The vegetative mycelium of *G. lucidum* strain yw-1-5 was cultured on PDA medium for 7 days at 28 °C in darkness. Fresh mycelium was transferred in 100 mL of modified Martin broth medium (20 g/L glucose, 5 g/L tryptone, 2 g/L yeast extract, 0.5 g/L $MgSO_4$ and 1 g/L $KH_2PO_4$) and cultured for 6 days at 28 °C and 120 rpm. Tween80 at a concentration of 2.5% (*v/v*) was added to the medium on day 0. Additionally, reagent-grade Tween80 was purchased from Sangon Biotech Co., Ltd. (Shanghai, China). Mycelium and EPS from the fermentation broth were collected by centrifugation. Mycelium of the Tween80-treated group and control group were then frozen in liquid nitrogen and stored at −80 °C. The content of EPS was assayed by the phenol sulfuric acid method [12]. Data on the content of EPS are presented as the mean ± standard deviation of three replicates. Statistical analysis with GraphPad Prism5 software was performed using Student's *t*-test.

For revealing the transcriptomic landscape of the Tween80-treated group and control group, total RNA was extracted by RNA Purification Reagent. The concentration and purity of the extracted RNA were evaluated by Nanodrop2000. The quality evaluation was determined by 1.5% agarose gels and Agilent 2100 Bioanalyzer. RNA sequencing libraries were then generated with the Illumina TruseqTM RNA sample prep kit. Next-generation sequencing of these libraries was performed on the Illumina Novaseq 6000 platform at Shanghai Major Biomedical Technology Co., Ltd. (Shanghai, China). Standard bioinformatic analysis was performed using Majorbio Cloud Platform (www.majorbio.com (accessed on 30 September 2022)). Briefly, clean reads were obtained after quality control using fastp v0.19.5 software [48]. The clean reads were then compared with a reference genome using Bowtie2 v2.4.1 software [49]. HISAT2 v2.1.0 [42] and TopHat2 v2.1.1 software [50] were applied to evaluate the results of comparison. All unigenes were functionally annotated based on the NCBI nonredundant protein (Nr) (2021.10), Protein Family (Pfam) v14.6, Swiss-Prot (2021.06), Kyoto Encyclopedia of Genes and Genomes (KEGG) (2021.09), GO (2021.0918) and EggNOG (2020.06) databases. For the quantification of gene-expression levels, FPKMs (fragments per kilobase of exon model per million mapped reads) were calculated using RSEM v1.3.3. The differential expression analysis was performed using DESeq R package v1.24.0 [51]. Genes with an adjusted $p$-value < 0.05 and absolute value of log2 (Fold change) > 1 were defined as differentially expressed genes (DEGs). For the functional annotation of DEGs, gene ontology (GO) enrichment analysis was implemented by GOATOOLS v0.6.5 [52] and the enrichment of differentially expression genes in KEGG pathways was performed by KOBAS v2.1.1 software [53].

*2.5. Real-Time Quantitative PCR*

In total, five differentially expressed genes were chosen to validate these RNA-seq data (Table S15). Fresh mycelium from a Tween80-treated group and control group were collected and ground into a powder with liquid nitrogen. Total RNA was extracted using a HiPure Fungal RNA Mini kit (Magen, Guangzhou, China) according to the manufacturer's instructions. First-strand cDNA was synthesized by HiScript® III All-in-one RT SuperMix Perfect for qRT-PCR (+gDNA wiper) (Vazyme Biotech, Nanjing, China). The qPCR was performed using a ChamQ Universal SYBR qPCR Master Mix (Vazyme Biotech, China). The 18S rRNA gene was used as the reference gene. The conditions for qPCR were as follows: initiation was conducted at 95 °C for 30 s; followed by 40 cycles at 95 °C for 10 s, 60 °C for 30 s. This experiment was conducted on an Applied Biosystems ABI 7500 (Applied Biosystems, Waltham, MA, USA) containing three technical replicates and three biological replicates. The relative expression of selected DEGs was calculated with the $2^{-\Delta\Delta CT}$ method [54]. The primers used are shown in Table S15.

*2.6. Data Availability*

The genome data of strain yw-1-5 had been submitted to NCBI (National Center for Biotechnology Information, Bethesda, MD, USA) with bioproject ID PRJNA886764. The raw data of the transcriptome generated in this study had deposited in NCBI associated with bioproject ID PRJNA886761.

## 3. Results

*3.1. Sequencing and De Novo Assembly of Ganoderma lucidum Genome*

The monokaryon strain yw-1-5, generated from the basidiospore of the dikaryon strain yw-1, was selected for this study (Figures S1 and S2). After filtering for adaptor sequences, low-quality reads, and shorter fragments (<2 kb), 7.23 Gb clean data yielded from the Nanopore sequel platform were obtained and assembled into a 58.16 Mb genome with a GC content of 55.91% for *G. lucidum* yw-1-5 (Table S1 and Table 1). The assembly genome contained 95 contigs (>1 kb), with a contig N50 and N90 of 2.49 Mb and 0.37 Mb, respectively (Table 1). Furthermore, Hi-C technique was used to scaffold the contig assembly and 19.32 million paired reads with 5.78 Gb clean data were obtained (Table S2). Chromosome-scale de novo assemblies of the yw-1-5 genome were achieved by combining the 95 contig sequences and genome-wide chromatin-interaction data sets. A total of 55.52 Mb (95.46%) of the genome sequences were assigned into twelve chromosome groups and there was 100% accuracy in ordering and orienting contigs within chromosome groups (Figure 1A, Table S2). The scaffold N50 value was 4.78 Mb (Table 1). Compared with previous reported *G. lucidum* genomes, the size of yw-1-5 was larger, with fewer scaffold numbers and a larger N50 value (Table S3), indicating the high quality of the genome sequence assembly. In addition, 95.05% of the clean reads from next-generation sequencing could be mapped to the assembly genome (Table S4). The high degree of completeness of the genome was further supported by Benchmarking Universal Single-Copy Ortholog (BUSCO) analysis as 279 BUSCOs (96.21%) were identified and 276 of them were complete (Table S4).

**Table 1.** Statistics for genome assembly.

| | **Contig** | **Scaffold** |
|---|---|---|
| Number | 95 | 58 |
| Total length (bp) | 58,157,106 | 58,160,806 |
| N50 (bp) | 2,485,691 | 4,775,195 |
| N90 (bp) | 365,330 | 3,060,068 |
| Maximum length (bp) | 5,247,307 | 8,872,950 |
| GC content (%) | 55.91 | 55.91 |

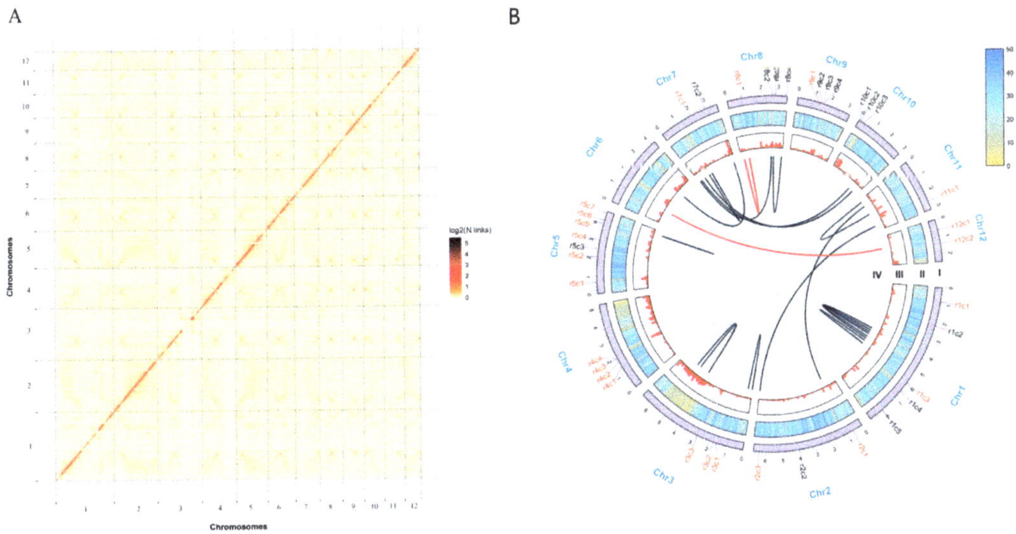

**Figure 1.** Hi-C interaction heat map and genomic features of the *Ganoderma lucidum* yw-1-5. (**A**), Hi-C interaction heat map of *Ganoderma lucidum* yw-1-5 haploid genome. (**B**), Global view of the yw-1-5 genomic features. The names marked outside the circle were identified as gene clusters and the cluster names in red represent the terpene clusters. The detailed information of these gene clusters was listed in Table S6. Circle I represents the 12 pseudo-chromosomes of *G. lucidum* (Mb). Circle II represents gene density on every pseudo-chromosome. Circle III represents repetitive sequence density on each pseudo-chromosome. Circle IV represents large segmental duplications: regions sharing more than 90% sequence similarity over 10 kb were connected by red lines; those over 5 kb were connected by dark blue lines.

### 3.2. Gene Prediction of *Ganoderma lucidum* Yw-1-5 Genome

A total of 14,849 genes were mined in yw-1-5 genome based on the homology and de novo methods, with an average gene length of 1824.07 bp, and each predicted gene contained 5.57 exons (Table S5). Using antiSMASH, we identified 38 gene clusters comprising types of terpene, non-ribosomal peptide synthetase cluster (NRPS), NRPS-like fragment, type I polyketide synthase (T1PKS), and beta-lactone containing protease inhibitor (Figure 1B, Table S6). Among these gene clusters, 60.53% are a type of terpene, which is in agreement with the abundant diversity of terpenoid products in *G. lucidum*.

In addition to the protein-encoding gene models, 272 tRNAs were predicted and 10 of them are putative pseudo-genes and the others correspond to the 20 common amino acid and selenium cysteine (SeC) codons. Furthermore, 1,133,073 bases were predicted to encode 319 pseudo-genes and the average length was 3551.95 bp. Moreover, repetitive sequences accounted for 17.45% of the genome, and were more abundant than 8.15% in the haploid *G. lucidum* strain 260125-1 [18]. The most abundant type was the LTR/Gypsy, representing 6.03% of the genome (Table S7).

### 3.3. Functional Annotation of *Ganoderma lucidum* Yw-1-5 Genome

The genome of *G. lucidum* yw-1-5 had 13,957 genes that could be annotated using a sequence-similarity search in public protein databases (Table S8), including 6699 (45.11%), 3681 (24.79%), 6264 (42.18%), 8599 (57.91%), 6943 (46.76%), 13,768 (92.72%) and 13,921 (93.75%) genes were successfully annotated according to GO, KEGG, KOG, Pfam, Swissprot, TrEMBL and NCBI Nr databases, respectively (Table 2). In addition to these general databases, specific functional databases such as the Carbohydrate-Active enZYmes Database (CAZy), Cytochrome P450 Engineering Database (CYPED) and Transporter Classification Database (TCDB) were used to annotate the function of the identified genes in yw-1-5. A total of 1427 genes could be assigned to these three databases, accounting for 9.61% of the predicted genes of assembled genome (Table 2).

**Table 2.** Functional annotation of deduced proteins by sequence-similarity search.

| Annotated Database | Annotated Number | $100 \leq$ length < 300 | Length $\geq 300$ |
|---|---|---|---|
| GO_Annotation | 6699 | 1752 | 4827 |
| KEGG_Annotation | 3681 | 1046 | 2561 |
| KOG_Annotation | 6264 | 1509 | 4698 |
| Pfam_Annotation | 8599 | 2102 | 6409 |
| Swissprot_Annotation | 6943 | 1617 | 5248 |
| TrEMBL_Annotation | 13,768 | 3926 | 9646 |
| nr_Annotation | 13,921 | 3999 | 9718 |
| All_Annotated | 13,957 | 4020 | 9733 |
| CAZy_Annotation | 643 | | |
| TCDB_Annotation | 99 | | |
| CYPED_Annotation | 685 | | |

In the GO analysis, genes classified into the biological process were mostly found to participate in the metabolic, cellular and single-organism processes. Within the molecular function category, genes having catalytic and binding activities accounted for 49.1% and 39.2%, respectively (Figure 2). In KOG analysis, most genes were involved in metabolism and 21.36% of them were assigned to secondary metabolite biosynthesis, transport and catabolism, which is well in accordance with the great variety of secondary metabolites in *G. lucidum* (Figure 3). In addition, 2014 genes were found to be involved in cellular processes and signaling, including post-translational modification, protein turnover, chaperones and signal transduction mechanisms. Similar to the KOG annotation results, most genes in the KEGG pathway annotation were involved in metabolism, especially the biosynthesis of amino acids and carbon metabolism, followed by genetic information processing such as RNA transport, ribosome and spliceosome (Figure 2).

It has been reported that cytochrome P450s (CYP450s) are involved in many essential cellular processes and play diverse roles in fungi [55]. In the yw-1-5 genome, we found a total of 685 candidate CYP450 sequences (Table S9), which could be classified into 43 families, and the CYP51 family contained highest counts (135), followed by the CYP620 (88), CYP53 (78) and CYP79 (63) families. The CYP51 family is found across all kingdoms which participate in membrane-ergosterol biosynthesis (sterol 14 alpha-demethylase cytochrome P450 (CYP51), a P450 in all biological kingdoms). The expression profile of one CYP51 member (GL26139) in *G. lucidum* was highly correlated with that of the lanosterol synthase (correlation coefficient (r) = 0.9920). Therefore, we could not exclude the possibility that some of these CYP51 genes in yw-1-5 might be involved in triterpenoid biosynthesis. Furthermore, we found that there are only two CYP51 members have been identified in the haploid *G. lucidum* strain 260125-1 [18], and the *Antrodia cinnamomea* genome contains only one putative CYP51 gene [55]; the increased quantity of CYP51 gene identification and annotation might be a consequence of the improvement in sequencing and annotation pipelines.

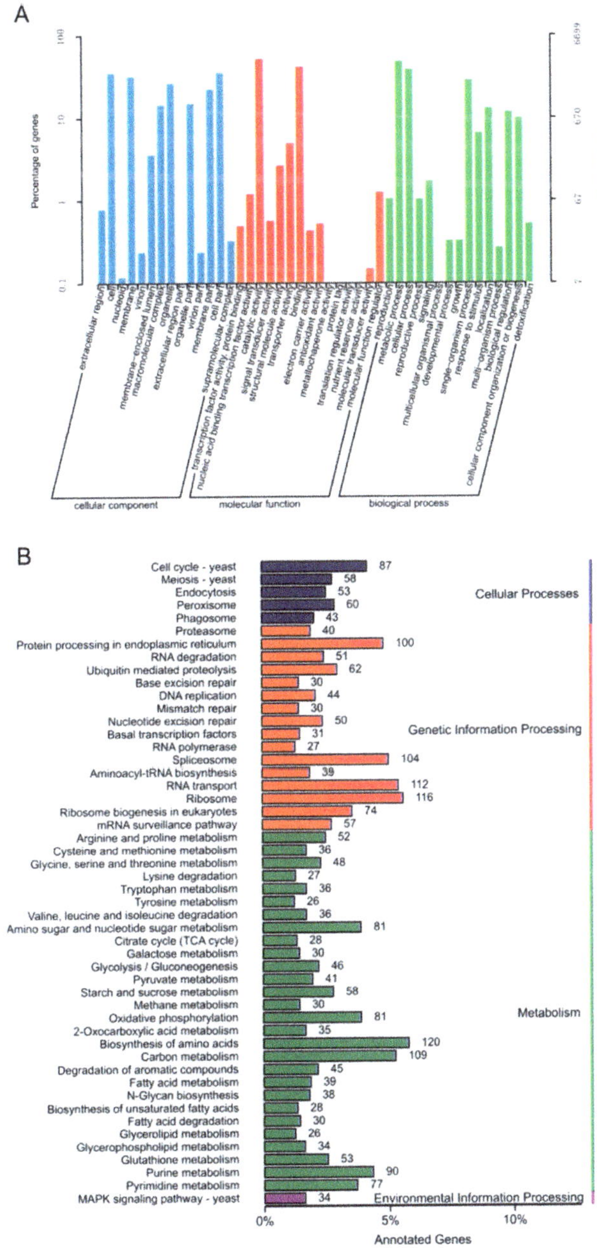


**Figure 2.** Gene ontology (GO) and Kyoto Encyclopedia of Genes and Genomes (KEGG) functional annotation of the *Ganoderma lucidum* yw-1-5 genome. GO functional enrichment (**A**) and KEGG pathway (**B**) enrichment of putative genes in yw-1-5.

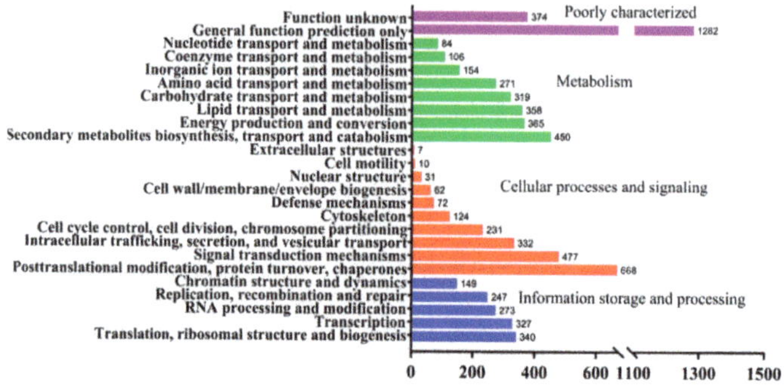


**Figure 3.** Genes functional classification of yw-1-5 by KOG.

### 3.4. Annotation in CAZy Databases

CAZy enzymes (CAZymes) that degrade, modify, or create glycosidic bonds confer wood-decay capability to *G. lucidum*, and play important roles in *G. lucidum* growth and development. Therefore the Carbohydrate-Active enZYmes Database (CAZy) was used to map the genome of *G. lucidum*. In total, 643 candidate CAZymes were identified in the yw-1-5 genome including 101 auxiliary activities (AAs), 15 polysaccharide lyases (PLs), 90 carbohydrate-binding modules (CBMs), 114 carbohydrate esterases (CEs), 313 glycoside hydrolases (GHs) and 83 glycosyl transferases (GTs). The most abundant type was the glycoside hydrolases. For revealing the evolution of CAZymes, a comparison between *G. lucidum* and 10 selected Basidiomycete species was conducted (Table S10). Interestingly, we found that the abundance of CAZymes in yw-1-5 was much more than the 489 in the previous *G. lucidum* genome [19] and was comparable with 614 candidate CAZymes in *Ganoderma leucocontextum* [56].

### 3.5. Genes Involved in Synthesis of Polysaccharides from G. lucidum Genome

Polysaccharides identified from *G. lucidum* fermentation are mainly divided into exopolysaccharides and cell-wall polysaccharides. Generally, the biosynthesis of exopolysaccharides includes the synthesis of nucleotide-activated sugars, the linking and modification of sugar chains, and extracellular export. Exopolysaccharides from *G. lucidum* fermentation are mainly composed of glucose, galactose, mannose, arabinose and rhamnose [7]. Based on the KEGG pathway, the identification of genes associated with the synthesis of nucleotide-activated sugars was performed in our genome data. In the amino sugar and nucleotide sugar metabolism (00520) KEGG pathway, genes related to the synthesis of UDP-Glc, UDG-Gal, GDP-man, GDP-Fuc, GDP-Arb, UDP-GlcNAc, UDP-D-Xyl and UDP-GlcA have been found (Table S11, Figure S3). For instance, the metabolic pathway of the most important nucleotide-activated sugar, UDP-Glc, containing phosphoglucomutase (PGM), glucose phosphate isomerase (PGI) and UDP-glucose pyrophosphorylase (UGP), had been predicted in *G. lucidum* yw-1-5 genome (Table S11). Additionally, for the synthesis of GDP-man, genes encoding mannose-1-phosphate guanylyltransferase (GMPP), phosphomannomutase (PMM) and hexokinase (HK) were also obtained from the *G. lucidum* yw-1-5 genome (Table S11).

Previous studies showed that reactions linking and modifying sugar chains were catalyzed by glycosyltransferases and glycoside hydrolases. Glycosyltransferases are a response for the sugar chains, while glycoside hydrolases are considered as transglycosides in polysaccharide modification. Additionally, the cell wall of fungi mainly consists of chitin and glucans (β-1,3-(1,6)-glucan, β-1,6-glucan). Key enzymes participating in these polysaccharides are glycosyltransferases and glycoside hydrolases. Genes encoding glycosyltransferases and glycoside hydrolases had been predicted through functional annotation in the CAZy database. Because the synthesis of β-1,3-glucan, β-1,6-glucan, chitin had been investigated fully, genes encoding glycosyltransferases and glycoside hydrolases associated with these polysaccharides had been predicted for the *G. lucidum* yw-1-5 genome, as showed in Table S11. In particular, we obtained four genes encoding 1,3-beta-glucan synthase FKS and eight genes encoding beta-glucan synthesis-associated protein KRE6. Finally, there are no reports on the extracellular export of polysaccharides in fungi. New tools should be developed to reveal the mechanism of extracellular export of polysaccharides in *G. lucidum*. Genes annotated in the MAPK (04011) KEGG pathway were also identified in the yw-1-5 genome (see Figure S4).

### 3.6. Transcriptomic Analysis of G. lucidum Mycelium Treated with Tween80

Our previous research showed that a wide range of concentrations of Tween80 had the capacity to increase *G. lucidum* exopolysaccharide production in submerged fermentation [12]. In this study, 2.5% (v/v) Tween80 also increased the production of exopolysaccharide (Figure 4). However, the mechanism of the effect of Tween80 on enhanced exopolysaccharide production in *G. lucidum* is still unknown. Recently, RNA-seq was performed in a Tween80-treated group and control group for the identification of genes associated with this effect (Figure S5). In total, 30.8 Gb clean data were generated from six samples (Tween80-treated group and control group, 3 replicates) (Table S12). There were an average of 5.1 Gb clean data per sample. The Q20 value of each sample was above 97%, the GC of which was above 59%. A total of 10,030 transcripts were obtained after assembly. The transcripts were annotated with the GO, KEGG, NR, Swiss-Prot, Pfam and EggNOG databases for the prediction of possible functions (see Figure S6). In total, 9316 transcripts were annotated and numbers of transcripts annotated in the GO and KEGG were 7303, 3910.

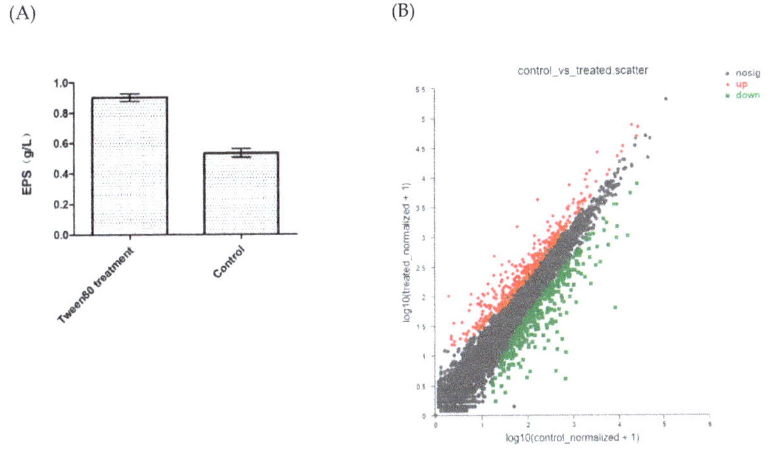


**Figure 4.** The content of exopolysaccharides (**A**) and the comparison of DEG expression between Tween80-treated group and control group (**B**).

The expression level of each gene was analyzed with the FPKM method (Figure S7). Genes which $|\log2FC| \geq 1$ and padjust < 0.05 were defined as differentially expressed genes (DEGs). In this study, a total of 655 genes were classified as differentially expressed genes (DEGs), including 341 up-regulated DEGs and 314 down-regulated. The functional annotation of these DEGs with GO and KEGG databases was performed. In GO enrichment analysis, 655 DEGs were categorized into 219 GO terms (Figure S8). Most of the DEGs belonged to the carbohydrate metabolic process, acyl-CoA dehydrogenase activity, interspecies interaction between organisms and carbon–oxygen lyase activity, acting on polysaccharide GO terms. Moreover, KEGG analysis revealed that 655 DEGs were found to map to 78 KEGG pathways (Figure S9). The KEGG pathways with most of the DEGs were peroxisome, steroid biosynthesis and starch, sucrose and galactose metabolism. The results of the KEGG analysis showed that most of the DEGs were categorized into metabolism, suggesting the addition of Tween80 majorly changed the metabolic processes of *G. lucidum* mycelium. Finally, in total, five DEGs were selected to perform qRT-PCR to validate the results obtained from RNA-seq. For each selected DEG, the results of qRT-PCR exhibited similar expression patterns between groups compared with the FPKM values of RNA-seq data, suggesting that the results of RNA-seq were reliable (see Table S17).

### *3.7. Differentially Expressed Genes Involved in the Enhanced Effect of Tween80 on Exopolysaccharide Production*

MAPK pathway is essential for the formation and regulation of fungal cell walls [57]. Further analysis of DEGs showed that nine DEGs were mapped in the MAPK signaling pathway—yeast KEGG pathway, including six up-regulated DEGs (*fks2*, *rsp5*, *stt4*, *bmh1,2*, *ptp2,3*, *mkk1,2*) and three down-regulated DEGs (*yck1,2*, *sln1*, *ctt1*). Importantly, gene *fks2* encode glycosyltransferase-family 48 protein, which is a key enzyme of the synthesis of 1,3-β-glucan. These results suggest that these DEGs participated in cell-wall modeling and the regulation of polysaccharide synthesis (Table S13).

There are few exopolysaccharides purified from submerged cultured *G. lucidum*. For instance, GLEP-2 is a exopolysaccharide from submerged cultured *G. lucidum* which is mainly composed of glucose [7]. Besides this, polysaccharides from intracellular parts and the cell wall of *G. lucidum* mycelium include chitin, β-1,3-glucan and β-1,6-glucan. The mechanism of exopolysaccharide synthesis was discussed from three perspectives (nucleotide-activated sugars, linking and modification of sugar chains and extracellular export) based on KEGG annotation. Nucleotide sugars serve as donors for glycosyltransferases that are key enzymes of polysaccharides biosynthesis. In the metabolic pathway of synthetic sugar nucleotides, amino sugar and nucleotide sugar metabolism (00520) KEGG pathway, three genes including *nagB* (encoding glucosamine-6-phosphate deaminase), *nagA* (encoding N-acetyl glucosamine-6-phosphate deacetylase), and gene encoding HEXA_B (hexosaminidase) were up-regulated. In addition, genes encoding chitinase (a glycoside-hydrolase-family 18 protein), xylan 1,4-beta-xylosidase (XYL4), chitin deacetylase, mannose-1-phosphate guanylyltransferase (ManC) were down-regulated (Table S13). In addition, glycosyltransferases and glycoside hydrolases are two important enzymes of polysaccharide synthesis in *G. lucidum*. In this study, we obtained six DEGs encoding glycosyltransferases, including five up-regulated DEGs and one down-regulated DEG. Among them, two DEGs were found to encode β-1,3-glucan synthase and three DEGs participated in the biosynthesis of chitin (Table S14). In addition, eight DEGs encoding glycoside hydrolases were found, along with three up-regulated DEGs and five down-regulated DEGs (Table S14). These results suggest that Tween80 had an effect on the process of exopolysaccharide synthesis in *G. lucidum*.

Previous studies have shown that glucose was consumed more rapidly and the ATP level increased in the Tween80-treated group [12]. They suggested that the carbohydrate-related energy system could participate in the enhanced effect of Tween80 on exopolysaccharide production in *G. lucidum*. In the glycolysis/gluconeogenesis KEGG pathway (00010), *pckA* was found to be up-regulated in the Tween80-treated group. On the other hand, genes encoding pyruvate carboxylase (PYC), phosphoenolpyruvate carboxykinase (PCKA) and fumarate hydratase (FH) from the tricarboxylic acid cycle KEGG pathway (TCA cycle, 00020) were up-regulated. In the starch and sucrose metabolism KEGG pathway (00500), genes encoding glucan 1,3-beta-glucosidase and beta-glucosidase (BglX) were up-regulated, suggesting that the capability for degrading carbohydrates in a medium increased when providing more glucose. In conclusion, starch and sucrose metabolism, TCA cycle and glycolysis/gluconeogenesis KEGG pathways produced energy and materials for polysaccharide synthesis.

Reactive oxygen species (ROS) is an essential part of the abiotic stress response and development in plants [58]. The Tween80 treatment of mycelium in *G. lucidum* led to the accumulation of ROS [12]. The peroxisome KEGG pathway participated in the metabolism of ROS and stress response. In the present study, we found that in the peroxisome KEGG pathway (04146), genes encoding acyl-CoA oxidase (ACOX) and acetyl-CoA acyltransferase 1 (ACAA1) from fatty acid oxidation (β-oxidation) section, PDCR and ABCD from unsaturated fatty acid β-oxidation section, CRAT from other oxidation PTS1-type section, PEX11 and PMP34 were up-regulated. Moreover, gene-encoding CAT was down-regulated. Interestingly, genes encoding ACOX and ACAA1 were also involved in the fatty acid degradation (00071) KEGG pathway.

Previous studies have shown that fatty acid degradation and the accumulation of ROS induces autophagy [59]. The ubiquitin-mediated proteolysis and autophagy play important roles in cellular homeostasis [60]. In the autophagy-related KEGG pathways autophagy—yeast (04138) and autophagy—other (04139), two genes encoding ATG4 and ATG7 were up-regulated. Additionally, in the KEGG pathway ubiquitin-mediated proteolysis (04120), two up-regulated genes (encoding NEDD4 and ARF-BP1) from "HECT type E3" section and one up-regulated gene (encoding Apc1) from the "target recognizing other subunits" section were found. These results suggest that protein degradation through autophagy and ubiquitin-mediated proteolysis might be an important type of regulation during exopolysaccharide synthesis and Tween80 treatment.

Transcription factors are types of proteins which play important roles in the regulation of transcription through interaction with the promoter region of protein-coding genes [61]. Genes encoding transcription factors associated with the effect of Tween80 on enhanced exopolysaccharide production in *G. lucidum* had never been reported. In this work, we obtained 19 DEGs encoding transcription factors from RNA-seq data, including 14 up-regulated and 5 down-regulated (see Table S16). Among them, 11 DEGs encoding zinc-finger-type transcription factors, especially the HMG-box domain type and fungal-specific type. The DEGs encoding transcription factors need to be investigated by further knock-down and overexpression experiments. Genes related to the synthesis of exopolysaccharides and Tween80 treatment above are listed in Tables S13, S14 and S16.

## 4. Discussion

*G. lucidum* is one of most important medicinal mushrooms in Asia countries. The mycelium of *G. lucidum* yw-1 strain has been used as raw material for Lingzhi products in Guangdong Yuewei Edible Fungi Technology Co., (Guangzhou, China). In addition, polysaccharides have been considered one of main bioactive components of *G. lucidum* for more than twenty years. Due to the medicinal value of *G. lucidum*, we had developed a submerged cultivation technique supplied with Tween80 for enhanced exopolysaccharide production in *G. lucidum* [12]. Since the first report on the genome of *G. lucidum* appeared in 2012, several efforts had been made to achieve a more complete genome [18]. For instance, a more complete reference genome sequence of *G. lucidum* laid the foundation for the breeding of new strains with a high content of bio-active products [19]. Herein, we report a 58.16 Mb reference genome of *G. lucidum* monokaryon strain yw-1-5 containing 14,849 protein-coding genes. Moreover, the strain yw-1-5 genome sequence has been applied to revealing the mechanism of Tween80-enhanced exopolysaccharide synthesis in this study.

Previous studies have reported that long reads, which are generated by the single-molecule Oxford Nanopore technology sequencing platform, have been assembled to a high-quality reference human genome [62]. Besides this, Hi-C is a current approach, based on chromosome conformation capture to detect genome-wide chromatin interaction, and which can be used to widely assist genome assembly [63]. A sequencing strategy combining Oxford Nanopore technology sequence, Illumina data and Hi-C data to generate a chromosome-scale *de novo* assembly of *Lentinus edodes* has been reported [20]. In the present study, we carried out de novo genome sequencing of *G. lucidum* using this sequencing strategy. To our knowledge, the evaluation of assembly by the number of Scaffold and N50 of *G. lucidum* yw-1-5 genome in this work is the best compared with *G. lucidum* genomes from NCBI, proving that Hi-C is an effective method to assist genome assembly in *G. lucidum*. Hi-C data has been mapped to 12 pseudo-chromosomes, whereas the number of chromosomes was thirteen in a previous study [18]. Additional experiments are needed to solve this controversial issue.

Glycosyltransferases are key enzymes which catalyze the formation of glycosidic bonds during the synthesis of polysaccharides [16]. Glycoside hydrolases can degrade, modify and create glycosidic bonds, and are responsible for the modification of sugar chains [17]. In total, 313 genes encoding glycoside hydrolases and 83 glycosyl transferases were obtained from the *G. lucidum* yw-1-5 genome. Genes from the amino sugar and nucleotide sugar metabolism KEGG pathway (00520) had been proved to be involved in the synthesis of nucleotide-activated sugars [64]. Research papers have shown that genes which are involved in the MAPK signaling pathway—yeast KEGG pathway participate in the regulation of exopolysaccharide synthesis [65]. In this study, we also found genes mapped to these two KEGG pathways in the *G. lucidum* genome. Therefore, our *G. lucidum* yw-1-5 genome sequence laid the foundation for researching the synthesis of exopolysaccharides.

Many efforts (e.g., culturing in fine wheat-bran powder [66], the limitation of nitrogen [67], the optimization of pH [6]) have been made to improve exopolysaccharide production in *G. lucidum*. Our previous work showed that the content of exopolysaccharides in *G. lucidum* increased when the mycelium was treated with Tween80 [12], but the mechanism of Tween80-enhanced exopolysaccharide production is unclear. In this research, the transcriptome sequencing of a Tween80-treated group and control group was conducted based on next-generation technology. In total, 655 differentially expressed genes (DEGs) were found in our transcriptome data, including 341 up-regulated DEGs and 314 down-regulated.

Our transcriptomic study showed that six genes in the MAPK signaling pathway—yeast KEGG pathway were found to be up-regulated, including *fks2* and *mkk1,2*. The increase in the content of exopolysaccharides might be due to the up-regulation of *fks2* which encodes β-1,3-glucan synthase, a key enzyme in β-1,3-glucan synthesis. Another piece of research on *Schizophyllum commune* showed that the up-regulation of *fks2* correlated with changes in the content of the exopolysaccharide schizophyllan after Tween80 treatment [11]. Therefore, we proposed that perhaps Tween80 increases exopolysaccharide production through the enhancement of β-1,3-glucan synthesis in *G. lucidum* mycelium. Previous studies have shown that cell-wall integrity kinase (Mkk1,2) was involved in cell-wall integrity and remodeling [68]. By increasing the expression level of *mkk1,2*, Tween80 regulated cell-wall integrity and cell-wall remodeling in *G. lucidum* mycelium.

Nucleotide sugars are blocks for the synthesis of polysaccharides. The amino sugar and nucleotide sugar metabolism (00520) KEGG pathway is involved in the metabolic pathway of synthetic sugar nucleotides. Our transcriptomic results showed that seven genes from this KEGG pathway were differentially expressed. Surprisingly, genes encoding phosphoglucomutase (PGM) and UDP-glucose pyrophosphorylase (UGP) were not differentially expressed. Our previous work showed that genes encoding phosphoglucomutase and UDP-glucose pyrophosphorylase were up-regulated in *G. lucidum* mycelium treated with 0.25% Tween 80 on day 8 [12]. Controversial results in our transcriptome data might be due to the synthesis of nucleotide-activated sugar UDP-Glc in *G. lucidum* mycelium was dynamic during fermentation and Tween80 treatment. In addition, glycosyltransferases and glycoside hydrolases are two key types of enzyme in polysaccharide synthesis in *G. lucidum*. Importantly, we obtained six DEGs encoding glycosyltransferases, two of which encoded β-1,3-glucan synthase. Therefore, Tween80 might increase the content of exopolysaccharide in *G. lucidum* mycelium through the enhancement of β-1,3-glucan synthesis. Three DEGs encoding glycosyltransferases participated in the biosynthesis of chitin were also obtained, suggesting that the change in cell-wall component and cell-wall integrity is adapted to Tween80 treatment. Eight DEGs encoding glycoside hydrolases were identified and will be functionally investigated further.

In *Arabidopsis*, RBOH-mediated ROS facilitated lateral root emergence by promoting the remodeling of the cell wall of overlying parental tissues [69], indicating that ROS are associated with changes in cell-wall components. Surfactant CTAB induced the accumulation of ROS and enhanced mannatide production [70]. Previous studies have reported that the Tween80 treatment of *G. lucidum* mycelium lead to the accumulation of ROS [12]. It is suggested that Tween80 induced ROS might play a specific role in growth, the excretion of exopolysaccharides and cell-wall remodeling, resulting in an increase in the content of exopolysaccharides in *G. lucidum*. In our transcriptome data, seven up-regulated genes in the peroxisome KEGG pathway were achieved and the expression of these genes might contribute to the metabolism of ROS. Above all, Tween80 could induce ROS accumulation through the peroxisome pathway, which might lead to cell-wall remodeling and the excretion of exopolysaccharides. However, proteins linking ROS accumulation with the excretion of exopolysaccharides are needed to investigate this.

ROS-induced autophagy and ubiquitin-mediated proteolysis are two types of protein degradation, participating in cellular homeostasis [60]. ATG4 protease, considered an essential component for the biogenesis of autophagosome, was regulated by reactive oxygen species in *Chlamydomonas reinhardtii* [71]. In this study, one gene encoding ATG4 in the autophagy-related KEGG pathway was up-regulated, indicating that ATG4 protease might be a link of Tween80-induced ROS and autophagy. It then promotes the production of exopolysaccharides. In addition, two up-regulated genes encoding NEDD4 and ARF-BP1 from the ubiquitin-mediated proteolysis KEGG pathway were identified. Other studies have shown that as a post-translational modification, ubiquitination controls a few of the steps in autophagy [71], that is, there is cross-talk between autophagy and ubiquitin-mediated proteolysis. In conclusion, autophagy and ubiquitin-mediated proteolysis might be crucial for the Tween80-enhanced production of exopolysaccharides in *G. lucidum*. Previous studies showed transcription factors such as the blue light receptor–white collar complex (WC-1, WC-2) regulate fruiting-body formation and cell-wall development [72]. We found 19 DEGs encoding transcription factors from RNA-seq data, which might play important roles in Tween80 treatment and exopolysaccharide production. Interestingly, 11 DEGs encode zinc-type transcription factors. These 19 DEGs had not been studied in *G. lucidum* and need to be investigated by functional experiments.

Finally, according to the results of our genome and transcriptome, we propose a model for explaining the enhanced effect of Tween80 on exopolysaccharide production, as shown in Figure 5. Abstractly, through the regulation of the expression level of genes belonging to KEGG pathways (e.g., the MAPK signaling pathway) and genes encoding transcription factors, glycosyltransferases and glycoside hydrolases, Tween80 enhances exopolysaccharide production in *G. lucidum*. This model should be further investigated by knock-down (e.g., RNAi), knock-out (e.g., CRISPR-Cas9 technology) and over-expression experiments.

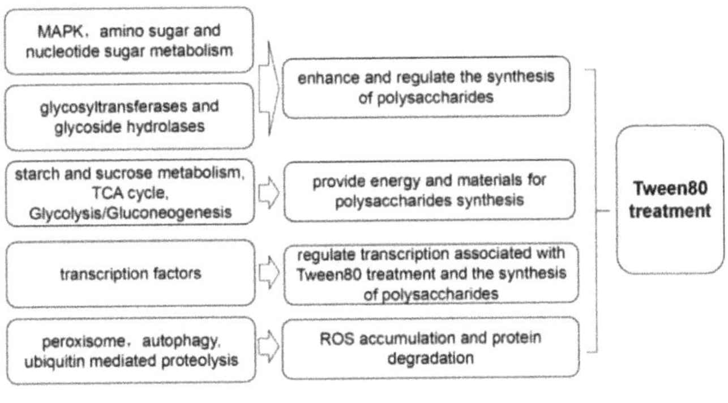


**Figure 5.** A model of the enhanced effect of Tween80 on exopolysaccharide production.

## 5. Conclusions

Firstly, the addition of Tween80 under submerged fermentation is an effective approach to increase exopolysaccharide production in *G. lucidum*. We obtained a high quality de novo assembly of *G. lucidum* strain yw-1-5 combining Oxford Nanopore technology sequencing, Illumina data and Hi-C data. The elucidation of this genome provided crucial genomic information to investigate the synthesis of exopolysaccharides. We also obtained genes encoding glycosyltransferases and glycoside hydrolases, which are potential key enzymes of the synthesis of exopolysaccharides. Moreover, our study is the first to show that genes from MAPK, amino sugar and nucleotide sugar metabolisms, autophagy, ubiquitin-mediated proteolysis, peroxisome, starch and sucrose metabolism, TCA cycle and glycolysis/gluconeogenesis KEGG pathways and genes encoding glycosyltransferases and glycoside hydrolases might be involved in the enhanced effect of Tween80 on exopolysaccharide production as revealed by RNA-seq technology. To summarize, this study provides new insights into the synthesis of exopolysaccharides in an important medicinal mushroom, *G. lucidum*, and the addition of Tween80 can be used to enhance exopolysaccharide production under submerged fermentation.

**Supplementary Materials:** The following supporting information can be downloaded at: https://www.mdpi.com/article/10.3390/jof8101081/s1. Table S1: Statistics of the raw data and clean reads; Table S2: Statistics of the Hi-C assembly; Table S3: Comparisons of *G. lucidum* assembled genome; Table S4: Mapping rate of the Illumina data and Benchmarking Universal Single-Copy Ortholog analysis; Table S5: Statistics of the gene model; Table S6: Gene clusters in the yw-1-5 genome; Table S7: Statistics of the repetitive sequences in the *G. lucidum* genome; Table S8: Statistics of the gene functional annotation in different databases; Table S9: Cytochrome P450s genes in the *G. lucidum* genome. Table S10: The CAZyme spectra were compared between *G. lucidum* and 8 selected Basidiomycete species; Table S11: The putative genes involved in polysaccharide biosynthesis of *G. lucidum*; Table S12: Summary of sequencing and assembly of RNA-seq; Table S13: Putative genes involved in Tween80 treatment obtained from KEGG pathway analysis; Table S14: Genes encoding glycosyltransferase and glycoside hydrolase associated with Tween80 treatment; Table S15: Oligonucleotide primer sets for qRT-PCR; Table S16: Genes encoding transcription factors involved in Tween80 treatment; Table S17: The relative expression levels of selected DEGs. Supplemental Figures, Figure S1: Fruiting body of *G. lucidum* dikaryon strain yw-1; Figure S2: *G. lucidum* dikaryon strain yw-1 and monokaryon strain yw-1-5 grown on PDA plate; Figure S3: Genes annotated in amino sugar and nucleotide sugar metabolism (00520) KEGG pathway from genome; Figure S4: Genes annotated in MAPK (04011) KEGG pathway from genome. Figure S5: Tween80-treated group and control group; Figure S6: Functional annotation and classification of transcripts; Figure S7: The distribution of expression level in Tween80-treated group and control group; Figure S8: Gene ontology analysis of DEGs between Tween80-treated group and control group; Figure S9: KEGG pathway distribution of DEGs between Tween80-treated group and control group.

**Author Contributions:** Conceptualization, T.W. and M.C.; methodology, T.W.; software, J.C. and Y.L.; C.X., X.L., X.G. and S.C. collected the samples; writing—original draft preparation, T.W. and M.C.; writing—review and editing, Z.Z.; supervision, H.H. and C.J.; project administration, Q.W. and Y.X.; funding acquisition, Q.W. and Y.X. All authors have read and agreed to the published version of the manuscript.

**Funding:** This work was supported by the GDAS Project of Science and Technology Development (2020GDASYL-20200103022) and Key-Area Research and Development Program of Guangdong Province (Grant No.2018B0202050010).

**Institutional Review Board Statement:** Not applicable.

**Informed Consent Statement:** Not applicable.

**Data Availability Statement:** The data that support the findings of this study had deposited in NCBI associated with bioproject PRJNA886764 and PRJNA886761. In addition, data are also available from the corresponding author upon reasonable request. All data generated or analyzed during this study are included in this published article and its supplementary additional files.

**Acknowledgments:** We would like to thank Biomarker Technologies (Beijing, China) and Xiaoxian Wu from Institute of Microbiology, Guangdong Academy of Sciences for providing technological assistance. We also thank Guangdong Yuewei Edible Fungi Technology Co. For providing the place to culture *G. lucidum*.

**Conflicts of Interest:** The authors declare that there are no conflicts of interest about this work. The Guangdong Yuewei Edible Fungi Technology Co., has no conflict of interest about the work.

## References


1. Zhang, K.; Liu, Y.; Zhao, X.; Tang, Q.; Dernedde, J.; Zhang, J.; Fan, H. Anti-inflammatory properties of GLPss58, a sulfated polysaccharide from *Ganoderma lucidum*. *Int. J. Biol. Macromol.* **2018**, *107*, 486–493. [CrossRef] [PubMed]
2. Xiao, C.; Wu, Q.; Xie, Y.; Tan, J.; Ding, Y.; Bai, L. Hypoglycemic mechanisms of *Ganoderma lucidum* polysaccharides F31 in db/db mice via RNA-seq and iTRAQ. *Food Funct.* **2018**, *9*, 6495–6507. [CrossRef] [PubMed]
3. Zhang, Y.; Li, H.; Song, L.; Xue, J.; Wang, X.; Song, S.; Wang, S. Polysaccharide from *Ganoderma lucidum* ameliorates cognitive impairment by regulating the inflammation of the brain-liver axis in rats. *Food Funct.* **2021**, *12*, 6900–6914. [CrossRef] [PubMed]
4. Lu, J.; He, R.; Sun, P.; Zhang, F.; Linhardt, R.J.; Zhang, A. Molecular mechanisms of bioactive polysaccharides from *Ganoderma lucidum* (Lingzhi), a review. *Int. J. Biol. Macromol.* **2020**, *150*, 765–774. [CrossRef] [PubMed]
5. Jan, J.; Cheng, T.R.; Juang, Y.; Ma, H.; Wu, Y.; Yang, W.; Cheng, C.; Chen, X.; Chou, T.; Shie, J.; et al. Identification of existing pharmaceuticals and herbal medicines as inhibitors of SARS-CoV-2 infection. *Proc. Natl. Acad. Sci. USA* **2021**, *118*, e2021579118. [CrossRef]
6. Papinutti, L. Effects of nutrients, pH and water potential on exopolysaccharides production by a fungal strain belonging to *Ganoderma lucidum* complex. *Bioresour Technol.* **2010**, *101*, 1941–1946. [CrossRef]
7. Zhou, H.; Liu, G.; Huang, F.; Wu, X.; Yang, H. Improved production, purification and bioactivity of a polysaccharide from submerged cultured *Ganoderma lucidum*. *Arch. Pharmacal Res.* **2014**, *37*, 1530–1537. [CrossRef]
8. Zhang, W.; Tang, Y.-J. A Novel Three-Stage Light Irradiation Strategy in the Submerged Fermentation of Medicinal Mushroom *Ganoderma lucidum* for the Efficient Production of Ganoderic Acid and Ganoderma Polysaccharides. *Biotechnol. Prog.* **2008**, *24*, 1249–1261. [CrossRef]
9. Liang, Y.; Zhu, L.; Gao, M.; Zheng, Z.; Wu, J.; Zhan, X. Influence of Tween-80 on the production and structure of water-insoluble curdlan from *Agrobacterium* sp. *Int. J. Biol. Macromol.* **2018**, *106*, 611–619. [CrossRef]
10. Li, Q.; Lei, Y.; Hu, G.; Lei, Y.; Dan, D. Effects of Tween 80 on the liquid fermentation of *Lentinus edodes*. *Food Sci. Biotechnol.* **2018**, *27*, 1103–1109. [CrossRef]
11. Meng, Q.; Chuai, S.; Chen, L.; Wang, L.; Cai, G.; Mao, J.; Gu, Z.; Shi, G.; Ding, Z. Effect of surfactants on the production of polysaccharides from *Schizophyllum commune* through submerged fermentation. *Int. J. Biol. Macromol.* **2021**, *192*, 210–218. [CrossRef] [PubMed]
12. Yang, X.; Yang, Y.; Zhang, Y.; He, J.; Xie, Y. Enhanced exopolysaccharide production in submerged fermentation of *Ganoderma lucidum* by Tween 80 supplementation. *Bioprocess Biosyst. Eng.* **2021**, *44*, 47–56. [CrossRef] [PubMed]
13. Schmid, J.; Sieber, V. Enzymatic transformations involved in the biosynthesis of microbial exo-polysaccharides based on the assembly of repeat units. *Chembiochem* **2015**, *16*, 1141–1147. [CrossRef] [PubMed]
14. Xu, J.-W.; Ji, S.-L.; Li, H.-J.; Zhou, J.-S.; Duan, Y.-Q.; Dang, L.-Z.; Mo, M.-H. Increased polysaccharide production and biosynthetic gene expressions in a submerged culture of *Ganoderma lucidum* by the overexpression of the homologous α-phosphoglucomutase gene. *Bioprocess Biosyst. Eng.* **2015**, *38*, 399–405. [CrossRef]
15. Li, M.; Chen, T.; Gao, T.; Miao, Z.; Jiang, A.; Shi, L.; Ren, A.; Zhao, M. UDP-glucose pyrophosphorylase influences polysaccharide synthesis, cell wall components, and hyphal branching in *Ganoderma lucidum* via regulation of the balance between glucose-1-phosphate and UDP-glucose. *Fungal Genet. Biol.* **2015**, *82*, 251–263. [CrossRef]
16. Lairson, L.L.; Henrissat, B.; Davies, G.J.; Withers, S.G. Glycosyltransferases: Structures, functions, and mechanisms. *Annu. Rev. Biochem.* **2008**, *77*, 521–555. [CrossRef]
17. Thanh Nguyen, H.; Zhang, R.; Inokawa, N.; Oura, T.; Chen, X.; Iwatani, S.; Niimi, K.; Niimi, M.; Holmes, A.R.; Cannon, R.D.; et al. *Candida albicans* Bgl2p, Ecm33p, and Als1p proteins are involved in adhesion to saliva-coated hydroxyapatite. *J. Oral Microbiol.* **2021**, *13*, 1879497. [CrossRef]
18. Chen, S.; Xu, J.; Liu, C.; Zhu, Y.; Nelson, D.R.; Zhou, S.; Li, C.; Wang, L.; Guo, X.; Sun, Y.; et al. Genome sequence of the model medicinal mushroom *Ganoderma lucidum*. *Nat. Commun.* **2012**, *3*, 913. [CrossRef]
19. Tian, Y.-Z.; Wang, Z.-F.; Liu, Y.-D.; Zhang, G.-Z.; Li, G. The whole-genome sequencing and analysis of a *Ganoderma lucidum* strain provide insights into the genetic basis of its high triterpene content. *Genomics* **2021**, *113*, 840–849. [CrossRef]
20. Yu, H.; Zhang, L.; Shang, X.; Peng, B.; Li, Y.; Xiao, S.; Tan, Q.; Fu, Y. Chromosomal genome and population genetic analyses to reveal genetic architecture, breeding history and genes related to cadmium accumulation in *Lentinula edodes*. *BMC Genom.* **2022**, *23*, 120. [CrossRef]
21. Yang, L.; Tang, J.; Chen, J.-J.; Peng, A.-Y.; Wang, Q.-M.; Rao, L.-Q.; Yang, H.; Zhang, X.-W.; Yang, H.-Z.; Zhang, C.; et al. Transcriptome analysis of three cultivars of *Poria cocos* reveals genes related to the biosynthesis of polysaccharides. *J. Asian Nat. Prod. Res.* **2019**, *21*, 462–475. [CrossRef] [PubMed]
22. Zhang, N.; Tang, Z.; Zhang, J.; Li, X.; Yang, Z.; Yang, C.; Zhang, Z.; Huang, Z. Comparative transcriptome analysis reveals the genetic basis underlying the biosynthesis of polysaccharides in *Hericium erinaceus*. *Bot. Stud.* **2019**, *60*, 15. [CrossRef] [PubMed]
23. Li, Q.; Chen, J.; Liu, J.; Yu, H.; Zhang, L.; Song, C.; Li, Y.; Jiang, N.; Tan, Q.; Shang, X.; et al. *De novo* Sequencing and Comparative Transcriptome Analyses Provide First Insights Into Polysaccharide Biosynthesis During Fruiting Body Development of *Lentinula edodes*. *Front. Microbiol.* **2021**, *12*, 627099. [CrossRef] [PubMed]

24. Walker, B.J.; Abeel, T.; Shea, T.; Priest, M.; Abouelliel, A.; Sakthikumar, S.; Cuomo, C.A.; Zeng, Q.; Wortman, J.; Young, S.K.; et al. Pilon: An integrated tool for comprehensive microbial variant detection and genome assembly improvement. *PLoS ONE* **2014**, *9*, e112963. [CrossRef]
25. Belton, J.-M.; McCord, R.P.; Gibcus, J.H.; Naumova, N.; Zhan, Y.; Dekker, J. Hi—C: A comprehensive technique to capture the conformation of genomes. *Methods* **2012**, *58*, 268–276. [CrossRef]
26. Li, H.; Durbin, R. Fast and accurate short read alignment with Burrows—Wheeler transform. *Bioinformatics* **2009**, *25*, 1754–1760. [CrossRef]
27. Simão, F.A.; Waterhouse, R.M.; Ioannidis, P.; Kriventseva, E.V.; Zdobnov, E.M. BUSCO: Assessing genome assembly and annotation completeness with single-copy orthologs. *Bioinformatics* **2015**, *31*, 3210–3212. [CrossRef]
28. Xu, Z.; Wang, H. LTR_FINDER: An efficient tool for the prediction of full-length LTR retrotransposons. *Nucleic Acids Res.* **2007**, *35*, W265–W268. [CrossRef]
29. Han, Y.; Wessler, S.R. MITE-Hunter: A program for discovering miniature inverted-repeat transposable elements from genomic sequences. *Nucleic Acids Res.* **2010**, *38*, e199. [CrossRef]
30. Price, A.L.; Jones, N.C.; Pevzner, P.A. *De novo* identification of repeat families in large genomes. *Bioinformatics* **2005**, *21* (Suppl. S1), i351–i358. [CrossRef]
31. Edgar, R.C.; Myers, E.W. PILER: Identification and classification of genomic repeats. *Bioinformatics* **2005**, *21* (Suppl. S1), i152–i158. [CrossRef] [PubMed]
32. Chen, N. Using RepeatMasker to identify repetitive elements in genomic sequences. *Curr. Protoc. Bioinform.* **2004**, *5*, 4–10. [CrossRef] [PubMed]
33. Lowe, T.M.; Eddy, S.R. tRNAscan-SE: A program for improved detection of transfer RNA genes in genomic sequence. *Nucleic Acids Res.* **1997**, *25*, 955–964. [CrossRef] [PubMed]
34. Nawrocki, E.P.; Eddy, S.R. Infernal 1.1: 100-fold faster RNA homology searches. *Bioinformatics* **2013**, *29*, 2933–2935. [CrossRef]
35. Medema, M.H.; Blin, K.; Cimermancic, P.; de Jager, V.; Zakrzewski, P.; Fischbach, M.A.; Weber, T.; Takano, E.; Breitling, R. antiSMASH: Rapid identification, annotation and analysis of secondary metabolite biosynthesis gene clusters in bacterial and fungal genome sequences. *Nucleic Acids Res.* **2011**, *39*, W339–W346. [CrossRef]
36. Burge, C.; Karlin, S. Prediction of complete gene structures in human genomic DNA. *J. Mol. Biol.* **1997**, *268*, 78–94. [CrossRef]
37. Stanke, M.; Waack, S. Gene prediction with a hidden Markov model and a new intron submodel. *Bioinformatics* **2003**, *19* (Suppl. S2), i215–i225. [CrossRef]
38. Majoros, W.H.; Pertea, M.; Salzberg, S.L. TigrScan and GlimmerHMM: Two open source ab initio eukaryotic gene-finders. *Bioinformatics* **2004**, *20*, 2878–2879. [CrossRef]
39. Blanco, E.; Parra, G.; Guigó, R. Using geneid to identify genes. *Curr. Protoc. Bioinform.* **2007**, *18*, 3–4.
40. Korf, I. Gene finding in novel genomes. *BMC Bioinform.* **2004**, *5*, 59. [CrossRef]
41. Keilwagen, J.; Wenk, M.; Erickson, J.L.; Schattat, M.H.; Grau, J.; Hartung, F. Using intron position conservation for homology-based gene prediction. *Nucleic Acids Res.* **2016**, *44*, e89. [CrossRef] [PubMed]
42. Sirén, J.; Välimäki, N.; Mäkinen, V. Indexing Graphs for Path Queries with Applications in Genome Research. *IEEE/ACM Trans. Comput. Biol. Bioinform.* **2014**, *11*, 375–388. [CrossRef] [PubMed]
43. Kovaka, S.; Zimin, A.V.; Pertea, G.M.; Razaghi, R.; Salzberg, S.L.; Pertea, M. Transcriptome assembly from long-read RNA-seq alignments with StringTie2. *Genome Biol.* **2019**, *20*, 278. [CrossRef] [PubMed]
44. Haas, B.J.; Delcher, A.L.; Mount, S.M.; Wortman, J.R.; Smith, R.J.; Hannick, L.I.; Maiti, R.; Ronning, C.M.; Rusch, D.B.; Town, C.D.; et al. Improving the *Arabidopsis* genome annotation using maximal transcript alignment assemblies. *Nucleic Acids Res.* **2003**, *31*, 5654–5666. [CrossRef] [PubMed]
45. Haas, B.J.; Salzberg, S.L.; Zhu, W.; Pertea, M.; Allen, J.E.; Orvis, J.; White, O.; Buell, C.R.; Wortman, J.R. Automated eukaryotic gene structure annotation using EVidenceModeler and the Program to Assemble Spliced Alignments. *Genome Biol.* **2008**, *9*, R7. [CrossRef]
46. Conesa, A.; Götz, S.; García-Gómez, J.M.; Terol, J.; Talón, M.; Robles, M. Blast2GO: A universal tool for annotation, visualization and analysis in functional genomics research. *Bioinformatics* **2005**, *21*, 3674–3676. [CrossRef]
47. Eddy, S.R. Profile hidden Markov models. *Bioinformatics* **1998**, *14*, 755–763. [CrossRef]
48. Chen, S.; Zhou, Y.; Chen, Y.; Gu, J. fastp: An ultra-fast all-in-one FASTQ preprocessor. *Bioinformatics* **2018**, *34*, i884–i890. [CrossRef]
49. Langmead, B.; Salzberg, S.L. Fast gapped-read alignment with Bowtie 2. *Nat. Methods* **2012**, *9*, 357–359. [CrossRef]
50. Kim, D.; Pertea, G.; Trapnell, C.; Pimentel, H.; Kelley, R.; Salzberg, S.L. TopHat2: Accurate alignment of transcriptomes in the presence of insertions, deletions and gene fusions. *Genome Biol.* **2013**, *14*, R36. [CrossRef]
51. Anders, S.; Huber, W. Differential expression analysis for sequence count data. *Genome Biol.* **2010**, *11*, R106. [CrossRef] [PubMed]
52. Klopfenstein, D.V.; Zhang, L.; Pedersen, B.S.; Ramírez, F.; Vesztrocy, A.W.; Naldi, A.; Mungall, C.J.; Yunes, J.M.; Botvinnik, O.; Weigel, M.; et al. GOATOOLS: A Python library for Gene Ontology analyses. *Sci. Rep.* **2018**, *8*, 10872. [CrossRef] [PubMed]
53. Xie, C.; Mao, X.; Huang, J.; Ding, Y.; Wu, J.; Dong, S.; Kong, L.; Gao, G.; Li, C.-Y.; Wei, L. KOBAS 2.0: A web server for annotation and identification of enriched pathways and diseases. *Nucleic Acids Res.* **2011**, *39*, W316–W322. [CrossRef] [PubMed]
54. Schmittgen, T.D.; Livak, K.J. Analyzing real-time PCR data by the comparative C(T) method. *Nat. Protoc.* **2008**, *3*, 1101–1108. [CrossRef] [PubMed]

55. Lu, M.-Y.J.; Fan, W.-L.; Wang, W.-F.; Chen, T.; Tang, Y.-C.; Chu, F.-H.; Chang, T.-T.; Wang, S.-Y.; Li, M.-Y.; Chen, Y.-H.; et al. Genomic and transcriptomic analyses of the medicinal fungus *Antrodia cinnamomea* for its metabolite biosynthesis and sexual development. *Proc. Natl. Acad. Sci. USA* **2014**, *111*, E4743–E4752. [CrossRef]
56. Liu, Y.; Huang, L.; Hu, H.; Cai, M.; Liang, X.; Li, X.; Zhang, Z.; Xie, Y.; Xiao, C.; Chen, S.; et al. Whole-genome assembly of *Ganoderma leucocontextum* (Ganodermataceae, Fungi) discovered from the Tibetan Plateau of China. *G3 Genes Genomes Genet.* **2021**, *11*, jkab337. [CrossRef]
57. González-Rubio, G.; Sastre-Vergara, L.; Molina, M.; Martín, H.; Fernández Acero, T. Substrates of the MAPK Slt2: Shaping Yeast Cell Integrity. *J. Fungi* **2022**, *8*, 368. [CrossRef]
58. Considine, M.J.; Foyer, C.H. Oxygen and reactive oxygen species-dependent regulation of plant growth and development. *Plant Physiol.* **2021**, *186*, 79–92. [CrossRef]
59. Singh, R.; Kaushik, S.; Wang, Y.; Xiang, Y.; Novak, I.; Komatsu, M.; Tanaka, K.; Cuervo, A.M.; Czaja, M.J. Autophagy regulates lipid metabolism. *Nature* **2009**, *458*, 1131–1135. [CrossRef]
60. Kwon, Y.T.; Ciechanover, A. The Ubiquitin Code in the Ubiquitin-Proteasome System and Autophagy. *Trends Biochem. Sci.* **2017**, *42*, 873–886. [CrossRef]
61. Li, R.; Wang, X.; Zhang, S.; Liu, X.; Zhou, Z.; Liu, Z.; Wang, K.; Tian, Y.; Wang, H.; Zhang, Y.; et al. Two zinc-finger proteins control the initiation and elongation of long stalk trichomes in tomato. *J. Genet. Genom.* **2021**, *48*, 1057–1069. [CrossRef] [PubMed]
62. Jain, M.; Koren, S.; Miga, K.H.; Quick, J.; Rand, A.C.; Sasani, T.A.; Tyson, J.R.; Beggs, A.D.; Dilthey, A.T.; Fiddes, I.T.; et al. Nanopore sequencing and assembly of a human genome with ultra-long reads. *Nat. Biotechnol.* **2018**, *36*, 338–345. [CrossRef] [PubMed]
63. Burton, J.N.; Adey, A.; Patwardhan, R.P.; Qiu, R.; Kitzman, J.O.; Shendure, J. Chromosome-scale scaffolding of *de novo* genome assemblies based on chromatin interactions. *Nat. Biotechnol.* **2013**, *31*, 1119–1125. [CrossRef] [PubMed]
64. Orlean, P. Architecture and Biosynthesis of the *Saccharomyces cerevisiae* Cell Wall. *Genetics* **2012**, *192*, 775–818. [CrossRef]
65. Zan, X.-Y.; Zhu, H.-A.; Jiang, L.-H.; Liang, Y.-Y.; Sun, W.-J.; Tao, T.-L.; Cui, F.-J. The role of Rho1 gene in the cell wall integrity and polysaccharides biosynthesis of the edible mushroom *Grifola frondosa*. *Int. J. Biol. Macromol.* **2020**, *165*, 1593–1603. [CrossRef]
66. Liu, S.-R.; Zhang, W.-R. Hyperproduction of exopolysaccharides by submerged mycelial culture of *Ganoderma lucidum* using a solid seed grown in fine-powder of wheat bran and in vitro evaluation of the antioxidant activity of the exopolysaccharides produced. *Food Sci. Biotechnol.* **2018**, *27*, 1129–1136. [CrossRef]
67. Hsieh, C.; Tseng, M.-H.; Liu, C.-J. Production of polysaccharides from *Ganoderma lucidum* (CCRC 36041) under limitations of nutrients. *Enzym. Microb. Technol.* **2006**, *38*, 109–117. [CrossRef]
68. Levin, D.E. Regulation of cell wall biogenesis in *Saccharomyces cerevisiae*: The cell wall integrity signaling pathway. *Genetics* **2011**, *189*, 1145–1175. [CrossRef]
69. Orman-Ligeza, B.; Parizot, B.; de Rycke, R.; Fernandez, A.; Himschoot, E.; Van Breusegem, F.; Bennett, M.J.; Périlleux, C.; Beeckman, T.; Draye, X. RBOH mediated ROS production facilitates lateral root emergence in *Arabidopsis*. *Development* **2016**, *143*, 3328–3339. [CrossRef]
70. Wang, L.; Sha, Y.; Wu, D.; Wei, Q.; Chen, D.; Yang, S.; Jia, F.; Yuan, Q.; Han, X.; Wang, J. Surfactant induces ROS-mediated cell membrane permeabilization for the enhancement of mannatide production. *Process Biochem.* **2020**, *91*, 172–180. [CrossRef]
71. Pérez-Pérez, M.E.; Lemaire, S.D.; Crespo, J.L. Control of Autophagy in *Chlamydomonas* Is Mediated through Redox-Dependent Inactivation of the ATG4 Protease. *Plant Physiol.* **2016**, *172*, 2219–2234. [CrossRef] [PubMed]
72. Ohm, R.A.; de Jong, J.F.; de Bekker, C.; Wösten, H.A.B.; Lugones, L.G. Transcription factor genes of *Schizophyllum commune* involved in regulation of mushroom formation. *Mol. Microbiol.* **2011**, *81*, 1433–1445. [CrossRef] [PubMed]

Journal of
*Fungi*

MDPI

*Article*

# Effects of Oleic Acid Addition Methods on the Metabolic Flux Distribution of Ganoderic Acids R, S and T's Biosynthesis

**Meng-Qiu Yan, Xiao-Wei Su, Yan-Fang Liu, Chuan-Hong Tang, Qing-Jiu Tang, Shuai Zhou, Yi Tan, Li-Ping Liu, Jing-Song Zhang * and Jie Feng ***

Institute of Edible Fungi, Shanghai Academy of Agricultural Sciences, National Engineering Research Center of Edible Fungi, Key Laboratory of Edible Fungi Resources and Utilization (South), Ministry of Agriculture, Shanghai 201403, China; yanmengqiu@sina.com (M.-Q.Y.); susu1822615@163.com (X.-W.S.); liuyanfang@saas.sh.cn (Y.-F.L.); tangchuanhong123@163.com (C.-H.T.); tangqingjiu@saas.sh.cn (Q.-J.T.); simonzsz@163.com (S.Z.); judy_1989317@126.com (Y.T.); l1228llp@163.com (L.-P.L.)

* Correspondence: sytufengjie@163.com (J.F.); syja16@saas.sh.cn (J.-S.Z.)

**Abstract:** The effects of oleic acid addition methods on the metabolic flux distribution of ganoderic acids R, S and T's biosynthesis from *Ganoderma lucidum* were investigated. The results showed that adding filter-sterilized oleic acid in the process of submerged fermentation and static culture is of benefit to the synthesis of ganoderic acids R, S and T. The metabolic fluxes were increased by 97.48%, 78.42% and 43.39%, respectively. The content of ganoderic acids R, S and T were 3.11 times, 5.19 times and 1.44 times higher, respectively, than they were in the control group, which was without additional oleic acid. Ganoderic acids R, S and T's synthesis pathways (GAP), tricarboxylic acid cycles (TCA), pentose phosphate pathways (PP) and glycolysis pathways (EMP) were all enhanced in the process. Therefore, additional oleic acid can strengthen the overall metabolic flux distribution of *G. lucidum* in a submerged fermentation-static culture and it can reduce the accumulation of the by-product mycosterol. This study has laid an important foundation for improving the production of triterpenes in the submerged fermentation of *G. lucidum*.

**Keywords:** *Ganoderma lucidum*; submerged fermentation-static culture; oleic acid; metabolic flux analysis; ganoderic acid



**Citation:** Yan, M.-Q.; Su, X.-W.; Liu, Y.-F.; Tang, C.-H.; Tang, Q.-J.; Zhou, S.; Tan, Y.; Liu, L.-P.; Zhang, J.-S.; Feng, J. Effects of Oleic Acid Addition Methods on the Metabolic Flux Distribution of Ganoderic Acids R, S and T's Biosynthesis. *J. Fungi* **2022**, *8*, 615. https://doi.org/10.3390/jof8060615

Academic Editors: Mingwen Zhao, Gen Zou and Jing Zhu

Received: 18 May 2022
Accepted: 7 June 2022
Published: 9 June 2022



## 1. Introduction

*Ganoderma lucidum*, a traditional medicinal mushroom included in Chinese Pharmacopoeia, has a history of more than 6800 years. It contains a variety of active components, such as polysaccharides, triterpenes and nucleotides. Among them, triterpenes have a wide range of pharmacological activities, such as antioxidant, anti-aging, anti-inflammatory and anti-tumor [1]. Triterpenes are lanostane derivatives, and lanosterol is their precursor. Lanostane-type triterpenes can be divided into tetracyclic triterpenes and pentacyclic triterpenes, according to their chemical structure; according to carbon atoms, they can be divided into C24, C27 and C30. In addition, they can be divided into ganoderic acid, ganoderol, ganoderal and ganolactone, according to different functional groups and side chains, with ganoderic acid and ganoderol as the main ones [2]. It has been found that ganoderic acid R, ganoderic acid S [3] and ganoderic acid T [4] are the main triterpenes in *Ganoderma lucidum* mycelium, which exhibit anti-liver cancer, anti-cervical cancer and anti-prostate cancer effects [5]. Triterpenes can be obtained by artificial cultivation and liquid submerged fermentation; liquid submerged fermentation is considered to be an effective way to accumulate triterpenes from *Ganoderma lucidum* due to its advantages of short cultivation time and the ease of controlling the culture conditions [6].

In recent years, there have been many studies on the acquisition of triterpenes from *G. lucidum* by submerged fermentation. However, due to a lack of understanding of the biosynthesis and the metabolism of triterpenes from *G. lucidum*, the application of key technologies, such as the metabolic regulation in the fermentation of triterpenes from *G. lucidum*, is limited. The anabolism of triterpenes in *G. lucidum* is complex, and the subtle differences in the fermentation process may cause great changes in the metabolic flux of products. Therefore, fermentation regulation is an important way to obtain a high yield of triterpenes. There have been reports that the addition of exogenous substances in the fermentation system could interfere with triterpene synthesis, for example, methyl jasmonate, phenobarbital, aspirin, copper ions and calcium ions could induce triterpene synthesis [7,8]. In addition, due to the aerobic fermentation of *G. lucidum* mycelium, in order to improve the oxygen concentration in the culture medium during its liquid fermentation, we found that adding monounsaturated fatty acid oleic acid in the fermentation process of *G. lucidum* mycelium could greatly increase the content of triterpenes in the mycelium [9]. Through the addition method of oleic acid in the fermentation process of *G. lucidum*, the mycelium was systematically optimized. The study showed that two ways of adding oleic acid, high-temperature-sterilized oleic acid and filter-sterilized oleic acid, had a great influence on the synthesis of triterpenes from *G. lucidum* [10].

However, there are few and insufficient studies on fungal triterpene anabolism at present. The existing studies on triterpene anabolism have mainly focused on plants. For many years, terpenes biosynthesis has been considered to start from the mevalonate pathway (MVP) [11]. It has been confirmed that the triterpenes of *G. lucidum*, like other terpenoids, were synthesized by the mevalonic acid biosynthesis pathway [12,13]. The upstream process from acetyl-coenzyme A to mevalonic acid and then to lanosterol is basically clear, but the further pathways of forming various triterpenes are still unclear. That is to say, after the synthesis of lanosterol, the carbocyclic skeleton must undergo a complex modification, and little is known about the specific steps of this process. The unclear downstream pathway of triterpenes synthesis greatly limits the application of metabolic engineering or pathway engineering in the submerged fermentation regulation of triterpenes. However, metabolic flux analysis may provide a good solution.

Metabolic flux analysis quantitatively describes the flow distribution of each branch in the metabolic network, which is helpful in analyzing the bottleneck in the synthesis process of target products, and it has great significance for understanding the intracellular metabolic regulation mechanism [14]. When using metabolic flux analysis, the metabolic reactions without branches can be combined into one reaction in the process of constructing a metabolic network. This makes it possible to combine the unknown reactions from lanosterol to triterpenes of different structures into one reaction by using this principle.

On the basis of the above analysis and of previous studies, metabolic flux analysis was used to analyze the effects of oleic acid addition methods on triterpenes' metabolic biosynthesis during the liquid fermentation of *G. lucidum* mycelium. It provides guidance for the following modification of *G. lucidum* strains and the optimization of the fermentation strategy to promote the triterpenes yield.

## 2. Materials and Methods

### *2.1. Strains*

Strain: The strain used in this study was *Ganoderma lucidum* G0023, which was preserved by the Agricultural Culture Collection of China (the edible fungi branch). It was maintained on potato dextrose agar (PDA) slants. The culture was inoculated and incubated at 26 °C for 10 days, then stored at 4 °C for approximately two months.

### *2.2. Medium and Culture Conditions*

The seed medium contained (g/L) the following: glucose 30, yeast extract 4, $KH_2PO_4$ 1.5, $MgSO_4 \cdot 7H_2O$ 1.5 and pH 5.5. For the seed culture, three loops of agar culture, in pea-size, from a slant culture were inoculated into a 250 mL flask, with 100 mL seed medium and were cultivated for nine days in a reciprocal shaker at 120 rpm and 26 °C. The seed culture (10%, $v/v$) was inoculated into the fermentation medium.

The fermentation-static medium contained (g/L) the following: glucose 30, yeast extract 3, $KH_2PO_4$ 2, $MgSO_4 \cdot 7H_2O$ 2 and pH 5.5. For liquid submerged fermentation, the seed culture was inoculated at 10% into a 250 mL flask, with 100 mL seed medium. Culture conditions of temperature and agitation rate were fixed at 26 °C and 120 rpm, respectively. According to the literature and preliminary experimental data, 1.21% high-temperature-sterilized oleic acid (121 °C, 30 min) was added to the flask at 32 h; 1.35% filter-sterilized oleic acid (filtered with 0.22 μm microfiltration membrane) was added at 7 h; after 5 days submerged fermentation, the broth was transferred to static culture at 26 °C, and growth period after static culture for 15 days. The culture without oleic acid was used as control. All experiments were performed in triplicate.

### 2.3. Analysis Methods

Determination of dry weight of mycelial samples: 100 mL of fermentation broth were centrifuged at 15,000× g for 15 min, reserved supernatant, and the pellet was washed with distilled water 3 times and dried at 60 °C, until a constant weight was achieved. We calculated the dry weight of mycelium per liter of fermentation broth after precise weighing, measured in g/L. Three parallel experiments were set in each group, and the average value was measured [15].

Determination of glucose: The supernatant prepared in the determination of the dry weight of the mycelium was diluted and then determined by DNS method to calculate the content of glucose [16].

Determination of ergosterol: The dried mycelium was extracted with absolute ethanol, at the ratio of solid to liquid of 1:10. Ultrasonic extraction for 1 h, centrifuged at 8000× g for 5 min, and the supernatant was taken for high-performance liquid chromatography (HPLC) analysis. Chromatographic conditions: Agilent ZORBAX SB-Aq (150 × 4.6 mm, 5 μm) column was selected. The mobile phase was acetonitrile-acetic acid (0.01%) to water, with a ratio of 95:5, the flow rate was 1.0 mL/min, the column temperature was 30 °C, the injection volume was 10 μL and the analysis wavelength was 282 nm [17].

Determination of ganoderic acids R, S and T: The dried mycelium was extracted with absolute ethanol at the ratio of solid to liquid of 1:20, and then centrifuged at 8000× g for 5 min, with ultrasound for 1 h, and the supernatant was taken for detection. Chromatographic conditions: Agilent ZORBAX Eclipse C18 (4.6 × 250 mm, 5 μm) column was selected. The mobile phase was acetonitrile-acetic acid (0.01%) to water, with a ratio of 55:45, the flow rate was 1.0 mL/min, the column temperature was 30 °C, the injection volume was 10 μL and the analysis wavelength was 240 nm [18].

### 2.4. Statistical Analysis

The experimental data were analyzed by MATLAB R2018a software, plotted by Origin 2021 software and analyzed by Microsoft Excel 2019. The data averaged from three individual experiments are expressed as the mean ± standard error (SE). Error bars present the standard deviations from the means of triplicates.

## 3. Results

### 3.1. Effects of Oleic Acid Addition Methods on Ganoderic Acids R, S and T's Biosynthesis

The consumption of glucose, the synthesis of the by-product ergosterol and the target product (ganoderic acids R, S and T) in the two-stage fermentation culture of *G. lucidum* were analyzed during the fermentation process. As shown in Figure 1, the consumption of glucose under the conditions of adding high-temperature-sterilized oleic acid and filter-sterilized oleic acid were consistent with that of the control group, and the glucose was basically consumed on the 20th day of static culture. However, the synthesis of ergosterol was quite different. Under the condition of adding high-temperature-sterilized oleic acid, the synthesis of ergosterol increased in advance and then decreased, with a maximum value of 1.03 mg/g on the 10th day of static culture. The content of ergosterol was positively correlated with the culture time under the condition of adding filter-sterilized oleic acid, and reached the maximum value of 1.47 mg/g on the 20th day of culture. The ergosterol content in the control group was the highest on the 5th day of static culture, which was 0.99 mg/g. Compared with the control group, the content of ganoderic acids R, S and T was quite different. Under the condition of adding high-temperature-sterilized oleic acid, the ganoderic acids R, S and T reached the maximum of 2.00 mg/g, 17.26 mg/g and 9.33 mg/g on the 10th day of static culture, which were 1.44 times, 3.55 times and 0.38 times higher than the control of 0.82 mg/g (the 10th day of static culture), 3.79 mg/g (the 10th day of static culture) and 6.77 mg/g (the 20th day of static culture), respectively. The highest results of the ganoderic acids R, S and T in the filter-sterilized oleic acid group were 3.38 mg/g (15 days), 23.46 mg/g (20 days) and 16.49 mg/g (20 days), respectively, which were 3.11 times, 5.19 times and 1.44 times higher, respectively, than those of the control group.

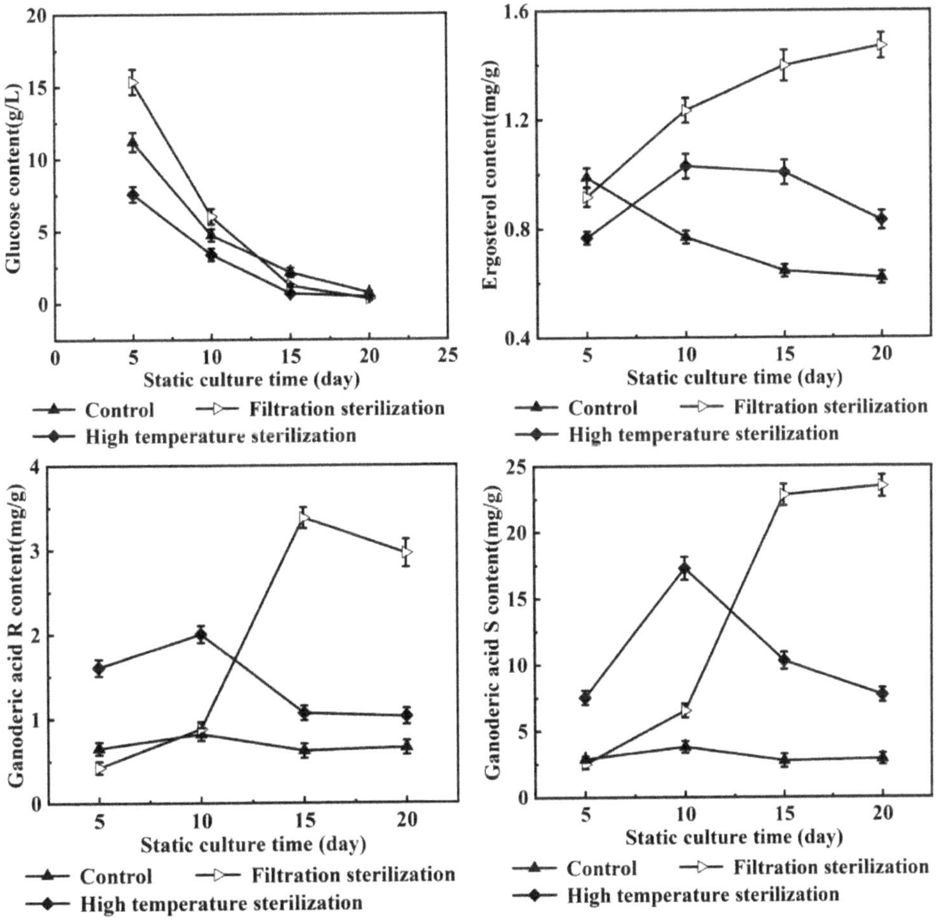


**Figure 1.** *Cont.*

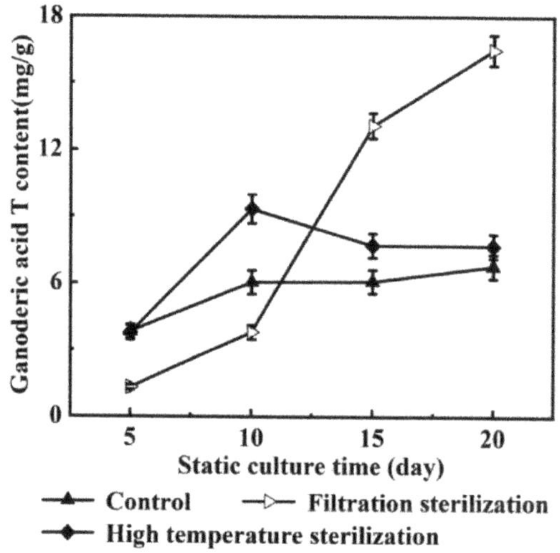


**Figure 1.** Substrate and product contents of *G. lucidum* submerged fermentation-static culture under different oleic acid addition methods.

It can be seen from Figure 1, that both methods of oleic acid addition are beneficial in increasing ganoderic acids R, S and T's content. In order to further interpret the effects of oleic acid on these ganoderic acids' biosynthesis under two different addition methods, the specific regulation methods of oleic acid on the liquid fermentation metabolism of *G. lucidum* mycelium were analyzed through a metabolic flux analysis.

### 3.2. Metabolic Flux Balance Model of Ganoderic Acids R, S and T

Based on the relevant literature [19–22], a metabolic network of the synthesis of the ganoderic acids R, S and T in a submerged fermentation-static culture was established (Figure 2). The metabolic network consisted of a glycolysis pathway (EMP), a tricarboxylic acid cycle (TCA), a pentose phosphate pathway (PP) and a ganoderic acids (R, S and T) synthesis pathway (GAP). Detailed metabolic reaction equations and metabolic node reaction rate equations are shown in Table 1. The metabolic network included 26 unknown reaction rates and 24 reaction equations with a degree of freedom of two; therefore linear equations can be solved merely by measuring just two-step reaction rates. In this study, the main variables measured were as follows: the glucose consumption rate and the ergosterol production rate.

**Table 1.** Metabolic reaction equations and metabolic node reaction rate equations of ganoderic acids R, S and T's synthesis from submerged fermentation-static culture of *G. lucidum*.

| Metabolic Reaction Equations | Metabolic Node Reaction Rate Equations |
|---|---|
| Glycolysis pathway (EMP): | (1) GLC: GLC(e) − r1 = 0 |
| r1: GLC + ATP → G6P + ADP | (2) G6P: r1 − r2 − r10 = 0 |
| r2: G6P → F6P | (3) F6P: r2 − r3 + r13 + r14 = 0 |
| r3: F6P + ATP → 2GA3P + ADP | (4) GA3P: 2r3 − r4 + r14 = 0 |
| r4: GA3P + NAD + PI + 2ADP → PYR + NADH + 2ATP | (5) PYR: r4 − r5 − r6 = 0 |
| Tricarboxylic acid cycle (TCA): | (6) AcCoA: r5 − r7 − 2r15 − r16 = 0 |
| r5: PYR + CoA + NAD → AcCoA + NADH + $CO_2$ | (7) ICI: r7 − r8 = 0 |
| r6: PYR + ATP + $CO_2$ → OAA + ADP + PI | (8) α-KG: r8 − r9 = 0 |
| r7: OAA + AcCoA → ICI + CoA | (9) OAA: r6 − r7 + r9 = 0 |
| r8: ICI + NADP → α − KG + NADPH + $CO_2$ | (10) Ru5P: r10 − r11 − r12 = 0 |
| r9: α − KG + 2NAD + FAD + ADP + PI → OAA + 2NADH + $FADH_2$ + ATP + $CO_2$ | (11) X5P: r11 − r13 − r14 = 0 |

Table 1. Cont.

| Metabolic Reaction Equations | Metabolic Node Reaction Rate Equations |
| --- | --- |
| Pentose phosphate pathway (PP): | (12) R5P: r12 − r13 = 0 |
| r10: G6P + 2NADP → Ru5P + 2NADPH + $CO_2$ | (13) E4P: r13 − r14 = 0 |
| r11: Ru5P → X5P | (14) CoA: -r5 + r7 + r15 + 2r16 = 0 |
| r12: Ru5P → R5P | (15) AcAcCoA: r15 − r16 = 0 |
| r13: X5P + R5P → F6P + E4P | (16) IPP: r16 − r17 − r18 − r19 − r20 = 0 |
| r14: X5P + E4P → F6P + GA3P | (17) DMAPP: r17 − r18 = 0 |
| Ganoderic acids R, S, T synthesis pathway (GAP): | (18) GPP: r18 − r19 = 0 |
| r15: 2AcCoA → AcAcCoA + CoA | (19) FPP: r19 − r20 − 2r21 = 0 |
| r16: AcAcCoA + AcCoA + 2NADPH + $2H^+$ + 3ATP + $H_2O$ → IPP + 2CoA + $2NADP^+$ + 3ADP + PI + $CO_2$ | (20) SQ: r21 − r22 = 0 |
| r17: IPP → DMAPP | (21) OS: r22 − r23 = 0 |
| r18: DMAPP + IPP → GPP + PPI | (22) LAN: r23 − r24 − r25 − r26 = 0 |
| r19: GPP + IPP → FPP + PPI | (23) ERG: r24 − ERG(e) = 0 |
| r20: FPP + IPP → GGPP + PPI | (24) GaR: r26 − GaR(e) = 0 |
| r21: 2FPP + NADPH + $H^+$ → SQ + $NADP^+$ + 2PPI | GaS: r26 − GaS(e) = 0 |
| r22: SQ + $O_2$ + NADPH + $H^+$ → OS + $NADP^+$ + $H_2O$ | GaT: r26 − GaT(e) = 0 |
| r23: OS → LAN | |
| r24: LAN → ERG | |
| r25: LAN → MY | |
| r26: LAN → GaR/GaS/GaT | |

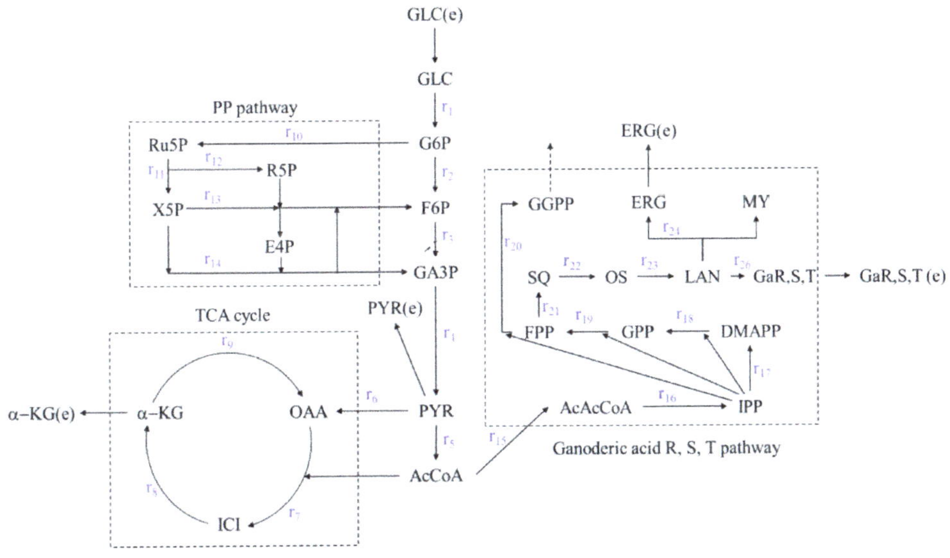


**Figure 2.** The metabolic network of ganoderic acids R, S and T synthesis by *G. lucidum*, submerged fermentation-static culture. AcAcCoA—acetoacetyl coenzyme A; AcCoA—Acetyl Coenzyme A; DMAPP—Dimethylallyl pyrophosphate; E4P—Erythrose-4-phosphate; ERG—Ergosterol; ERG (e)—Extracellular ergosterol; F6P—fructose-7-phosphate; FADH2—Reduced Flavin Dinucleotide; FPP—Farnesylpyrophosphate; G6P—Glucose-6-phosphate; GA3P—glyceraldehyde 3-phosphate; Ga R, S, T—Ganoderic acid R, Ganoderic acid S, Ganoderic acid T; GGPP—geranylgeranyl pyrophosphate; GLC—Glucose; GLC (e)—Extracellular glucose; GPP—Geranyl pyrophosphate; ICI—Isocitric acid; IPP—Isopentenyl pyrophosphate; LAN—Lanosterol; MY—Mycosterol; NADH—Nicotinamide adenine dinucleotide; OAA—Oxaloacetic acid; OS—2,3-Oxidosqualene; PYR—Pyruvate; PYR (e)—Extracellular pyruvate; R5P—Ribose-5-phosphate; Ru5P—ribulose 5-phosphate; SQ—Squalene; X5P—Xylulose phosphate; α-KG—α-ketoglutaric acid; α-KG (e)—Extracellular α-ketoglutaric acid.

### 3.3. Effects of Oleic Acid Addition Methods on Metabolic Flux Distribution of Ganoderic Acids R, S and T's Biosynthesis

The metabolic flux of the ganoderic acids R, S and T–synthesized from *G. lucidum* with two oleic acid addition methods and no addition group in two stages of submerged fermentation-static culture, under the highest target products yields–were analyzed. The concentration of glucose and ergosterol in the fermentation were determined. The metabolic flow distribution of the ganoderic acids R, S and T were obtained by linear programming with MATLAB software. The results are shown in Figures 3–5.

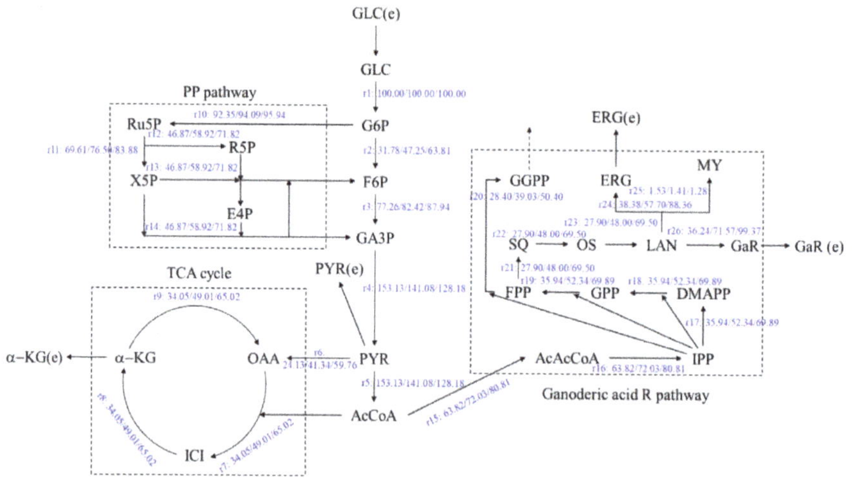


**Figure 3.** Metabolic flux distribution of ganoderic acid R, synthesized by submerged fermentation-static culture under two methods of oleic acid addition. Left: Control (the 10th day of static culture); Middle: Filter-sterilized oleic acid (15th day of static culture); Right: High-temperature-sterilized oleic acid (10th day of static culture).

As shown in Figure 3, the flux at the acetoacetyl-CoA (a key metabolic node to enter into the ganoderic acid synthesis) to the GARP (r15) of under filter-sterilized and high-temperature-sterilized oleic acid addition, were 72.03 and 80.81, which increased by 12.86% and 26.62%, respectively, compared with the control fermentation process of 63.82. The flux to the lanosterol nodes(r23) were 48.00 and 69.50, which increased by 72.05% and 149.15%, respectively, compared with the control of 27.90. The flux to the ergosterol node(r24) were 57.70 and 88.36, which were enhanced by 50.33% and 130.22%, respectively, compared with the control of 38.38. The increasement of the ergosterol flux promoted the synthesis of ergosterol, which was consistent with the ergosterol contents in Figure 2 (due to the enhancement of this node, the synthesis of ergosterol was promoted). The flux (r26) to the ganoderic acid R node were 71.57 and 99.37, which increased by 97.48% and 174.19%, respectively, compared with the control (36.24). Therefore, the addition of oleic acid can improve the metabolic flux through the above key nodes and can promote the synthesis of ganoderic acid R.

As shown in Figure 4, the flux at the acetoacetyl-CoA to the GASP (r15) of under filter-sterilized and high-temperature-sterilized oleic acid addition, were 89.24 and 95.24, increased by 19.67% and 27.72%, respectively, compared with the control fermentation process of 74.57. The flux to the lanosterol nodes (r23) were 90.07 and 104.71, which increased by 66.13% and 93.14%, respectively, compared with the control (54.21). The flux to the ergosterol node(r24) were 83.67 and 97.11, which were enhanced by 47.60% and 71.32%, respectively, compared with the control of 56.69. The increasement of the ergosterol flux promoted the synthesis of ergosterol, which also corresponded to the ergosterol contents in Figure 2. The flux (r26) to the ganoderic acid S node were 159.67 and 185.78, which increased by 78.42% and 107.59%, respectively, compared with the control (89.49). Therefore, the addition of oleic acid can improve the metabolic flux through the above key nodes and can promote the synthesis of ganoderic acid S.

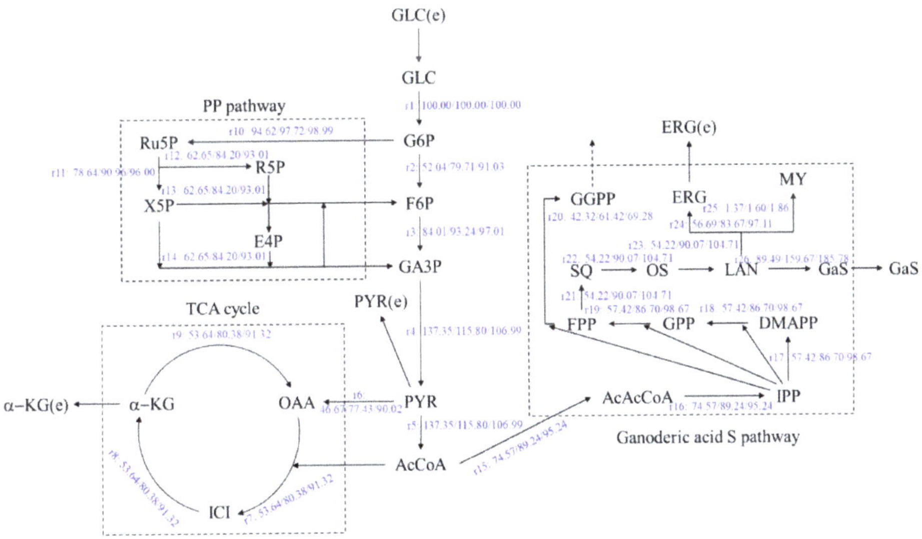


**Figure 4.** Metabolic flow distribution of ganoderic acid S, synthesized by submerged fermentation-static culture under two methods of oleic acid addition. Left: Control (the 10th day of static culture); Middle: Filtration-sterilization oleic acid (20 days of static culture); Right: High-temperature-sterilization oleic acid (10 days of static culture).

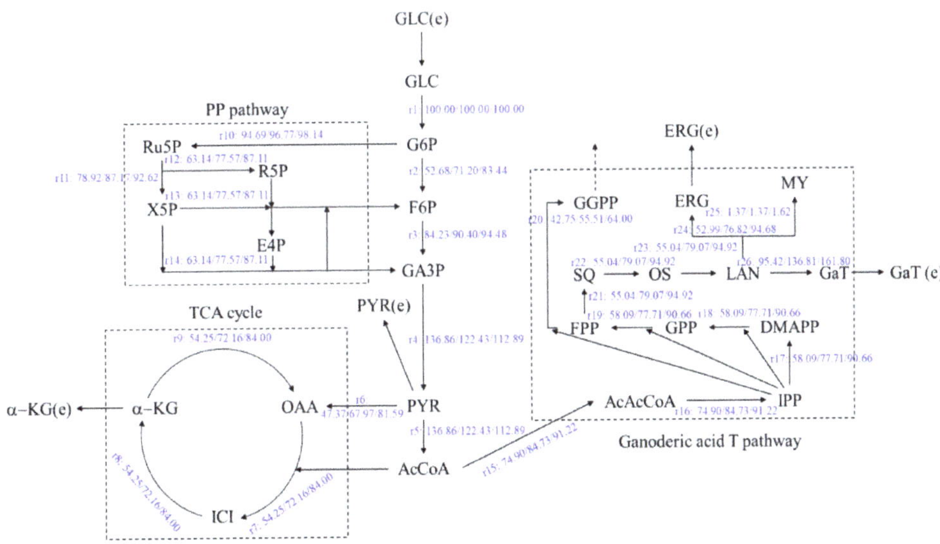


**Figure 5.** Metabolic flow distribution of ganoderic acid T, synthesized by submerged fermentation-static culture under two methods of oleic acid addition. Left: Control (the 20th day of static culture); Middle: Filtration-sterilization oleic acid (20 days of static culture); Right: High-temperature-sterilization oleic acid (10 days of static culture).

As shown in Figure 5, the flux at the acetoacetyl-CoA to the GATP (r15) of under filter-sterilized and high-temperature-sterilized oleic acid addition, were 84.72 and 91.22, increased by 13.11% and 21.78%, respectively, compared with the control fermentation process of 74.90. The flux to the lanosterol node(r23) were 70.07 and 94.91, increased by 43.66% and 72.46%, respectively, compared with the control of 55.04. The flux to ergosterol node(r24) were 76.82 and 94.68, which were enhanced by 44.98% and 78.67%, respectively, compared with the control of 52.99. The increasement of the ergosterol flux promoted the synthesis of ergosterol, which also corresponded to the ergosterol contents in Figure 2. The flux (r26) to the ganoderic acid T node were 136.81 and 161.80, which increased by 43.38% and 69.56%, respectively, compared with the control (95.42). Therefore, the addition of oleic acid can improve the metabolic flux through the above key nodes and can promote the synthesis of ganoderic acid T.

### *3.4. Effects of Oleic Acid Addition Methods on Metabolic Pathway Flux in Ganoderic Acids R, S and T's Biosynthesis*

To further compare the effects of two oleic acid addition methods on the flux direction of the metabolic pathways in a submerged fermentation-static culture of *G. lucidum*, the distribution of the metabolic flux was calculated according to their metabolic pathways, respectively. The results are shown in Table 2. From these results, it can be seen that oleic acid added by filtration sterilization and high temperature sterilization promoted all of the four metabolic pathways in different degrees. Among them, the influence on the ganoderic acid R synthesis pathway was the greatest, which increased by 45.10% and 93.35%, respectively; it had the least effect on the glycolysis pathway, which increased by 2.37% and 4.91%, respectively. The improvement rates of the tricarboxylic acid cycle were 17.91% and 37.07%, respectively. The improvement rates of the pentose phosphate pathway were 14.80% and 30.64%, respectively. These results implied that the addition of oleic acid enhances the distribution of carbon metabolic flow as a whole, improves the flow direction in each pathway and contributes to enhancing the accumulation of ganoderic acid R. Further analysis of the metabolic flow distribution of ganoderic acid S and ganoderic acid T showed that the oleic acid added by two sterilization methods enhanced the metabolic flow of the four metabolic pathways in varying degrees. The enhancement of the metabolic flow in the biosynthesis of ganoderic acid R was increased by 80.18% and 165.97%, respectively; the enhancement of the metabolic flow in the biosynthesis of ganoderic acid S was increased by 102.84% and 144.89%, respectively; and the enhancement of the metabolic flow in the biosynthesis of ganoderic acid T was increased by 68.30% and 113.44%, respectively.

**Table 2.** Statistics of metabolic pathway flow direction of ganoderic acids R, S and T, synthesized by submerged fermentation-static culture under two methods of oleic acid addition.

| Metabolic Pathway | Control | Filter-Sterilized Oleic Acid | | High-Temperature-Sterilized Oleic Acid | |
|---|---|---|---|---|---|
| | | Flow | Increasing Rate (%) | Flow | Increasing Rate (%) |
| | | | Ganoderic acid R | | |
| EMP | 362.17 | 370.75 | 2.37 | 379.93 | 4.91 |
| TCA | 279.42 | 329.46 | 17.91 | 383.00 | 37.07 |
| PP | 302.57 | 347.36 | 14.80 | 395.28 | 30.64 |
| Ganoderic acid R synthesis pathway (GaRP) | 423.70 | 614.77 | 45.10 | 819.22 | 93.35 |
| | | | Ganoderic acid S | | |
| EMP | 373.41 | 388.75 | 4.11 | 395.02 | 5.79 |
| TCA | 344.94 | 434.39 | 25.93 | 470.98 | 36.54 |
| PP | 361.22 | 441.27 | 22.16 | 474.03 | 31.23 |
| Ganoderic acid S synthesis pathway (GaSP) | 673.91 | 1015.16 | 50.64 | 1154.66 | 71.34 |
| | | | Ganoderic acid T | | |
| EMP | 373.76 | 384.03 | 2.75 | 390.82 | 4.57 |
| TCA | 346.98 | 406.87 | 17.26 | 446.47 | 28.67 |
| PP | 363.05 | 416.64 | 14.76 | 452.09 | 24.53 |
| Ganoderic acid T synthesis pathway (GaTP) | 681.70 | 910.28 | 33.53 | 1061.26 | 55.68 |

## 4. Discussion

In the liquid submerged fermentation of *G. lucidum*, the downstream metabolic pathway of triterpenes synthesis is still unclear, it is difficult to improve the yield of triterpenes by means of genetic engineering or metabolic pathway engineering, and the low production of triterpenes has always been the bottleneck problem restricting the development of the industry. In view of this situation, we have cast a new light on the metabolic flux analysis method. Combined with previous studies, we found that adding oleic acid in fermentation can greatly improve the ganoderic acids R, S and T's contents. In order to further explore the effects of oleic acid, added by two sterilization methods, on the biosynthesis of ganoderic acids R, S and T, this study explained the flow direction of carbon metabolic and the changes of by-products in the synthesis of ganoderic acids R, S and T by means of metabolic flow distribution.

Metabolic flux analysis is a method of stoichiometry analysis with measured data. Based on the mass balance law and pseudo-steady-state assumption, it calculates the intracellular, microscopic and unmeasurable metabolic flux by using acquirable information, such as observable state variables and chemical reaction formulas in the metabolic network [23,24]. Metabolic flow distribution analysis has many applications in the metabolic synthesis of goal products by bacteria and fungi. Niu [25] constructed a metabolic network of *Escherichia coli* fermentation to promote glycerol to produce L-methionine and found that L-methionine was mainly obtained from the pathway of phosphoenolpyruvate (PEP) to oxaloacetic acid (OAA). Enhancing the HMP pathway can provide a large amount of NADPH, while increasing the dissolved oxygen level can promote the synthesis of goal products. Tomàs-Gamisans et al. [26] found that glycerol as a sole carbon source could promote amino acid production by *Pichia pastoris*, by metabolic flux analysis. Hayakawa et al. [27] conducted a metabolic model of adding ethanol to promote S-adenosyl-L-methionine production in *Saccharomyces cerevisiae* and found that the metabolic flux levels of the tricarboxylic acid cycle and glyoxylate shunt in the ethanol culture were significantly higher than that in the glucose culture, and more ATP was produced from ethanol by oxidative phosphorylation. However, to our knowledge, there are no reports in the existing literature on this type of research, due to the complexity of the metabolites of edible and medicinal fungi and the lack of in-depth research on metabolic networks.

In this study, we constructed a metabolic network of carbon metabolic flow, which consisted of a glycolysis pathway (EMP), a tricarboxylic acid cycle (TCA), a pentose phosphate pathway (PP) and ganoderic acids R, S and T synthesis pathways (GAP). Glucose was used as a substrate in the metabolic pathway, and the effects of oleic acid on each pathway were investigated before and after the oleic acid addition. The results showed that additional oleic acid promoted the pentose phosphate pathway and the tricarboxylic acid cycle and strengthened the energy supply. At the same time, it was found that oleic acid mainly enhanced the synthesis of lanosterol, weakened the synthesis of the by-product mycosterol, and then promoted the synthesis of ganoderic acid. It was found that although the metabolic pathway of lanosterol was strengthened, the metabolic flow of the by-product ergosterol was also strengthened under the two methods of oleic acid addition. The metabolic flow (r24) of ergosterol with the addition of high-temperature-sterilized oleic acid was much higher than that with the addition of filter-sterilized oleic acid. This is because 1.35% of filter-sterilized oleic acid was added at the 7th hour of fermentation, while 1.21% high-temperature-sterilized oleic acid was added at the 32nd hour of fermentation. Although the addition time and the amount of oleic acid were different, they promoted the differences in the flow direction of the mycelium metabolism together, which led to the differences in the ganoderic acids' content. According to the correlation analysis from the results in Table 2 and the results of the contents of the three ganoderic acids in Figure 1, under the condition of adding filter-sterilized oleic acid, the increasing rate of the total metabolic flow was less than that of adding high-temperature-sterilized oleic acid. However, the content of ganoderic acids R, S and T in the former was higher than that in the latter, and the content of the by-product ergosterol was much higher than that in the latter. The reason for this might be due to the influence of the other by-products (except ergosterol) on the metabolism of the ganoderic acids R, S and T. In the follow-up study, if the metabolic flow of ergosterol and the other by-products' synthesis are further blocked by strain transformation or process optimization, then the metabolic flow of its by-product synthesis would be inhibited. It will be beneficial to strengthen the metabolic flow of ganoderic acid synthesis and then increase the yield of ganoderic acid.

**Author Contributions:** Conceptualization, M.-Q.Y. and J.F.; methodology, Q.-J.T.; software, S.Z.; validation, J.F.; formal analysis, Y.-F.L.; investigation, L.-P.L.; resources, C.-H.T.; data curation, J.F.; writing—original draft preparation, M.-Q.Y.; writing—review and editing, J.F. and J.-S.Z.; visualization, Y.T.; supervision, X.-W.S.; project administration, J.F. and J.-S.Z.; funding acquisition, J.F. and J.-S.Z. All authors have read and agreed to the published version of the manuscript.

**Funding:** This work was supported by the National Natural Science Foundation of China (No. 31801919), Natural Science Foundation of Shanghai, China (No. 20ZR1418700), the development fund for Shanghai Talents (2019052), the Shanghai Academy of Agricultural Sciences' Climbing Project (2022–2024) and by the project of the Shanghai Academy of Agricultural Sciences' excellent team(2022A-03).

**Institutional Review Board Statement:** Not applicable.

**Informed Consent Statement:** Not applicable.

**Data Availability Statement:** Not applicable.

**Acknowledgments:** We thank the Institute of Edible Fungi, the Shanghai Academy of Agricultural Sciences, the National Engineering Research Center of Edible Fungi, the Key Laboratory of Edible Fungi Resources and Utilization (South), the Ministry of Agriculture of P. R. China, the Key Laboratory of Agricultural Genetics and the Breeding of Shanghai.

**Conflicts of Interest:** The authors declare that the research was conducted in the absence of any commercial or financial relationships that could be construed as a potential conflict of interest.

## References

1. Wei, Z.H.; Liu, L.; Guo, X.F.; Li, Y.J.; Hou, B.C.; Fan, Q.L.; Wang, K.X.; Luo, Y.; Zhong, J.J. Sucrose fed-batch strategy enhanced biomass, polysaccharide, and ganoderic acids production in fermentation of *Ganoderma lucidum* 5.26. *Bioprocess Biosyst. Eng.* **2016**, *39*, 37–44. [CrossRef] [PubMed]
2. Xia, W.; Zhang, H.; Sun, X.; Zhao, H.; Wu, L.; Zhu, D.; Yang, G.; Shao, Y.; Zhang, X.; Mao, X.; et al. A comprehensive review of the structure elucidation and biological activity of triterpenoids from *Ganoderma* spp. *Molecules* **2014**, *19*, 17478–17535. [CrossRef] [PubMed]
3. Liu, R.M.; Zhong, J.J. Ganoderic acid Mf and S induce mitochondria mediated apoptosis in human cervical carcinoma HeLa cells. *Phytomedicine* **2011**, *18*, 349–355. [CrossRef] [PubMed]
4. Tang, W.; Liu, J.W.; Zhao, W.M.; Wei, D.Z.; Zhong, J.J. Ganoderic acid T from *Ganoderma lucidum* mycelia induces mitochondria mediated apoptosis in lung cancer cells. *Life Sci.* **2006**, *80*, 205–211. [CrossRef] [PubMed]
5. Hirotani, M.; Ino, C.; Furuya, T.; Shiro, M. Ganoderic acids T, S and R, new triterpenoids from the cultured mycelia of *Ganoderma lucidum*. *Chem. Pharm. Bull.* **1986**, *34*, 2282–2285. [CrossRef]
6. Xiao, H.; Zhong, J.J. Production of useful terpenoids by higher-fungus cell factory and synthetic biology approaches. *Trends Biotechnol.* **2016**, *34*, 242–255. [CrossRef]
7. Ren, A.; Li, X.B.; Miao, Z.G.; Shi, L.; Jaing, A.L.; Zhao, M.W. Transcript and metabolite alterations increase ganoderic acid content in *Ganoderma lucidum* using acetic acid as an inducer. *Biotechnol. Lett.* **2014**, *36*, 2529–2536. [CrossRef]
8. Ren, A.; Li, M.J.; Shi, L.; Mu, D.S.; Jiang, A.L.; Han, Q.; Zhao, M.W. Profiling and quantifying differential gene transcription provide insights into ganoderic acid biosynthesis in *Ganoderma lucidum* in response to methyl jasmonate. *PLoS ONE* **2013**, *8*, e65027. [CrossRef]
9. Feng, J.; Zhang, J.S.; Feng, N.; Yan, M.Q.; Yang, Y.; Jia, W.; Lin, C.C. A novel *Ganoderma lucidum* G0119 fermentation strategy for enhanced triterpenes production by statistical process optimization and addition of oleic acid. *Eng. Life Sci.* **2016**, *17*, 430–439. [CrossRef]
10. Su, X.W.; Tang, Q.J.; Zhang, J.S.; Feng, N.; Wang, J.Y.; Zhou, S.; Feng, J.; Yu, L. Synthesis of triterpenes in liquid submerged fermentation of *Ganoderma lingzhi* promoted by oleic acid. *Mycosystema* **2021**, *40*, 2445–2460. (In Chinese)
11. Wang, X.L.; Yang, H.; Liu, G.Q. Enhanced triterpene acid production by *Ganoderma lucidum* using a feeding stimulus integrated with a two-stage pH-control strategy. *J. Chem. Technol. Biotechnol.* **2016**, *91*, 974–982.
12. Wang, Q.; Cao, R.; Zhang, Y.; Qi, P.; Wang, L.; Fang, S. Biosynthesis and regulation of terpenoids from basidiomycetes: Exploration of new research. *AMB Express* **2021**, *11*, 150. [CrossRef] [PubMed]

13. Chen, S.; Xu, J.; Liu, C.; Zhu, Y.; Nelson, D.R.; Zhou, S.; Li, C.; Wang, L.; Guo, X.; Sun, Y.; et al. Genome sequence of the model medicinal mushroom *Ganoderma lucidum*. *Nat. Commun.* **2012**, *3*, 913. [CrossRef] [PubMed]
14. Goel, A.; Ferrance, J.; Jeong, J.; Ataai, M.M. Analysis of metabolic fluxes in batch and continuous cultures of *Bacillus subtilis*. *Biotechnol. Bioeng.* **1993**, *42*, 686–696. [CrossRef]
15. Feng, J.; Feng, N.; Tang, Q.J.; Liu, Y.F.; Tang, C.H.; Zhou, S.; Wang, J.Y.; Tan, Y.; Zhang, J.S.; Lin, C.C. Development and optimization of the triterpenoid and sterol production process with Lingzhi or Reishi medicinal mushroom, *Ganoderma lucidum* strain G0017 (agaricomycetes), in liquid submerged fermentation at large scale. *Int. J. Med. Mushrooms* **2021**, *23*, 43–53. [CrossRef]
16. Sharma, C.; Bhardwaj, N.; Sharma, A.; Tuli, H.S.; Batra, P.; Beniwal, V.; Gupta, G.K.; Sharma, A.K. Bioactive metabolites of *Ganoderma lucidum*: Factors, mechanism and broad spectrum therapeutic potential. *J. Herb. Med.* **2019**, *17–18*, 100268. [CrossRef]
17. Wang, J.Y.; Feng, N.; Liu, Y.F.; Yan, P.L.; Tang, Q.J. Ergosterol and fingerprint of liposoluble constituent in the *Ganoderma lingzhi* basidiospore. *Mycosystema* **2018**, *37*, 1215–1223. (In Chinese)
18. Zhao, N.; Feng, N.; Jia, W.; Feng, J.; Liu, Y.F.; Zhang, J.S. Quantitative analysis of ganoderic acid T, S and R in Ganoderma mycelium by HPLC. *Sci. Technol. Food Ind.* **2014**, *35*, 71–75. (In Chinese)
19. Krömer, J.O.; Wittmann, C.; Schröder, H.; Heinzle, E. Metabolic pathway analysis for rational design of L-methionine production by Escherichia coli and Corynebacterium glutamicum. *Metab. Eng.* **2006**, *8*, 353–369. [CrossRef]
20. Lin, S.; Liu, Z.Q.; Baker, P.J.; Yi, M.; Wu, H.; Xu, F.; Teng, Y.; Zheng, Y.G. Enhancement of cordyceps polysaccharide production via biosynthetic pathway analysis in *Hirsutella sinensis*. *Int. J. Biol. Macromol.* **2016**, *92*, 872–880. [CrossRef]
21. Wushensky, J.A.; Youngster, T.; Mendonca, C.M.; Aristilde, L. Flux connections between gluconate pathway, glycolysis, and pentose-phosphate pathway during carbohydrate metabolism in *Bacillus megaterium* QM B1551. *Front. Microbiol.* **2018**, *9*, 2789. [CrossRef] [PubMed]
22. Sacco, S.A.; Young, J.D. $^{13}$C metabolic flux analysis in cell line and bioprocess development. *Curr. Opin. Chem. Eng.* **2021**, *34*, 100718. [CrossRef]
23. Vallino, J.J.; Stephanopoulos, G. Metabolic flux distributions in *Corynebacterium glutamicum* during growth and lysine overproduction. *Biotechnol. Bioeng.* **1993**, *41*, 633–646. [CrossRef] [PubMed]
24. Zhang, Y.; Li, X.; Wang, Z.; Wang, Y.; Ma, Y.; Su, Z. Metabolic flux analysis of simultaneous production of vitamin B12 and propionic acid in a coupled fermentation process by *Propionibacterium freudenreichii*. *Appl. Biochem. Biotechnol.* **2021**, *193*, 3045–3061. [CrossRef]
25. Niu, K.; Xu, Y.Y.; Wu, W.J.; Zhou, H.Y.; Liu, Z.Q.; Zheng, Y.G. Effect of dissolved oxygen on L-methionine production from glycerol by Escherichia coli W3110BL using metabolic flux analysis method. *J. Ind. Microbiol. Biotechnol.* **2020**, *47*, 287–297. [CrossRef]
26. Tomàs-Gamisans, M.; Ødum, A.S.R.; Workman, M.; Ferrer, P.; Albiol, J. Glycerol metabolism of *Pichia pastoris* (*Komagataella* spp.) characterised by $^{13}$C-based metabolic flux analysis. *New Biotechnol.* **2019**, *50*, 52–59. [CrossRef]
27. Hayakawa, K.; Matsuda, F.; Shimizu, H. $^{13}$C-metabolic flux analysis of ethanol-assimilating *Saccharomyces cerevisiae* for S-adenosyl-L-methionine production. *Microb. Cell Fact.* **2018**, *17*, 82. [CrossRef]

Journal of *Fungi*

MDPI

*Article*

# Two Strains of *Lentinula edodes* Differ in Their Transcriptional and Metabolic Patterns and Respond Differently to Thermostress

**Yuan Guo [1], Qi Gao [1], Yangyang Fan [1], Shuang Song [1], Dong Yan [1], Jing Zhao [2], Yulin Chen [3], Yu Liu [1,*] and Shouxian Wang [1,*]**

[1] Beijing Engineering Research Center for Edible Mushroom, Institute of Plant Protection, Beijing Academy of Agriculture and Forestry Sciences, Beijing 100097, China
[2] College of Horticulture and Plant Protection, Inner Mongolia Agricultural University, Hohhot 010018, China
[3] College of Agriculture and Food Engineering, Baise University, Baise 533000, China
* Correspondence: liuyu@ipepbaafs.cn (Y.L.); wangshouxian@ipepbaafs.cn (S.W.)

**Abstract:** Temperature type is one of the key traits determining the cultivation regime of *Lentinula edodes*. However, the molecular and metabolic basis underling temperature type remain unclear. Here, we investigated the phenotypic, transcriptomic, and metabolic features of *L. edodes* with different temperature types under both control (25 °C) and high (37 °C) temperature conditions. We found that under the control condition, the high- and low-temperature types of *L. edodes* harbored distinct transcriptional and metabolic profiles. The high-temperature (H-)-type strain had a higher expression level of genes involved in the toxin processes and carbohydrate binding, while the low-temperature (L-)-type strain had a high expression level of oxidoreductase activity. Heat stress significantly inhibited the growth of both H- and L-type strains, while the latter had a higher growth inhibition rate. Upon exposure to heat, the H-type strain significantly up-regulated genes associated with the components of the cellular membrane, whereas the L-type strain markedly up-regulated genes involved in the extracellular region and carbohydrate binding. Metabolome data showed that thermostress altered purine and pyrimidine metabolism in the H-type strain, whereas it altered cysteine, methionine, and glycerophospholipid metabolism in the L-type strain. Transcriptome and metabolome integrative analysis was able to identify three independent thermotolerance-related gene–metabolite regulatory networks. Our results deepen the current understanding of the molecular and metabolic basis underlying temperature type and suggest, for the first time, that thermotolerance mechanisms can be temperature-type-dependent for *L. edodes*.

**Keywords:** *Lentinula edodes*; temperature type; heat stress; multi-omics; transcriptome; metabolome



**Citation:** Guo, Y.; Gao, Q.; Fan, Y.; Song, S.; Yan, D.; Zhao, J.; Chen, Y.; Liu, Y.; Wang, S. Two Strains of *Lentinula edodes* Differ in Their Transcriptional and Metabolic Patterns and Respond Differently to Thermostress. *J. Fungi* **2023**, *9*, 179. https://doi.org/10.3390/jof9020179

Academic Editors: Mingwen Zhao, Gen Zou and Jing Zhu

Received: 30 December 2022
Revised: 20 January 2023
Accepted: 23 January 2023
Published: 29 January 2023



## 1. Introduction

*Lentinula edodes* (Berk.) Pegler is a white-rot fungus broadly distributed in the subtropical to temperate regions of the of North and South America, Asia, and Australia [1], particularly in China and Japan [2,3]. It is known as "shiitake" in Japan and "Xianggu" in China. It is the second most cultivated and the most popular edible fungus in the world [4,5]. Except for its important role as food, *L. edodes* possesses huge potential for therapeutic applications due to its abundant bioactive components including polysaccharides, sulfur-compounds, phenolics, flavonoids, etc. [6–8].

It is estimated that there are approximately 500 *L. edodes* cultivars present in China [9]. Based on the optimal temperature range for fruiting body formation, *L. edodes* can be classified into different temperature types including the high-temperature type (H-type, fruiting at 15–25 °C), medium-temperature type (M-type, fruiting at 10–20 °C), low-temperature type (L-type, fruiting at 5–15 °C) and broad-temperature type (B-type, fruiting at 5 to 25 °C) [10,11]. Temperature type is the key trait in determining the regional cultivation regimes of *L. edodes* [10,11]. Temperature type is also among one of the most important traits for breeding purposes. The temperature type of *L. edodes* can be characterized by different strain-typing approaches including amplified fragment length polymorphism (AFLP) markers and inter-simple sequence repeat markers (ISSR), suggesting that strains with the same temperature type may share the same gene pool [10,12]. However, the genetic basis underlying temperature type remains largely unknown [11,13]. More recently, it was reported that temperature is the key environmental factor involved in the genetic divergence and phenotypic differentiation of *L. edodes* from the aspect of population genomics [14]. Therefore, it is of great value to uncover the molecular basis behind temperature type.

Heat stress is a major abiotic constraint in mycelial growth and fruiting body productivity for macrofungi [15,16]. Direct injuries that result from high temperatures include cell wall damage [17], increased membrane fluidity and permeability [18,19], nuclear condensation, mitochondrial dysfunction, and DNA fragmentation [20,21]. Thermostress also causes cell metabolism disorder as microorganisms typically respond to stress through the reprogramming of metabolic profiles [16]. In macrofungi, thermostress has been found to alter metabolic pathways including the tricarboxylic acid cycle, glucose metabolism, sphingolipid metabolism, and some amino acid metabolism [21–23]. Indirect impacts include increased susceptivity to fungal pathogens [24]. In addition, heat stress could also inhibit the growth and development of *L. edodes* by reducing carbon availability [25,26]. The molecular mechanisms by which macrofungi address thermostress comprise activation of the antioxidant defense system, the synthesis of trehalose and auxins, the expression of heat shock proteins (HSPs), etc. [16]. In *Pleurotus eryngii*, Kong et al. found that nitric oxide may function as a signaling molecule to mitigate HS-induced oxidative damage [27]. Upon exposure to heat, heat response factors (HSFs) can boost the transcription and accumulation of heat-stimulated gene products by interacting with heat stress-related genes [28]. In *L. edodes*, it was reported that HSPs could enhance thermotolerance by regulating IAA biosynthesis [15,25,29–31]. To date, however, the differences in gene expression and metabolic adaptation for different temperature types of *L. edodes* in response to high temperature stress have not yet been studied.

In this study, we compare the phenotypic, gene expression, and metabolic profiles of two widely cultivated H- and L-type *L. edodes* strains in China under normal (25 °C) and high (37 °C) temperature conditions using RNA-Seq and liquid chromatography–mass spectrometry (LC-MS)-based metabolomics analysis. We found that different temperature types of *L. edodes* differ in their transcriptional and metabolic profiles. Upon exposure to heat, H- and L-type strains responded differently at the level of the transcriptome and metabolome. These results deepen our understanding of the molecular and metabolic basis underlying the temperature type of *L. edodes*, and suggest that the thermotolerance mechanisms might be temperature-type-dependent. The data also indicate that the temperature types might be associated with the thermotolerance of mycelium.

## 2. Materials and Methods

### *2.1. Fungal Strains and Cultivation*

The *L. edodes* strains JZB2102217 (H-type) and JZB2102031 (L-type) used in this study were supplied by the Beijing Germplasm Resource Bank for Edible Fungi. The two selected strains are among the typical H- and L-type strains widely cultivated in northern and southern China. Mycelia were punched out using a cork borer (1 cm diameter), and then, were inoculated in Petri dishes (10 cm diameter) containing 35 mL of potato dextrose agar (PDA) medium as described previously [32]. Prior to inoculation, a sterilized cellophane membrane was placed on the surface of the PDA medium for easier collection of the mycelium samples. The two strains were cultivated in a growth incubator at 25 °C and in permanent darkness. After 5 days of growth at 25 °C, the fungal cultures were divided into the control and treatment groups, where the latter were subjected to heat exposure at 37 °C for 24 h.

### 2.2. RNA Isolation, cDNA Library Construction, and Illumina Hiseq X Ten/Nova seq 6000 Sequencing

Total RNA was extracted in triplicate from the mycelia of *L. edodes* using TRIzol® Reagent according to the manufacturer's instructions (Invitrogen, Carlsbad, CA, USA). Genomic DNA was removed using DNase I (Takara, Kyoto, Japan). RNA quality was evaluated using the Agilent 2100 BioAnalyzer (Agilent Technologies, Palo Alto, CA, USA) and quantity was determined using ND-2000 (NanoDrop Technologies, Wilmington, DE, USA). High-quality RNA samples (OD260/280 = 1.8–2.2, OD260/230 $\geq$ 2, RIN $\geq$ 6.5, 28S:18S $\geq$ 1.0, > 1 μg) were used to construct a sequencing library.

RNA-seq transcriptome libraries were prepared following the instructions of the TruSeq™ RNA sample preparation Kit from Illumina (San Diego, CA, USA) using 1 μg of total RNA. Briefly, mRNA, which was enriched by poly A tail selection and chemically fragmented, was used for first-strand cDNA synthesis, followed by second-strand cDNA synthesis using a SuperScript double-stranded cDNA synthesis kit (Invitrogen, Carlsbad, CA, USA) with random hexamer primers (Illumina, San Diego, CA, USA). Then, the synthesized cDNA was subjected to end-repair, phosphorylation, and 'A' base addition according to Illumina's library construction protocol. The libraries were then size-selected for cDNA target fragments of 300 bp on 2% low-range ultra-agarose followed by PCR, amplified using Phusion DNA polymerase (NEB, Ipswich, MA, USA) for 15 PCR cycles. After quantification by TBS380, the paired-end RNA-seq sequencing library was sequenced using the Illumina HiSeq X Ten/NovaSeq 6000 sequencer (2 × 150 bp read length) (Illumina, San Diego, CA, USA).

### 2.3. Read Mapping

The raw paired-end reads were trimmed and quality-controlled using SeqPrep (https://github.com/jstjohn/SeqPrep (accessed on 25 March 2022)) and Sickle (https://github.com/najoshi/sickle (accessed on 25 March 2022)) with default parameters. Then, clean reads were separately aligned to the reference genome (GCA_015476405.1) in orientation mode using HISAT2 (http://ccb.jhu.edu/software/hisat2/index.shtml (accessed on 25 March 2022)) software [33]. The mapped reads of each sample were assembled using StringTie (https://ccb.jhu.edu/software/stringtie/index.shtml?t=example (accessed on 25 March 2022)) in a reference-based approach [34].

### 2.4. Differential Expression Analysis and Functional Enrichment

For gene expression analyses, fragments per kilobase per million reads (FPKM) values were calculated using RSEM software (Version 1.3.3, http://deweylab.biostat.wisc.edu/rsem/ (accessed on 25 March 2022)) [35]. Differentially expressed genes (DEGs) were identified using DESeq2 [36], in accordance with the following general criteria: $|\log2FC| > 1$ and padjust $\leq 0.05$. padjust was calculated using the BH (FDR correction with Benjamini–Hochberg) methods [37]. Gene ontology (GO) and Kyoto Encyclopedia of Genes and Genomes (KEGG) pathway functional enrichment analyses were performed via Goatools (Version 0.6.5, https://github.com/tanghaibao/Goatools (accessed on 25 March 2022)) and KOBAS (Version 2.1.1, http://kobas.cbi.pku.edu.cn/home.do (accessed on 25 March 2022)) [38].

### 2.5. Metabolite Extraction

The mycelium (50 mg) was homogenized in 2 mL of polypropylene in a tube under cryogenic conditions at −10 °C (50 Hz) for 6 min using a high-throughput tissue crusher Wonbio-96c (Shanghai Wanbo Biotechnology co., LTD, Shanghai, China). For extraction, 400 μL of methanol: water (4:1, *v*/*v*) extraction solvent mixture was added to the tube containing mycelium samples and 10 μL of internal standard (2-Chloro-L-Phenylalanine, HPLC hyper grade, Merck, Darmstadt, Germany, 0.02 mg/mL). Samples were sonicated at 40 kHz in an ultrasonic bath for 30 min at 5 °C, and then, kept at −20 °C for 30 min to precipitate proteins. The solution was centrifuged at 13,000× $g$ at 4 °C and the supernatant was recovered for further metabolomics analysis.

*2.6. Non-Target Metabolomics*

Metabolites were chromatographically separated using an ultra-high-performance liquid chromatography (UHPLC) system (Thermo Electron Corporation, San Jose, CA, USA). The UHPLC system was equipped with an ACQUITY BEH C18 column (100 mm × 2.1 mm i.d., 1.7 μm; Waters, Milford, MA, USA). The mobile phases consisted of 0.1% formic acid in water (solvent A) and 0.1% formic acid in acetonitrile: isopropanol (1:1, $v/v$) (solvent B). The gradient elution program was set as follows to equilibrate the systems: from 0 to 3 min, 95% (A): 5% (B) to 80% (A): 20% (B); from 3 to 9 min, 80% (A): 20% (B) to 5% (A): 95% (B); from 9 to 13 min, 5% (A): 95% (B) to 5% (A): 95% (B); from 13 to 13.1 min, 5% (A): 95% (B) to 95% (A): 5% (B); and from 13.1 to 16 min, 95% (A): 5% (B) to 95% (A): 5% (B). The sample injection volume was 2 μL, the flow rate was 0.4 mL $min^{-1}$, and the column temperature was kept at 40 °C throughout the chromatographic separation.

The mass spectra of the compounds were obtained using a Thermo UHPLC-Q Exactive Mass Spectrometer (MS/MS) equipped with an electrospray ionization (ESI) source (Thermo Electron Corporation, San Jose, CA, USA). The optimized parameters were set as follows: aus gas heater temperature, 400 °C; sheath gas flow rate, 40 psi; aus gas flow rate, 30 psi; ion-spray voltage floating (ISVF), −2800 V in negative mode and 3500 V in positive mode, respectively; and normalized collision energy, 20–40–60 V rolling for MS/MS. Data acquisition was performed in Data Dependent Acquisition (DDA) mode. The detection was carried out over a mass range of $m/z$ 70 to 1050. The samples were analyzed in both positive (+) and negative (−) ESI modes.

The obtained raw data were analyzed using Progenesis QI 2.3 software (Nonlinear Dynamics, Waters, USA) to perform the peak detection, alignments, integration, isotope filtering, and peak-grouping based on peak-area correlation [39]. The preprocessed data table contains the retention time (RT), mass-to-charge ratio ($m/z$) value, and peak intensity. The metabolic features detected to have values of at least 80% in any set of samples were retained. After filtering, the minimum metabolite values were imputed for specific samples in which the metabolite levels fell below the lower limit of quantitation, and each metabolic feature was normalized by sum. The internal standard was used for data QC (reproducibility). Metabolic features that had a relative standard deviation (RSD) of QC > 30% were discarded. Data were logarithmically (log10) transformed prior to conducting multivariate analysis [40]. Compounds were identified via comparison of accurate mass, MS/MS fragment spectra, and isotope ratio difference against the biochemical databases the Human Metabolome Database (HMDB) (http://www.hmdb.ca/ (accessed on 10 March 2022)) and the Metlin database (https://metlin.scripps.edu/ (accessed on 10 March 2022)). The mass tolerance between the measured $m/z$ values and the exact mass of the components of interest was ±10 ppm. Mass features without MS/MS spectra were tentatively annotated at the MS1 level using a 5.0 mDa tolerance. The metabolite abundances were quantified according to their peak areas.

*2.7. Differential Metabolite Analysis*

Orthogonal partial least squares discriminant analysis (OPLS-DA) was used to determine the differential expressed metabolites between pairwise groups. Differentially expressed metabolites (VIP $\geq 1$, $p \leq 0.05$) between groups were mapped into biochemical pathways through metabolic enrichment and pathway analysis based on a database search (KEGG, http://www.genome.jp/kegg/ (accessed on 10 March 2022)). scipy.stats (Python packages) (https://docs.scipy.org/doc/scipy/ (accessed on 10 March 2022)) was used to identify statistically significantly enriched pathway using Fisher's exact test.

### 2.8. Reactive Oxygen Species Detection and Growth Rate Measurement

Reactive oxygen species (ROS) production was measured according to the method described previously in [41,42]. For fluorescence detection, sterile glass coverslips were obliquely inserted into the Petri dishes just after the inoculation of fungal blocks, which we allowed the fungal mycelium grow on later. When the hyphae grew on the coverslips, the coverslips with hyphae were then incubated in 10 μmol/L of 2′,7′-dichlorodihydrofluorescein diacetate (DCHF-DA, Solarbio, Beijing, China) phosphate-buffered saline (PBS) solution for 25 min under dark conditions for ROS visualization. After staining, ROS production was visualized using a fluorescence microscope (Olympus, IX71, Tokyo, Japan). The mean fluorescence intensity, i.e., the mean pixel intensity over all the pixels in the region of fluorescence, was calculated using the ImageJ program (Version 1.53t, https://imagej.nih.gov/ij/index.html (accessed on 27 December 2022)) [43]. The inhibition rate was calculated as follows: growth inhibition rate $= \frac{G1-G2}{G2} \times 100\%$, where G1 and G2 denote the growth rate of the fungal colony after and before heat treatment. Growth rate was calculated according to the diameter change during the period of growth before and after heat treatment (mm/d).

### 2.9. Statistical Analysis

Orthogonal partial least squares regression discriminant analysis (OPLS-DA) was performed using SIMCA 14.1 (Umetrics, Umeå, Sweden). The robustness of the OPLS-DA models was assessed via seven-fold cross-validation [44]. The reliability of the predictive models was assessed via the analysis of variance testing of cross-validated predictive residuals (CV-ANOVA), R2 and Q2, which provide information on interpretability and predictability, respectively. The *p* value was estimated using a paired Student's t-test via single-dimensional statistical analysis. To test the overall associations between DEGs and DEMs, we performed Procrustes analysis in R using the 'vegan' package (version 4.2.0) [45,46]. The data were logarithmically (log10) transformed, centered, and Pareto-scaled prior to multivariate analyses [40]. For the integration and visualization of transcriptome × metabolome associations, the sparse partial least squares (sPLS) regression method [47] was implemented using the R package mixOmics [48].

## 3. Results

### 3.1. H- and L-Type L. edodes Strains Differ in Their Transcriptional and Metabolic Profiles

Transcriptomics and metabolomics were used to unearth the differences between *L. edodes* strains of different temperature type. A total of 7959 expressed genes (Table S1, RPKM $\geq$ 10) and 1594 metabolites (Table S2) were identified from the examined samples. PCA was used to visualize and evaluate the overall differences in gene expression and metabolics between H- and L-type *L. edodes* strains. For transcriptomics data, the first two PCs explained 86.3% of the variation; the first PC, which explained 74.5% of the total variation, was able to separate H- and L-type *L. edodes* strains (Figure 1a). A volcano plot shows the differentially expressed genes between H- and L-type strains (Figure 1b). Compared to the L-type strain, the up-regulated genes in the H-type strain were enriched mainly in toxin processes and carbohydrate binding, whereas the L-type strain had a high expression level mainly of oxidoreductase activity (Figure 1c,d).

The PCA analysis based on the detected metabolites was also able to discriminate H- and L-type *L. edodes* strains (Figure 1e). Compared to the L-type strain JZB2102031, the H-type strain JZB2102217 accumulated a significantly higher abundance of compounds enriched in starch and sucrose metabolism, and arginine and proline metabolism, while accumulating a lower abundance of compounds enriched in glycerophospholipid and purine metabolism (Figure 1f,h) based on the KEGG topology analysis (padjust $< 0.05$, VIP $\geq 1$, $|\log2FC| > 1$). The KEGG enrichment analysis of the up- and down-regulated genes is shown in Figure S1.

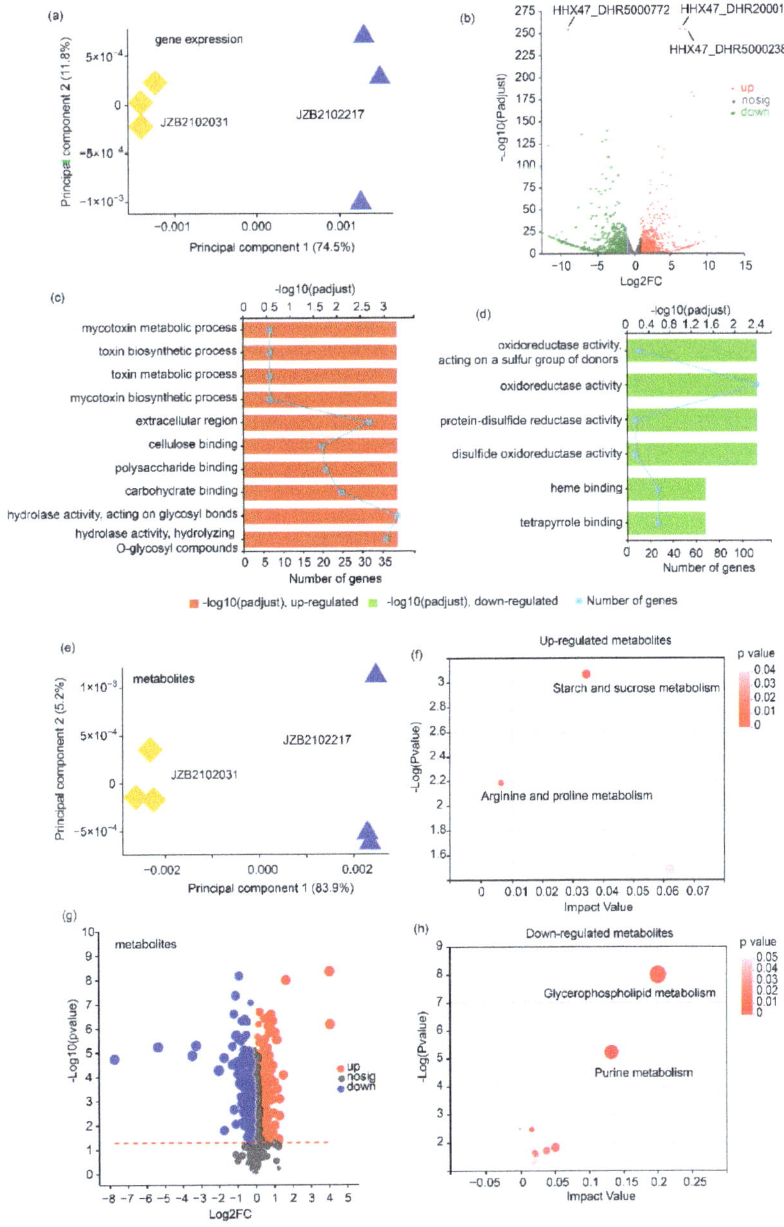


**Figure 1.** The transcriptional and metabolic profiling of H-type *L. edodes* strain JZB2102217 and L-type strain JZB2102031. (**a**) Principal component analysis (PCA) of transcriptome data of both strains; (**b**) volcano plot showing the up- and down-regulated genes; (**c**,**d**) the top 10 KEGG-enriched items of up- and down-regulated genes (padjust < 0.05), respectively; (**e**) PCA of the metabolome data of both strains; (**f**,**h**) the KEGG topology analysis of the up- and down-regulated metabolites, respectively; (**g**) volcano plot showing the up- and down-regulated metabolites (FDR < 0.05, VIP from OPLS-DA $\geq 1$, $|\log2FC| > 1$).

### 3.2. Heat Stress Inhibited the Growth and Induced the Production of ROS in Both Temperature Types of L. edodes

Figure 2a shows the morphology of the H-type strain JZB2102217 and the L-type strain JZB2102031 under control (25 °C) and HS (37 °C) conditions. We can observe that both strains could recover growth 2 d after heat treatment. According to the growth rate, heat stress significantly suppressed the growth of both the H- and L-type strains (Figure 2a,b), while the latter grew significantly slower as a result of heat stress. After the recovery of growth, the L-type strain had a higher growth inhibition rate than the H-type strain (Figure 2c). Under optimal growth conditions, the H- and the L-type strains had comparable levels of ROS production, which was significantly induced by heat stress (Figure 2d,e).

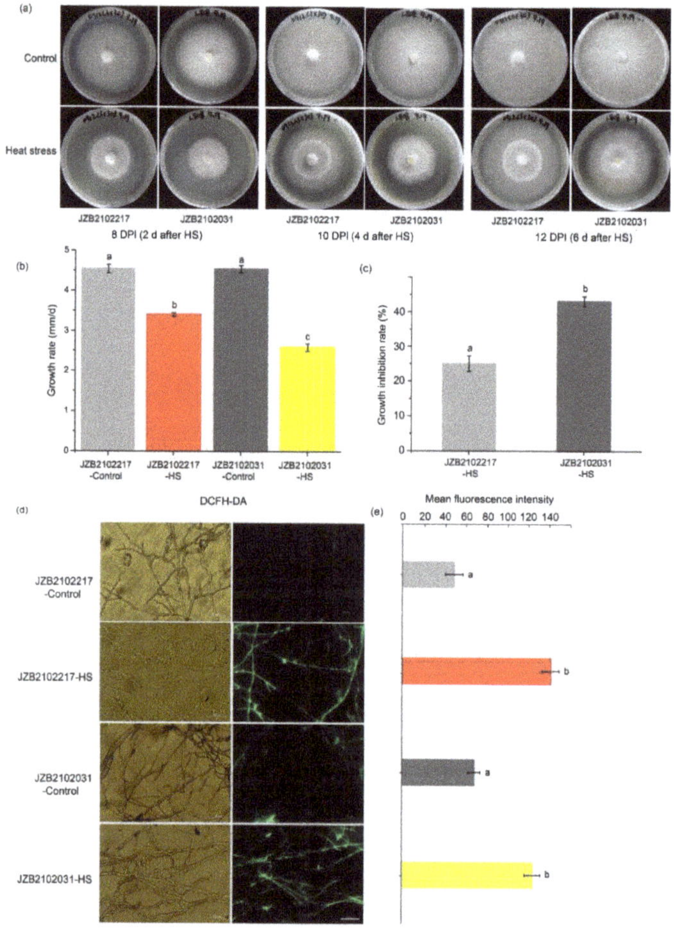


**Figure 2.** The phenotypic changes in *L. edodes* H-type strain JZB2102217 and L-type strain JZB2102031 under control (25 °C) and heat stress (HS, 37 °C) conditions. HS was conducted on the 5th day for 24 h. The fungal hyphae of two strains were able to recover growth 2 days after HS. (**a**) The morphology of cultures of H-type strain and L-type strain 2, 4, and 6 d after HS, i.e., 8, 10, and 12 days post-inoculation (DPI); (**b**,**c**) the growth rate and growth inhibition rate of two strains in response to heat stress, respectively; (**d**,**e**) the ROS production staining (Bar = 50 μm) and mean fluorescence intensity, respectively. Different letters indicate significant differences between groups (ANOVA, $p < 0.05$). Data are represented as mean ± se ($n = 4$).

### 3.3. H- and L-type L. edodes Strains Varied in Their Transcriptional Profiles in Response to Heat Stress

The sequencing data are summarized in Table S3. The pairwise correlation of all samples based on transcriptome data is showed in Figure S2. Transcripts with at least a 2 fold change in abundance and with a Benjamini–Hochberg-adjusted $p$ value < 0.05 were regarded as differentially expressed genes (DEGs). A total of 2329 (1030 upregulated and 1299 downregulated) and 2834 (1206 upregulated and 1628 downregulated) DEGs were identified in H- and L-type strains upon exposure to thermostress, respectively (Table S4). The OPLS-DA model (CV-ANOVA, $p < 0.05$) showed clear gene expression patterns separating the H- and L-type *L. edodes* strains (Figure 3a). The volcano plots show that the top five down- and up-regulated genes in the H-type strain were different to those top five genes in the L-type strain, except for gene with identified as HHX47_DHR5000174, which was up-regulated in both strains under heat stress (Figure 3b,e). All the identified up- and down-regulated DEGs were annotated using KEGG enrichment analysis. The results showed that the up-regulated genes in the H-type strain under heat stress were mostly enriched (padjust < 0.05) in the intrinsic component of the membrane, the cellular anatomical entity, the integral component of the membrane, and the cellular component, while the down-regulated genes were enriched in catalytic activity, oxidoreductase activity, monooxygenase activity, and DNA repair (Figure 3c,d). For the L-type strain, heat stress induced the up-regulation of genes associated with the extracellular region, cellulose, polysaccharide and carbohydrate binding, integral and intrinsic components of the membrane, etc., and down-regulated genes involved in enzyme activities, proteasome regulatory particles, base subcomplexes, and FAD binding (Figure 3f,g). The KEGG analysis of the DEGs of H- and L-type strains upon exposure to heat is shown in Figure S3.

### 3.4. H- and L-type L. edodes Strains Showed Different Metabolic Profiles in Response to Heat Stress

The OPLS-DA model based on the metabolomes of the H- and L-type strains showed both clear strain-derived and heat stress-derived metabolic variations (Figure 4a). The Top 30 metabolites shown in the loading plot (VIP > 1) were classified as carbohydrates, lipids, nucleic acids, and peptides using KEGG analysis (Figure 4a). For the H-type strain, the heat-induced DEMs (FDR < 0.05) were largely enriched in purine and pyrimidine metabolism (Figure 4b). Of the top 10 heat-induced metabolites in the H-type strain, 4 were peptides and 8 were up-regulated (Figure 4b). For the L-type strain under heat stress, the DEMs were mainly enriched in cysteine and methionine metabolism and in glycerophospholipid metabolism (Figure 4d). Most of the top 10 heat-induced compounds that discriminated the L-type strain under heat stress were down-regulated and unannotated based on the KEGG database (Figure 4e).

Upon exposure to heat, 394 DEMs were identified in the L-type strain, while only 24 DEMs were identified in the H-type strain when the selection criteria for DEMs were narrowed to FDR < 0.05, VIP of OPLS-DA > 1, and FC > 1.2 (Figure S4a,b). A Venn network shows the most unique and common DEMs for the H- and L-type strains in response to thermostress (Figure S4c).

### 3.5. Integrative Analysis of Transcriptome and Metabolome of L. edodes under Heat Stress

Sparse partial least squares (sPLS) regression was performed to examine the relationships between the transcriptome and metabolome data of all the samples. The results showed that the transcripts were strongly associated with metabolites (Figure 5a). Further, interaction networks were constructed to infer the substructures of highly correlated genes and metabolites using a threshold of 0.9 (Figure 5b). The results showed that two clear interaction substructures could be detected without any intermediate elements. Interestingly, most of the correlations in substructure I, consisting of nine genes and 35 metabolites, were positive, while most of the correlations in substructure II, containing four genes and 16 metabolites, were negative (Figure 5b).

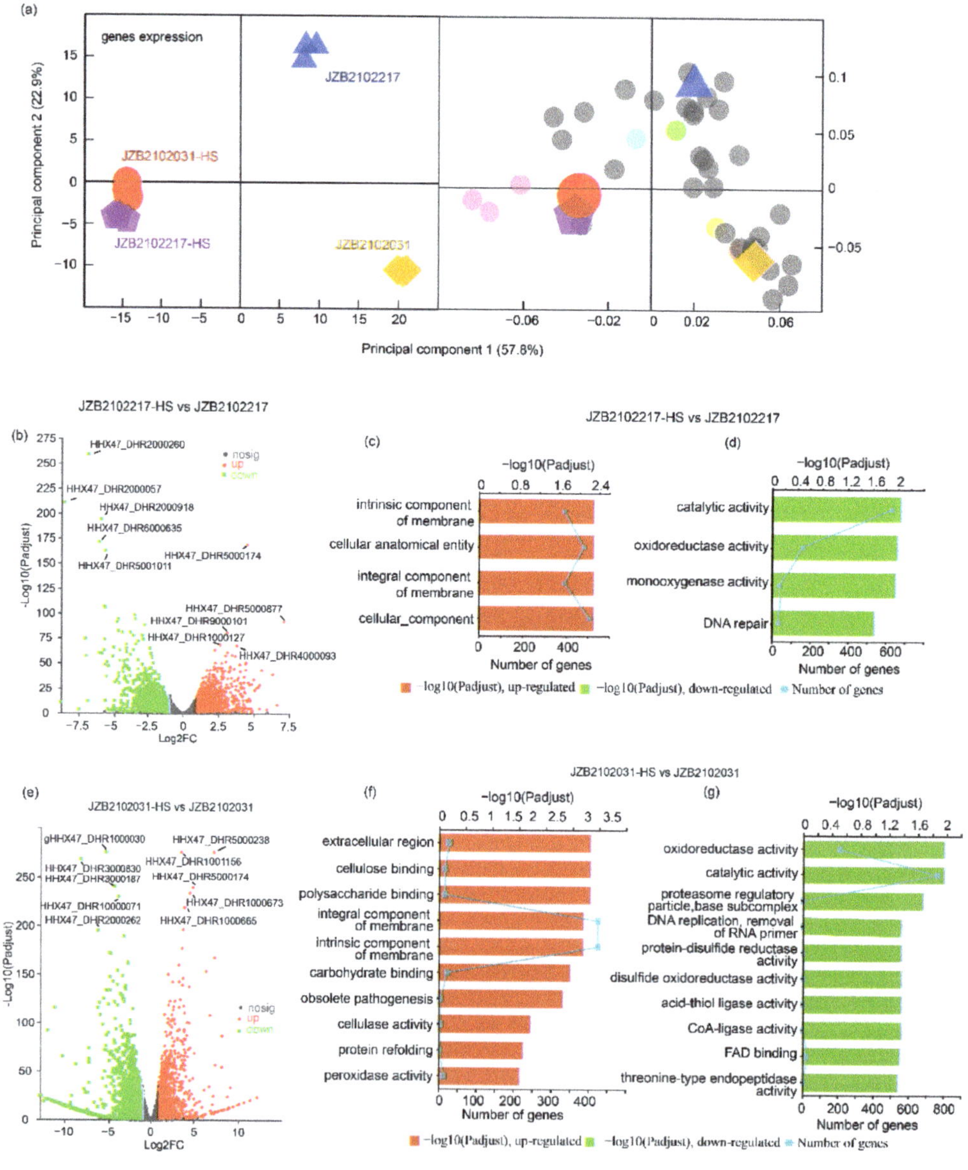


**Figure 3.** Gene expression analysis of H- and L-type *L. edodes* strains JZB2102217 and JZB2102031 in response to heat stress. (**a**) OPLS-DA analysis of transcriptome data of all samples (R2Xcum = 0.891, R2Ycum = 0.994, Q2cum = 0.931, CV-ANOVA, $p = 9.835 \times 10^{-7}$). OPLSDA loading showing the genes whose VIP > 4; color indicates the KEGG pathways (light bule—endocytosis; light orange—methane metabolism; light pink—protein processing in endoplasmic reticulum; light yellow—glycerophospholipid metabolism; grey—unannotated). (**b**,**e**) The volcano plot of expressed genes in H- and L-type strains upon exposure to heat, respectively. (**c**,**d**) The top 10 enriched GO terms of up- and down-regulated genes in H-type strains upon exposure to heat, respectively. (**f**,**g**) The top 10 enriched GO terms of up- and down-regulated genes in H-type strains upon exposure to heat, respectively.

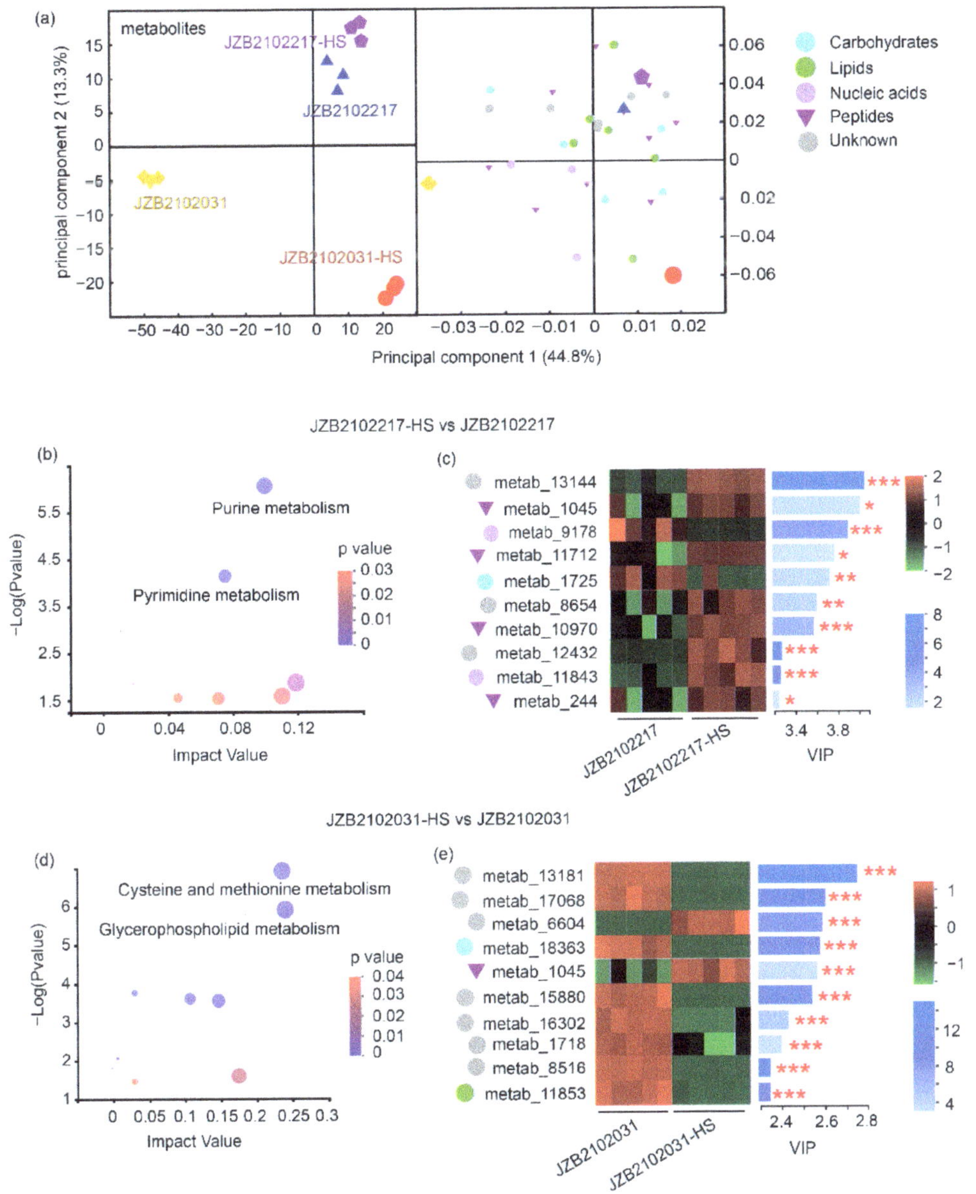


**Figure 4.** Metabolic profiling of H-type *L. edodes* strain JZB2102217 and L-type strain JZB2102031 in response to heat stress. (**a**) OPLS-DA analysis of the metabolome data of all samples (R2Xcum = 0.857, R2Ycum = 0.98, Q2cum = 0.899, CV-ANOVA, $p = 2.083 \times 10^{-11}$). Loading plot shows the top 30 metabolites (VIP > 1); (**b**,**c**) the KEGG topology analysis of DEMs and top 10 metabolites in H-type strain in response to heat stress (37 °C), respectively; (**d**,**e**) the KEGG topology analysis of DEMs and top 10 metabolites in L-type strain in response to heat stress (37 °C), respectively. The color of the blue bar indicates the value of -log10 (*p*-value); *, **, and *** indicate the abundance of compounds that were significant at *p* value < 0.05, <0.01, and <0.001, respectively.

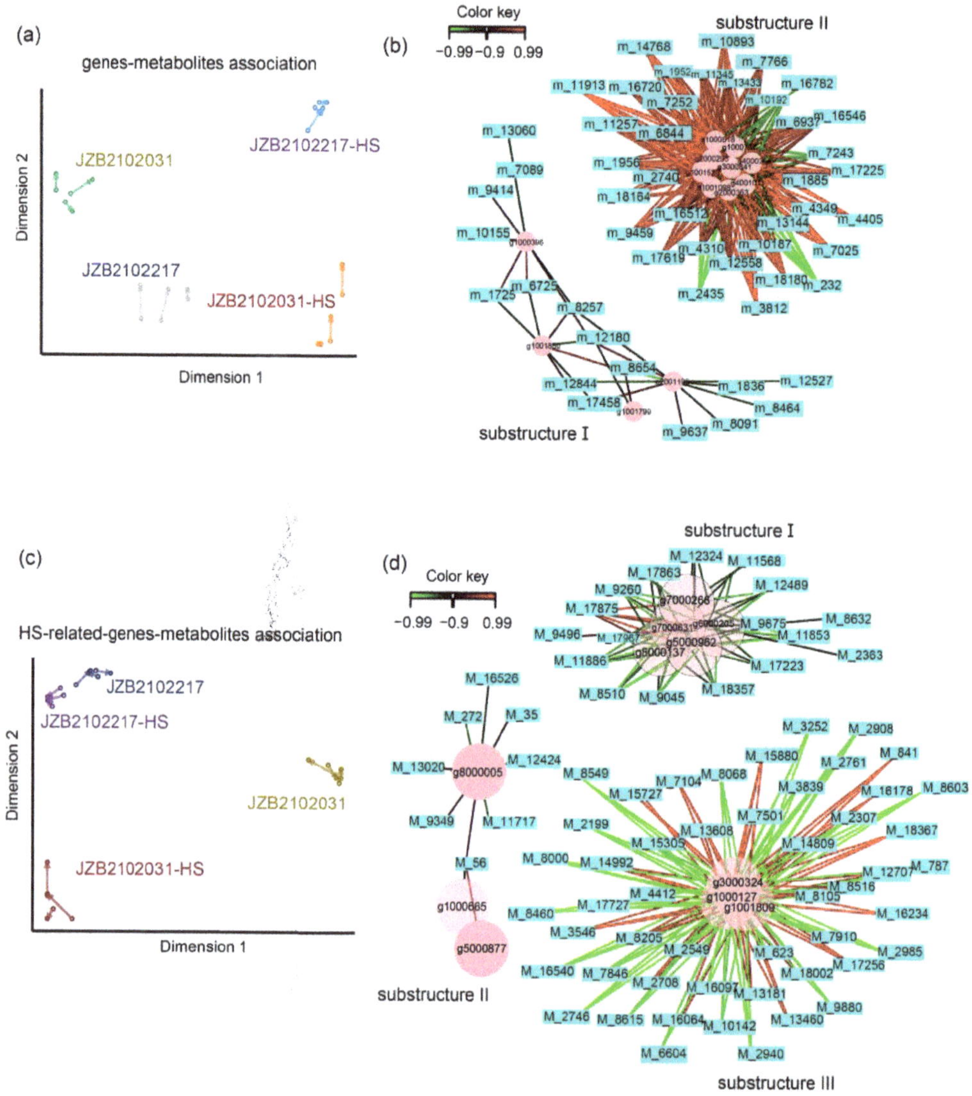


**Figure 5.** Transcriptome–metabolome-wide association analysis of all samples. (**a**) Sparse partial least squares (sPLS) association analysis between all genes and metabolites from all samples; (**b**) integration network of highly associated genes and metabolites at threshold of 0.9; (**c**) sPLS association analysis between thermotolerance-related genes and all metabolites from all samples; (**d**) integration network of highly associated thermotolerance-related genes and metabolites at threshold of 0.9. For the metabolite ID, "metab_" in Table S2 is abbreviated to "M_" for better readability.

To better understand the regulatory network of transcripts and metabolites in response to heat stress, 18 protentional heat-tolerance (HT)-related genes were selected based on either GO ontology, or KEGG or pfam annotations. These HT-related genes were further subjected to sPLS analysis with all the detected metabolites. Strong associations were found between HT-related genes and metabolites (Figure 5c). From the correlation network, we identified three clear HT–gene–metabolite regulatory substructures (Figure 5d). In substructure I, the gene HHX47_DHR7000266 related to HSP binding, the gene HHX47_DHR7000631 related to glutathione peroxidase activity, the gene HHX47_DHR6000205 related to cell redox homeostasis, the gene HHX47_DHR5000962 encoding HSP 90, and the gene HHX47_DHR8000137 encoding HSP 70 were highly correlated with 17 metabolites; in substructure II, HHX47_DHR8000005 encoding HSP70 was associated with eight compounds, and another HSP70-encoding gene, HHX47_DHR000665, and an HSP20-encoding gene, HHX47_DHR5000877, were solely associated with the monosaccharide compound M_56 (N-Acetyl-D-quinovosamine); substructure III was the largest gene–metabolite association, consisting of three HS-related genes and 50 metabolites (Figure 5d).

## 4. Discussion

Generally, based on the temperature for fruiting, *L. edodes*, including both the wild and cultivated strains, could be divided into the groups H (high-temperature)-type, M (medium temperature)-type, L (low-temperature)-type, and B (broad-temperature)-type [11,12]. The temperature type of *L. edodes* is usually obtained by observing the fruiting temperature during cultivations under natural climatic conditions [11,49]. The temperature type of *L. edodes* could, moreover, be characterized by other strain-typing approaches such as inter-simple sequence repeat (ISSR) analysis [12] and amplified fragment length polymorphism (AFLP) analysis [10]. These phenotypic divergence- and molecular marker-based results suggest strong genetic divergences underlying the temperature type of *L. edodes*.

Except for characterization from the aspect of the phenotype, the genetic basis underlying the temperature type of *L. edodes* is largely unknown. Here, using comparative transcriptomics and metabolomics, we show, for the first time, that the H- and L-type *L. edodes* strains possess distinct transcriptional and metabolic patterns under optimum temperature conditions, which underlie the genetic divergence of the temperature type. Similarly, a recent population genomics study shows that temperature is the key environmental factor involved in the genetic divergence and phenotypic differentiation of *L. edodes* [14]. Our data show that the H-type strain had a higher expression level in the genes involved in toxin processes and cellulose, polysaccharide, and carbohydrate binding, while it had a lower level in the genes involved in oxidoreductase activity, protein-disulfide reductase activity, etc. Oxidoreductase activity has been found to be involved in light-induced brown-film formation in *L. edodes* [50–52]. From the aspects of metabolism, our data show that the H-type *L. edodes* strains accumulated more metabolites in the pathways of starch and sucrose metabolism and arginine and proline metabolism, while the L-type strain accumulated more metabolites in glycerophospholipid and purine metabolism. Through we have found some interesting differences in gene expression and metabolism between different temperature types of the *L. edodes* strain, we are still a long way from building an explicit connection between genes and temperature type. Nevertheless, the current paper represents pioneering work that deepens our understanding of the genetic basis and metabolic processes regarding temperature type.

The optimum growth temperature for *L. edodes* mycelium ranges from 24 to 27 °C and the lethal temperature could be 38 °C or above, depending on the strain [26]. Heat stress appears to be the major abiotic constraint inhibiting mycelium growth, disease resistance, and fruiting body development, thereby seriously reducing fruiting body productivity and the quality of *L. edodes* [24,53–55]. Therefore, it is of great significance to decipher the mechanisms by which *L. edodes* addresses thermostress. For *L. edodes*, the optimum temperature required for fruiting body induction is different from that for vegetative growth. However, whether the temperature type is associated with the optimum mycelium growth temperature for *L. edodes* remains unknown. In this study, we compared the physiological and phenotypic responses of H- and L-type *L. edodes* strains under exposure to heat stress. We found that the growth rate of the L-type strain decreased more and had a higher growth inhibition rate after recovery than that of the H-type strain in response to heat stress, suggesting that the L-type strain might be more sensitive to heat stress. Though the temperature-type is defined by the fruiting temperature rather than the temperature at which mycelia grow [12], we found distinct responses of the H- and L-type *L. edodes* strains in their gene expression and metabolism under heat stress. Moreover, Wang et al. reported that the high-temperature type *L. edodes* strain could form a fruiting body at a higher temperature and had a higher fruiting body yield than the low-temperature type of *L. edodes* [56,57]. Taken together, it is interesting that the temperature types of *L. edodes* could be associated with the thermotolerance of mycelium. Thus, we suggest a large-scale evaluation on the thermotolerance of *L. edodes* strains with different temperature types.

A considerable number of studies have accumulated over the years addressing the topic of heat resistance in the mycelium of *L. edodes* [15,26,55]. Previous studies have documented that heat shock proteins (HSPs), indoleacetic acid (IAA), catalase, and trehalose play crucial roles in the thermotolerance of *L. edodes* [15,29,30]. HSPs are a family of conserved proteins that are up-regulated at high temperatures and play a crucial role in protein renaturation, enzyme and membrane stability, and cell homeostasis [58,59]. In this paper, we found that 13 *Hsp20* genes were significantly up-regulated in response to heat stress in both the H- and L-type *L. edodes* strains, highlighting the importance of HSPs for *L. edodes* to cope with heat stress. More recently, it was reported that overexpression of the *Agaricus bisporus Hsp20* gene enhanced the mycelial thermotolerance of *L. edodes* [55]. Moreover, the HSP40 protein LeDnaJ07 was also proven to be integral to conferring heat resistance on *L. edodes* [15,29,30].

The central outcome of our study is that the H- and L-type *L. edodes* strains respond differently upon exposure to heat stress. To guarantee a more representative assessment, we selected two typical H- and L-type strains which are widely cultivated in northern and southern China, respectively. However, we still suggest more comprehensive investigations to dig deeper in their temperature-type-dependent thermotolerance mechanisms.

Though the explicit mechanisms by which *L. edodes* copes with heat remain to be elucidated, our data suggest, for the first time, that the molecular strategy for addressing heat can be temperature-type-dependent and provide informative clues for further relevant studies.

**Supplementary Materials:** The following supporting information can be downloaded at: https://www.mdpi.com/article/10.3390/jof9020179/s1, Figure S1: KEGG enrichment analysis of DEGs between H-type strain JZB2102217 and L-type strain JZB2102031; Figure S2: Correlation of different samples based one RNA-seq data; Figure S3: KEGG enrichment of DEGs of H-type strain JZB2102217 and L-type strain JZB2102031 in response to thermostress; Figure S4: Volcano plot, Venn network, and KEGG enrichment for DEMS of H-type strain JZB2102217 and L-type strain JZB2102031 in response to thermostress; Table S1: Expression level and annotation of all expressed genes (RPKM > 10); Table S2: Abundance and annotations of all detected metabolites; Table S3: Overview of the RNA-seq data; Table S4: Number of DEGs in H-type strain JZB2102217 and L-type strain JZB2102031 in response to thermostress.

**Author Contributions:** Conceptualization, Y.G., Y.L. and S.W.; Data Curation, Y.F. and D.Y.; Formal Analysis, Y.G., Q.G., S.S., J.Z. and Y.C.; Funding Acquisition, S.W. and Y.L.; Investigation, S.S., J.Z. and Y.C.; Methodology, Y.G., Q.G., S.S., Y.L. and S.W.; Project Administration, Y.L. and S.W.; Resources, Y.G., Q.G., Y.F., Y.L. and S.W.; Software, Y.F. and D.Y.; Supervision, Y.L. and S.W.; Validation, S.S., Y.F., D.Y., J.Z. and Y.C.; Visualization, Y.G. Y.F. and Q.G.; Writing—Original Draft, Y.G.; Writing—Review and Editing, Y.G., Y.L. and S.W. All authors have read and agreed to the published version of the manuscript.

**Funding:** This work was funded by the China Agriculture Research System (CARS-20) and the Beijing Innovation Consortium of Agriculture Research System (BAIC03).

**Institutional Review Board Statement:** Not applicable.

**Informed Consent Statement:** Not applicable.

**Data Availability Statement:** The raw RNA-seq data and LC-MS data of the *L. edodes* strains are available at the National Genomics Data Center, China National Center for Bioinformation, under the BioProject IDs PRJCA014125 and PRJCA014136, respectively.

**Acknowledgments:** The authors thank ZJ song and XK Liu for their technical assistance.

**Conflicts of Interest:** The authors declare no conflict of interest.

## References

1. Xiang, Q.; Adil, B.; Chen, Q.; Gu, Y.; Zeng, X.; Li, X. Shiitake Mushroom (*Lentinula edodes* (Berk.) Sing. Breeding in China. In *Advances in Plant Breeding Strategies: Vegetable Crops*, 1st ed.; Al-Khayri, J.M., Jain, S.M., Johnson, D.V., Eds.; Springer: Cham, Switzerland, 2021; Volume 3, pp. 443–476.
2. Xiang, X.; Li, C.; Li, L.; Bian, Y.; Kwan, H.S.; Nong, W.; Cheung, M.K.; Xiao, Y. Genetic diversity and population structure of Chinese *Lentinula edodes* revealed by InDel and SSR markers. *Mycol. Prog.* **2016**, *15*, 37. [CrossRef]
3. Smith, C.A. Macrosynteny analysis between *Lentinula edodes* and *Lentinula novae-zelandiae* reveals signals of domestication in *Lentinula edodes*. *Sci. Rep.* **2021**, *11*, 9845. [CrossRef] [PubMed]
4. Yu, H.; Zhang, L.; Shang, X.; Peng, B.; Li, Y.; Xiao, S.; Tan, Q.; Fu, Y. Chromosomal genome and population genetic analyses to reveal genetic architecture, breeding history and genes related to cadmium accumulation in *Lentinula edodes*. *BMC Genom.* **2022**, *23*, 1–14. [CrossRef] [PubMed]
5. Li, C.; Gong, W.; Zhang, L.; Yang, Z.; Nong, W.; Bian, Y.; Kwan, H.S.; Cheung, M.K.; Xiao, Y. Association mapping reveals genetic loci associated with important agronomic traits in *Lentinula edodes*, shiitake mushroom. *Front. Microbiol.* **2017**, *8*, 237. [CrossRef] [PubMed]
6. Finimundy, T.C.; Dillon, A.J.P.; Henriques, J.A.P.; Ely, M.R. A review on general nutritional compounds and pharmacological properties of the *Lentinula edodes* mushroom. *Food Sci. Nutr.* **2014**, *5*, 1095–1105.
7. Nam, M.; Choi, J.Y.; Kim, M.S. Metabolic profiles, bioactive compounds, and antioxidant capacity in *Lentinula edodes* cultivated on log versus sawdust substrates. *Biomolecules* **2021**, *11*, 1654. [CrossRef]
8. Chowdhury, M.; Kubra, K.; Ahmed, S. Screening of antimicrobial, antioxidant properties and bioactive compounds of some edible mushrooms cultivated in Bangladesh. *Ann. Clin. Microbiol. Antimicrob.* **2015**, *14*, 8. [CrossRef]
9. Song, X.; Li, C.; Tan, Q.; Li, Q.; Ma, D.; Chen, M. The summary of resources of some cultivated strains of *Lentinula edodes* in China. *J. Fungal Res.* **2015**, *13*, 146–154. (In Chinese with an English abstract)
10. Terashima, K.; Matsumoto, T.; Hasebe, K.; Fukumasa-Nakai, Y. Genetic diversity and strain-typing in cultivated strains of *Lentinula edodes* (the shiitake mushroom) in Japan by AFLP analysis. *Rep. Tottori Mycol. Inst.* **2002**, *106*, 34–39.
11. Hasebe, K.; Ohira, I.; Arita, I. Genetic relationship between high-, medium-and low-temperature-type fruiting of *Lentinula edodes* in wood log culture. *Rep. Tottori Mycol. Inst.* **1998**, *36*, 21–28.
12. Zhang, R.; Huang, C.; Zheng, S.; Zhang, J.; Ng, T.B.; Jiang, R.; Zuo, X.; Wang, H. Strain-typing of *Lentinula edodes* in China with inter simple sequence repeat markers. *Appl. Microbiol. Biotechnol.* **2007**, *74*, 140–145. [CrossRef]
13. Ryu, S.R.; Bak, W.C.; Koo, C.D.; Lee, B.H. Studies on breeding and cultivation characteristics of *Lentinula edodes* strains for sawdust cultivation. *Kor. J. Mycol.* **2009**, *37*, 65–72. [CrossRef]
14. Zhang, J.; Shen, N.; Li, C.; Xiang, X.; Liu, G.; Gui, Y.; Patev, S.; Hibbett, D.S.; Barry, K.; Andreopoulos, W. Population genomics provides insights into the genetic basis of adaptive evolution in the mushroom-forming fungus *Lentinula edodes*. *J. Adv. Res.* **2022**, *38*, 91–106. [CrossRef]
15. Wang, G.; Luo, Y.; Wang, C.; Zhou, Y.; Mou, C.; Kang, H.; Xiao, Y.; Bian, Y.; Gong, Y. Hsp40 protein LeDnaJ07 enhances the thermotolerance of *Lentinula edodes* and regulates IAA biosynthesis by interacting LetrpE. *Front. Microbiol.* **2020**, *11*, 707. [CrossRef] [PubMed]
16. Luo, L.; Zhang, S.; Wu, J.; Sun, X.; Ma, A. Heat stress in macrofungi: Effects and response mechanisms. *Appl. Microbiol. Biotechnol.* **2021**, *105*, 7567–7576. [CrossRef]
17. Qiu, Z.; Wu, X.; Gao, W.; Zhang, J.; Huang, C. High temperature induced disruption of the cell wall integrity and structure in *Pleurotus ostreatus* mycelia. *Appl. Microbiol. Biotechnol.* **2018**, *102*, 6627–6636. [CrossRef] [PubMed]
18. Awasthi, R.; Bhandari, K.; Nayyar, H. Temperature stress and redox homeostasis in agricultural crops. *Front. Env. Sci-Switz.* **2015**, *3*, 1–24. [CrossRef]
19. Li, Z.; Yu, C.; Ren, Y.; Chen, M.; Cha, L.; Yang, H.; Song, X.; Zhao, Y. Effect of heat stress on fatty acids in *Stropharia rugosoan-nulata* mycelia. *Acta Edulis Fungi.* **2020**, *27*, 45–50. (In Chinese with an English abstract)
20. Yan, Z.; Zhao, M.; Wu, X.; Zhang, J. Metabolic response of *Pleurotus ostreatus* to continuous heat stress. *Front Microbiol.* **2020**, *10*, 3148. [CrossRef]
21. Zhao, X.; Yang, H.; Chen, M.; Song, X.; Yu, C.; Zhao, Y.; Wu, Y. Reference gene selection for quantitative real-time PCR of mycelia from *Lentinula edodes* under high-temperature stress. *Biomed. Res. Int.* **2018**, *2018*, 1670328. [CrossRef]
22. Tan, X.; Sun, J.; Ning, H.; Qin, Z.; Miao, Y.; Sun, T.; Zhang, X. De novo transcriptome sequencing and comprehensive analysis of the heat stress response genes in the basidiomycetes fungus *Ganoderma lucidum*. *Gene* **2018**, *661*, 139–151. [CrossRef] [PubMed]
23. Krah, F.; Hess, J.; Hennicke, F.; Kar, R.; Bässler, C. Transcriptional response of mushrooms to artificial sun exposure. *Ecol. Evol.* **2021**, *11*, 10538–10546. [CrossRef] [PubMed]
24. Cao, X.; Bian, Y.; Xiao, X.; Li, J.; Wang, G. Effect of heat stress on *Lentinula edodes* mycelial growth recovery and resistance to *Trichoderma harzianum*. *Acta Edulis Fungi.* **2015**, *22*, 81–85.

25. Xu, R.; Zhou, S.; Song, J.; Zhong, H.; Zhu, T.; Gong, Y.; Zhou, Y.; Bian, Y. Comparative transcriptome analysis provides insights into the mechanism by which 2, 4-Dichlorophenoxyacetic acid improves thermotolerance in *Lentinula edodes*. *Front. Microbiol.* **2022**, *13*, 910255. [CrossRef] [PubMed]
26. Zhao, X.; Chen, M.; Zhao, Y.; Zha, L.; Yang, H.; Wu, Y. GC–MS-based nontargeted and targeted metabolic profiling identifies changes in the *Lentinula edodes* mycelial metabolome under high-temperature stress. *Int. J. Mol. Sci.* **2019**, *20*, 2330. [CrossRef] [PubMed]
27. Kong, W.; Huang, C.; Chen, Q.; Zou, Y.; Zhang, J. Nitric oxide alleviates heat stress-induced oxidative damage in *Pleurotus eryngii* var. tuoliensis. *Fungal. Genet. Biol.* **2012**, *49*, 15–20. [CrossRef] [PubMed]
28. Vihervaara, A.; Duarte, F.M.; Lis, J.T. Molecular mechanisms driving transcriptional stress responses. *Nat. Rev. Genet.* **2018**, *19*, 385–397. [CrossRef] [PubMed]
29. Wang, G.; Zhou, S.; Luo, Y.; Ma, C.; Gong, Y.; Zhou, Y.; Gao, S.; Huang, Z.; Yan, L.; Hu, Y. The heat shock protein 40 LeDnaJ regulates stress resistance and indole-3-acetic acid biosynthesis in *Lentinula edodes*. *Fungal Genet. Biol.* **2018**, *118*, 37–44. [CrossRef]
30. Wang, G.; Ma, C.; Luo, Y.; Zhou, S.; Zhou, Y.; Ma, X.; Cai, Y.; Yu, J.; Bian, Y.; Gong, Y. Proteome and transcriptome reveal involvement of heat shock proteins and indoleacetic acid metabolism process in *Lentinula edodes* thermotolerance. *Cell. Physiol. Biochem.* **2018**, *50*, 1617–1637. [CrossRef] [PubMed]
31. Zhou, S.; Wang, G.; Luo, Y.; Ma, C.; Gong, Y.; Bian, Y.; Zhou, Y. Auxin and auxin analogues enhancing the thermotolerance of *Lentinula edodes*. *Mycosystema* **2018**, *37*, 1723–1730.
32. Chittaragi, A.; Kumar, A.; Muniraju, K. Evaluation of various lignocellulosic products for the cultivation of shiitake mushroom [*Lentinula edodes* (Berk.) Pegler]. *Int. J. Curr. Microbiol. App. Sci.* **2018**, *7*, 2199–2203. [CrossRef]
33. Kim, D.; Langmead, B.; Salzberg, S.L. HISAT: A fast spliced aligner with low memory requirements. *Nat. Methods* **2015**, *12*, 357–360. [CrossRef] [PubMed]
34. Pertea, M.; Pertea, G.M.; Antonescu, C.M.; Chang, T.-C.; Mendell, J.T.; Salzberg, S.L. StringTie enables improved reconstruction of a transcriptome from RNA-seq reads. *Nat. Biotechnol.* **2015**, *33*, 290–295. [CrossRef] [PubMed]
35. Li, B.; Dewey, C.N. RSEM: Accurate transcript quantification from RNA-Seq data with or without a reference genome. *BMC Bioinform.* **2011**, *12*, 323. [CrossRef] [PubMed]
36. Love, M.I.; Huber, W.; Anders, S. Moderated estimation of fold change and dispersion for RNA-seq data with DESeq2. *Genome Biol.* **2014**, *15*, 550. [CrossRef] [PubMed]
37. Benjamini, Y.; Hochberg, Y. Controlling the False Discovery Rate: A practical and powerful approach to multiple testing. *J R. Stat. Soc. Series. B Stat. Methodol.* **1995**, *57*, 289–300. [CrossRef]
38. Xie, C.; Mao, X.; Huang, J.; Ding, Y.; Wu, J.; Dong, S.; Kong, L.; Gao, G.; Li, C.Y.; Wei, L. KOBAS 2.0: A web server for annotation and identification of enriched pathways and diseases. *Nucleic Acids Res.* **2011**, *39*, W316–W322. [CrossRef]
39. Domingo-Almenara, X.; Montenegro-Burke, J.R. Annotation: A computational solution for streamlining metabolomics analysis. *Anal. Chem.* **2018**, *90*, 480–489. [CrossRef] [PubMed]
40. Worley, B.; Powers, R. Multivariate analysis in metabolomics. *Curr. Metab.* **2013**, *1*, 92–107.
41. Ren, A.; Liu, R.; Miao, Z.G.; Zhang, X.; Cao, P.F.; Chen, T.X.; Li, C.Y.; Shi, L.; Jiang, A.L.; Zhao, M.W. Hydrogen-rich water regulates effects of ROS balance on morphology, growth and secondary metabolism via glutathione peroxidase in *Ganoderma lucidum*. *Environ. Microbiol.* **2017**, *19*, 566–583. [CrossRef]
42. Mu, D.; Li, C.; Zhang, X.; Li, X.; Shi, L.; Ren, A.; Zhao, M. Functions of the nicotinamide adenine dinucleotide phosphate oxidase family in *Ganoderma lucidum*: An essential role in ganoderic acid biosynthesis regulation, hyphal branching, fruiting body development, and oxidative-stress resistance. *Environ. Microbiol.* **2014**, *16*, 1709–1728. [CrossRef] [PubMed]
43. Collins, T.J. ImageJ for microscopy. *Biotechniques* **2007**, *43*, S25–S30. [CrossRef] [PubMed]
44. Triba, M.N.; Le Moyec, L.; Amathieu, R.; Goossens, C.; Bouchemal, N.; Nahon, P.; Rutledge, D.N.; Savarin, P. PLS/OPLS models in metabolomics: The impact of permutation of dataset rows on the K-fold cross-validation quality parameters. *Mol. Biosyst.* **2015**, *11*, 13–19. [CrossRef] [PubMed]
45. Gower, J.C. Generalized procrustes analysis. *Psychometrika* **1975**, *40*, 33–51. [CrossRef]
46. Priya, S.; Burns, M.B.; Ward, T.; Mars, R.A.T.; Adamowicz, B.; Lock, E.F.; Kashyap, P.C.; Knights, D.; Blekhman, R. Identification of shared and disease-specific host gene-microbiome associations across human diseases using multi-omic integration. *Nat. Microbiol.* **2022**, *7*, 780–795. [CrossRef]
47. Lê Cao, K.A.; Rossouw, D.; Robert-Granié, C.; Besse, P. A sparse PLS for variable selection when integrating omics data. *Stat. Appl. Genet. Mol. Biol.* **2008**, *7*, 35. [CrossRef] [PubMed]
48. Rohart, F.; Gautier, B.; Singh, A.; Lê Cao, K.-A. mixOmics: An R package for 'omics feature selection and multiple data integration. *PLoS Comput. Biol.* **2017**, *13*, e1005752. [CrossRef]
49. Chang, S.; Kwan, H.; Kang, Y. Collection, characterization, and utilization of germ plasm of *Lentinula edodes*. *Canad. J. Bot.* **1995**, *73*, 955–961. [CrossRef]
50. Huang, X.; Zhang, R.; Qiu, Y.; Wu, H.; Xiang, Q.; Yu, X.; Zhao, K.; Zhang, X.; Chen, Q.; Penttinen, P. RNA-seq profiling showed divergent carbohydrate-active enzymes (CAZymes) expression patterns in *Lentinula edodes* at brown film formation stage under blue light induction. *Front. Microbiol.* **2020**, *11*, 1044. [CrossRef]
51. Tang, L.; Tan, Q.; Bao, D.; Zhang, X.; Jian, H.; Li, Y.; Wang, Y. Comparative proteomic analysis of light-induced mycelial brown film formation in *Lentinula edodes*. *Biomed. Res. Int.* **2016**, *2016*, 5837293. [CrossRef]

52. Kim, J.Y.; Kim, D.Y.; Park, Y.J.; Jang, M.J. Transcriptome analysis of the edible mushroom *Lentinula edodes* in response to blue light. *PLoS ONE* **2020**, *15*, e0230680. [CrossRef] [PubMed]
53. Zhang, X.; Ren, A.; Li, M.; Cao, P.; Chen, T.; Zhang, G.; Shi, L.; Jiang, A.; Zhao, M. Heat stress modulates mycelium growth, heat shock protein expression, ganoderic acid biosynthesis, and hyphal branching of *Ganoderma lucidum* via cytosolic $Ca^{2+}$. *Appl. Environ. Microbiol.* **2016**, *82*, 4112–4125. [CrossRef]
54. Guo, M.; Ye, Z.; Shen, G.; Bian, Y.; Xu, Z. Effects of mycovirus LeV-HKB on resistance of heat stress challenged *Lentinula edodes* mycelia against *Trichoderma atroviride*. *Acta Edulis Fungi.* **2020**, *27*, 143.
55. Ling, Y.; Ling, Z.; Zhao, R. Construction of a heat-resistant strain of *Lentinus edodes* by fungal Hsp20 protein overexpression and genetic transformation. *Front. Microbiol.* **2022**, *13*, 1009885. [CrossRef] [PubMed]
56. Wang, B.; Tang, L.; Xiong, Y.; Jiang, L.; Xian, L. Test on thermophilic stability of mycelium and fruitbody growth of *Lentinula edodes*. *J. Jilin Agric. Univ.* **2004**, *26*, 145–147.
57. Wu, X.; Deng, B.; Chen, W.; Xie, X. Breeding thermo-tolerant strains of *Lentinula edodes* by UV mutation. *Acta Agric. Zhejiangensis* **2017**, *29*, 2015–2022.
58. Park, C.J.; Seo, Y.S. Heat shock proteins: A review of the molecular chaperones for plant immunity. *Plant Pathol. J.* **2015**, *31*, 323. [CrossRef]
59. Tereshina, V. Thermotolerance in fungi: The role of heat shock proteins and trehalose. *Microbiology* **2005**, *74*, 247–257. [CrossRef]

Journal of *Fungi*

MDPI

*Article*

# The Influence of Different Pretreatment Methods of Highland Barley by Solid-State Fermentation with *Agaricus sinodeliciosus* var. Chaidam ZJU-TP-08 on Its Nutrient Content, Functional Properties and Physicochemical Characteristics

**Biao Liu [1], Hongyun Lu [2], Qin Shu [2], Qihe Chen [2,*] and Jinling Wang [1,*]**

[1] School of Forestry, Northeast Forestry University, Harbin 150040, China
[2] Department of Food Science and Nutrition, Zhejiang University, Hangzhou 310058, China
* Correspondence: chenqh@zju.edu.cn (Q.C.); wangjinling@nefu.edu.cn (J.W.); Tel.: +86-571-8698-4316 (Q.C.); +86-451-8219-0222 (J.W.)

**Abstract:** To enhance the nutritional value of highland barley (HB), this work investigated the effects of solid-state fermentation (SSF) by *Agaricus sinodeliciosus* var. Chaidam ZJU-TP-08 on nutrient content, phenolic components, antioxidant activities, and physicochemical characteristics of HB upon different pretreatments (germination, ultrasound and soaking). The results showed that germinated highland barley (GHB) exhibited higher levels of ergosterol ($0.19 \pm 0.01$ mg/g) in all fermentation groups. The content of β-glucan was higher in the SSF-GHB, with an increase of 24.21% compared to the control. The content of total amino acids, dietary fiber, total phenols and flavonoids were higher in the fermentation HB pretreated by ultrasound, increasing respectively by 5.60%, 61.50%, 25.10% and 65.32% compared to the control group. In addition, the colonized HB exhibited excellent physicochemical characteristics, including increased water solubility index and decreased pasting characteristics. Herein, the nutritional value and the biological activities were enriched in the pretreated HB through SSF, indicating its potential application for nutrition-enriched functional foods.

**Keywords:** highland barley; *Agaricus sinodeliciosus* var. Chaidam; solid-state fermentation; nutritional enrichment; biological activity



**Citation:** Liu, B.; Lu, H.; Shu, Q.; Chen, Q.; Wang, J. The Influence of Different Pretreatment Methods of Highland Barley by Solid-State Fermentation with *Agaricus sinodeliciosus* var. Chaidam ZJU-TP-08 on Its Nutrient Content, Functional Properties and Physicochemical Characteristics. *J. Fungi* **2022**, *8*, 940. https://doi.org/10.3390/jof8090940

Academic Editors: Mingwen Zhao, Gen Zou and Jing Zhu

Received: 24 July 2022
Accepted: 1 September 2022
Published: 7 September 2022



## 1. Introduction

Known for high nutritional value, edible macrofungi are favored by consumers worldwide for their unique flavor and organoleptic properties [1]. Especially, macrofungi-derived secondary metabolites, including polysaccharides, phenolic compounds, and triterpenes, have a variety of health-promoting biological activities, contributing to a wide-accept research hotspot [2]. Ergosterol, as a triterpenoid compound found in the cell walls of fungi, has a large number of biological activities, including antibacterial, anti-inflammatory, anti-cancer, and cholesterol-lowering effects [3]. *Agaricus sinodeliciosus* var. Chaidam, a novel underground edible macrofungus from the Qinghai-Tibetan Plateau [4], was reported to have plenty of nutritional and bioactive compounds in the fruiting bodies and mycelia [5–7]. Due to the immature cultivation technology and remote habitat, the utilization and development of *A. sinodeliciosus* var. Chaidam are rarely reported. Therefore, it is necessary to explore an efficient method to obtain ergosterol from the cultured mycelia of *A. sinodeliciosus* var. Chaidam and expand its application in food processing.

Besides the fruiting body, mycelia and culture media in macrofungi are being explored as potential sources of bioactive compounds. Cultured mycelia was emerging as a promising alternative source of macrofungal bioactive compounds, mainly due to its shorter incubation time and easier culture conditions [8]. Converging lines of evidence suggested that edible mycelia colonized in grains by SSF have resulted in fermented nutrient-rich grain flours enriched with functional ingredients [9,10], which could be potential ingredients for new food formulations. Hence, the SSF of grains using macrofungal colonization is an effective method to improve the utilization of *A. sinodeliciosus* var. Chaidam and agricultural products.

Highland barley (HB), known as Qingke in China, is a special variety of barley (*Hordeum vulgare* subsp. *vulgare*) and a staple food for the people in Qinghai and Tibet provinces. Recently, HB has attracted increasing attention due to its unique nutrition values and bioactive compounds [11], especially β-glucan, which is a non-starch polysaccharide and has a great positive effect on human health, including cholesterol level reduction, CVD disease inhibition, tumors prevention, and immune enhancement [12]. However, it is difficult for most people to accept HB as a dietary food due to its rough taste, because approximate 98% of barley is devoted to the production of animal feed and wine processing, while only 2% is employed to the development of barley-based foods [13]. As previously reported, the increase in amino acids and active flavor compounds can enrich the taste in SSF. Meanwhile, various released enzymes could lead to hydrolysis of starch, cellulose and hemicellulose, thus improving the taste of grains [10,14].

To better enhance nutritional value and utilization of grains, the pre-treatments such as germination, ultrasound, and soaking are used frequently in the food processing industry [15]. Germination of grains is a non-chemical and simple method, and has received much attention for its physiological benefits, which increases the content of bioactive compounds including β-glucan and γ-aminobutyric acid [16]. The previous studies exhibited that consumption of germinated grains could decrease the risk of diabetes, obesity, cardiovascular disease and cancers [16,17]. Furthermore, an extensive review focused on the application of ultrasound in various food processing technologies, including bioactive compound extraction, food fermentation and food degradation, and it was confirmed that rice treated by ultrasound could enhance nutrition and improve texture [18]. Therefore, we made an interesting attempt to use the germinated, ultrasonic and soaked HB as the substrates for *A. sinodeliciosus* var. Chaidam mycelia fermentation to increase the availability of nutrients and biological activity.

Herein, different pretreated HBs by germination, ultrasound and soaking were colonized by *A. sinodeliciosus* var. Chaidam with the aim to obtain HB with enriched nutrient, enhanced bioactivities and physicochemical properties. In this study, we firstly provided HB colonized with *A. sinodeliciosus* var. chaidam ZJU-TP-08 to obtain bioactive cereal materials, and further detected the fermented nutrients including ergosterol, phenolic compounds and antioxidant activity, which contribute to potential health-promoting and future food application.

## 2. Materials and Methods

### 2.1. Microorganism, Culture Media and Cultivation Conditions

The edible macrofungus was isolated, purified and cultured from the wild *A. sinodeliciosus* var. Chaidam fruiting body. It was identified by morphological observation, physiological, and biochemical experiments and molecular biology, named as *A. sinodeliciosus* var. Chaidam ZJU-TP-08 and stored at China Center for Type Culture Collection (CCTCC M 2021511) [19]. Before the fermentation experiment, *A. sinodeliciosus* var. Chaidam ZJU-TP-08 was grown in potato dextrose agar (PDA) solid slant medium and incubated at 25 °C. The seed liquid cultivation was conducted in Erlenmeyer flasks (250 mL), containing 100 mL of the sterile cultivation medium and inoculated with 3 mycelial discs (1.5 cm in diameter). The flasks were covered with a layer of gauze and sterile cotton and then incubated in a dark incubator at constant temperature (150 rpm, 25 °C) for 4 days. The composition of the liquid fermentation medium (g/L) included 30 g of fructose, 8 g of yeast extract, 1.79 g of $KH_2PO_4$ and 1.18 g of $MgSO_4$.

### 2.2. Pre-Treatment of HB

HB was purchased from a local farmers' markets (Xining, China). Next, HB was washed 6–8 times with distilled water. Drained HB was then soaked in distilled water for 12 h at 25 °C, as the soaked group (SHB). Then, the soaked HB was used for sonication and germination respectively. Soaked HB was sonicated in distilled water for 2 h with an ultrasonic cleaner (40 KHz, Skymen Cleaning Equipment Shenzhen Co., Ltd., Shenzhen, China), as the ultrasonic group (UHB). Soaked HB was carried out at a constant temperature and humidity incubator (SaFe Experimental Instrument Technology Co., Ltd., Ningbo, China) at 25 °C and 85% relative humidity in the dark for 36 h, as germinated group (GHB). Then, GHB, UHB and SHB samples (each of 50 g) were distributed separately in glass flasks of 250 mL, then 1% $CaCO_3$ and $CaSO_4$ were added. The water content of solid fermentation substrate was around 35%. All the flasks were sterilized at 121 °C for 30 min. After cooling, the flasks with the sterile grains were inoculated in a vertical laminar flow cabinet.

### 2.3. Solid-State Fermentation Process

GHB, UHB and SHB were separately inoculated with 16 mL of liquid medium of *A. sinodeliciosus* var. Chaidam ZJU-TP-08 (SSF-GHB, SSF-UHB and SSF-SHB as the colonized groups). At the same time, GHB, UHB and SHB were separately inoculated with 16 mL of sterile water and were kept under the same cultivation conditions as control groups. The flasks were covered with a layer of gauze and sterile cotton and then incubated in a dark incubator at constant temperature (25 ± 1 °C) for 25 d. The media colonization was monitored daily by visual inspection. After fermentation, SSF-GHB, SSF-UHB and SSF-SHB were dried in an oven at 60 °C until to a constant mass. Then, they were ground in a Multi-functional crusher to obtain the colonized HB flours (standardized in 60 mesh). Non-inoculated GHB, UHB and SHB were submitted to the same procedures and referred to as control groups. The flour was stored in polyethylene packages at 4 °C until determination.

### 2.4. Mycelial Biomass Evaluation

Mycelial biomass of each type of fermented samples was estimated by extraction and quantification of ergosterol, as described previously with slight modifications [10]. Firstly, the biomass of *A. sinodeliciosus* var. Chaidam ZJU-TP-08 was obtained by submerged cultivation. The selected colonies were inoculated into ready-made basic liquid medium in 100 mL/250 mL conical flasks, incubated at 25 °C, and shaken at 150 r/min for 34 d. The content of the flasks were collected every 4 d. The mycelial biomass was separated by centrifugation (3220× *g*, 20 min, 4 °C) and lyophilized for subsequent extraction and quantification of ergosterol.

The extraction and determination of ergosterol were based on described literature previously, with slight modifications [10]. The samples (60 mesh) were extracted with methanol at a ratio of 1:10 ($m/v$) for 1 h at 25 °C, followed by 1 h of ultrasound treatment (40 KHz, Skymen Cleaning Equipment Shenzhen Co., Ltd., Shenzhen, China). The ergosterol extracts were collected after centrifugation at 3220× *g* for 15 min at 4 °C and filtered through 0.22 μm nylon filters. The ergosterol extracted from the mycelial biomasses and the flour samples (colonized and control groups) were determined in high-performance liquid chromatography (HPLC) system (Agilent Technologies Inc., Santa Clara, CA, USA) with VWD detector and reverse phase column Cosmosil 5 C18-MS-II (Nacalai tesque, Kyoto, Japan, 250 × 4.6 mm and particle size 5 μm), according to the method described previously with slight modifications [20]. Then, 20 μL of ergosterol extracts was injected into the system at 35 °C with mobile phase methanol: water (98:2, $v/v$), with a flow rate of 1 mL/min and detection wave at 282 nm. The identification of ergosterol was based on comparing the retention time with the standard solutions (0.001–1 mg/mL). The data obtained for the production of ergosterol in submerged medium were used to assay the mycelial biomass.

### 2.5. Nutritional Composition Evaluation

The analysis of the proximal composition of moisture, ash, crude protein, crude fat, crude fiber and dietary fiber of colonized and control groups was determined according to the method described previously [21]: Moisture (925.09), ash (incineration in a muffle furnace for 24 h at 550 °C) (no. 923.03), crude protein (total N × 6.25) (no. 992.23), crude fat (Soxhlet extraction method) (no. 920.39), crude fiber (sample digestion with diluted acid and alkali) (no. 962.09), and dietary fiber (no. 985.29). Total sugar and reducing sugar content were determined by the phenol-sulfuric acid colorimetric and the 3,5-dinitrosalicylic acid (DNS) method, as described previously [22]. The γ-amino butyric acid (GABA) content was determined according to the method described previously with slight modifications [23]. Samples of 0.5 g were weighed into a test tube, and then 5 mL of distilled water was added and oscillated for 2 h. The GABA contained in supernatant layer was isolated by centrifugation (3200× *g*, 15 min). Then, 0.6 mL of the boric acid buffer (0.5 M, pH 9.0), 2 mL of 5% re-steamed phenol solution and 0.2 mL of 10% sodium hypochlorite were added to 1 mL of the collected supernatant. The resultant mixture was boiled for 10 min in a water bath, then cooled in ice bath and followed by oscillation for 20 min until appearance of blue-green. Finally, 2 mL of 60% ethanol was added and the absorbance at 645 nm was recorded. The final quantification was based on a GABA calibration. The β-glucan content was quantified from flour samples (colonized and control groups) with a mixed-linkage β-glucan content assay kit (Suzhou Grace Biotechnology Co., Ltd., Suzhou, China) following the standard protocols mentioned in the kits.

### 2.6. Amino Acids Composition Detection

Prior to analysis, each sample (100 mg) of colonized and control groups was hydrolyzed with 4 mL (6 mol/L) hydrochloric acid for 24 h at 110 °C and filtered through 0.22 μm nylon filters. The amino acid composition was profiled using an automatic amino acid analyzer with a Na+ type cation exchange column (4.6 mm ID × 60 mm, 3 μm particles) and detection with a UV-V detector. Proline was detected at 440 nm and other amino acids at 570 nm. The color developer was ninhydrin/sodium acetate buffer, the buffer system was citrate buffer B1 (pH 3.2), B2 (pH 3.0), B3 (pH 4.0), and B4 (pH 4.9); the buffer flow rate was 0.4 mL/min; the column temperature was 55 °C, and the reaction temperature was 135 °C. Free amino acid content in sample solution was determined by external standard method. The results of the 17 amino acids investigated (lysine, valine, leucine, isoleucine, threonine, methionine, phenylalanine, histidine, arginine, tyrosine, serine, glutamic acid, proline, glycine, alanine, cystine, aspartic acid) were expressed as milligram of amino acids per gram of dry HB samples.

### 2.7. Total Phenolic and Flavonoid Content Detection

Free and bound phenolic compounds were extracted according to the method described previously [24]. Methanolic extracts were subsequently used for the total phenols, total flavonoids and antioxidant activity assays. Then, 1 g HB flour samples (colonized and control groups) was extracted three times with 20 mL of 80% methanol (*v*/*v*). For each extraction, the mixture was shaken for 30 min at room temperature in a shaker, then centrifuged for 15 min (3500× *g*), then the supernatant was vacuum evaporated to dryness at 45 °C, and finally the free phenolic content was reconstituted in 10 mL of methanol. After the extraction of free phenolic, the residue was extracted with 20 mL 2M NaOH solution and shaken at room temperature for 2 h; afterward, the mixture was acidified with HCl to pH 1.5–2.0, then centrifuged at 3500× *g* for 15 min. Hexane was used to extract lipids in the supernatant. The remaining mixture was extracted three times with ethyl acetate. The ethyl acetate fractions were pooled and evaporated to dryness. The bound phenolic content was reconstituted in 10 mL of methanol. All phenolic extracts were stored at −20 °C before analysis. Total phenolic content (TPC) and flavonoids content (TFC) were determined according to the method described previously [24]. TPC was calculated and expressed as milligram gallic acid equivalents (mg GAE/g). TFC was calculated and expressed as milligram rutin equivalents (mg RE/g).

*2.8. Antioxidant Activity Assay*

The total antioxidant capacity was conducted using the Total Antioxidant Capacity (T-AOC) assay kit (Solarbio Biochemical Assay Division, Beijing, China). The results were expressed as micromoles of $Fe^{2+}$ equivalent antioxidant capacity per g of samples, using a $Fe^{2+}$ calibration curve. The DPPH• and ABTS+• radical scavenging activities were determined based on the method described previously [25]. The DPPH ethanol solution (4.5 mL, 0.1 mmol/L) was mixed with 1.0 mL of samples and incubated in darkness (30 min). The absorbance of the solution was read at 517 nm. The DPPH scavenging activity was calculated as follows: scavenging effect (%) = $(1 - A_1/A_0) \times 100\%$, where $A_1$ was the absorbance of the sample, and $A_0$ was the absorbance of the blank control, the results were expressed as micromoles of Trolox equivalent antioxidant capacity (μmol TEAC/g). ABTS•+ working solution (4.0 mL) was mixed with 200 μL of samples and incubated in darkness (30 min). The absorbance of the solution was read at 734 nm. The ABTS•+ scavenging activity was calculated as follows: scavenging effect (%) = $(1 - A_1/A_0) \times 100\%$, where $A_1$ was the absorbance of the sample, and $A_0$ was the absorbance of the blank control. The results were expressed as micromoles of Trolox equivalent antioxidant capacity (μmol TEAC/g), the hydroxyl radical scavenging rate was determined according to the method described previously [26]. The reaction mixture contained $FeSO_4$ (1 mL, 0.75 mM), 1,10-phenanthroline (1 mL, 0.75 mM), 1.5 mL of 0.15 M sodium phosphate buffer (pH 7.4), 1 mL of phenolic solution and 0.1 mL 0.3% $H_2O_2$ (1 mL, 0.01%, $v/v$) and 1 mL samples. After incubation at room temperature for 30 min, the absorbance of the mixture was measured at 536 nm. Hydroxyl radical scavenging (%) = $(1 - A_1/A_0) \times 100\%$, where $A_1$ was the absorbance of the sample, and $A_0$ was the absorbance of the blank control. The results were expressed as micromoles of Trolox equivalent antioxidant capacity (μmol TEAC/g).

*2.9. Analysis of Physicochemical Characteristics*

The bulk density (BD), water-holding capacity (WHC) and water solubility index (WSI) of the HB flour samples (colonized and control groups) were determined by a previously published method [27]. BD was calculated as the weight of sample per unit volume (g/mL). The pasting properties of the HB flour samples were analyzed according to previously reported method and used a Rapid Visco Analyzer (RVA Tecmaster, Perten, Hägersten, Sweden) [14]. Flour samples (3 g) in 25 mL of distilled water were added and measured with the RVA. The procedure was run as follows: 1 min at 50 °C, heated to 95 °C at a rate of 11.25 °C/min and held for 2 min, then cooled to 50 °C at 11.25 °C/min and maintained at 50 °C for 2 min. The peak viscosity (PV), tough viscosity (TV), breakdown viscosity (BV = PV−TV), final viscosity (FV), setback viscosity (SV = FV−TV) and peak temperature (PT) were recorded in this study.

*2.10. Scanning Electron Microscope (SEM) Observation Morphological Features*

The dehydrated samples were coated with gold–palladium in Hitachi Model E-1010 ion sputter for 4–5 min and observed in Hitachi Model SU-8010 SEM.

*2.11. Statistical Analysis*

All samples were conducted in triplicate. All experimental data were expressed in means ± SD and OriginPro software (2021 b, Southampton, MA, USA). One-way analysis of variance (ANOVA) followed by Duncan tests were conducted by SPSS statistics software (V.26.0, SPSS Inc., Chicago, IL, USA) to determine the significance level at $p < 0.05$. PCA analysis was implemented by using OriginPro software (2021 b, Southampton, MA, USA).

## 3. Results and Discussion

### 3.1. Effects of Different Pretreatments on Mycelial Biomass and Ergosterol Formation

*A. sinodeliciosus* var. Chaidam ZJU-TP-08 was investigated in this study, which was able to colonize in HB pretreated by germination, ultrasound, and soaking process. Mycelia were observed on the surface and interior of the different cultivation media (Figure 1A–C). According to the results of SEM (Figure 1D–I), *A. sinodeliciosus* var. Chaidam ZJU-TP-08 mycelium colonized firmly onto GHB, UHB and SHB by SSF. However, the surface of unfermented HB was relatively smooth compared to SSF groups.

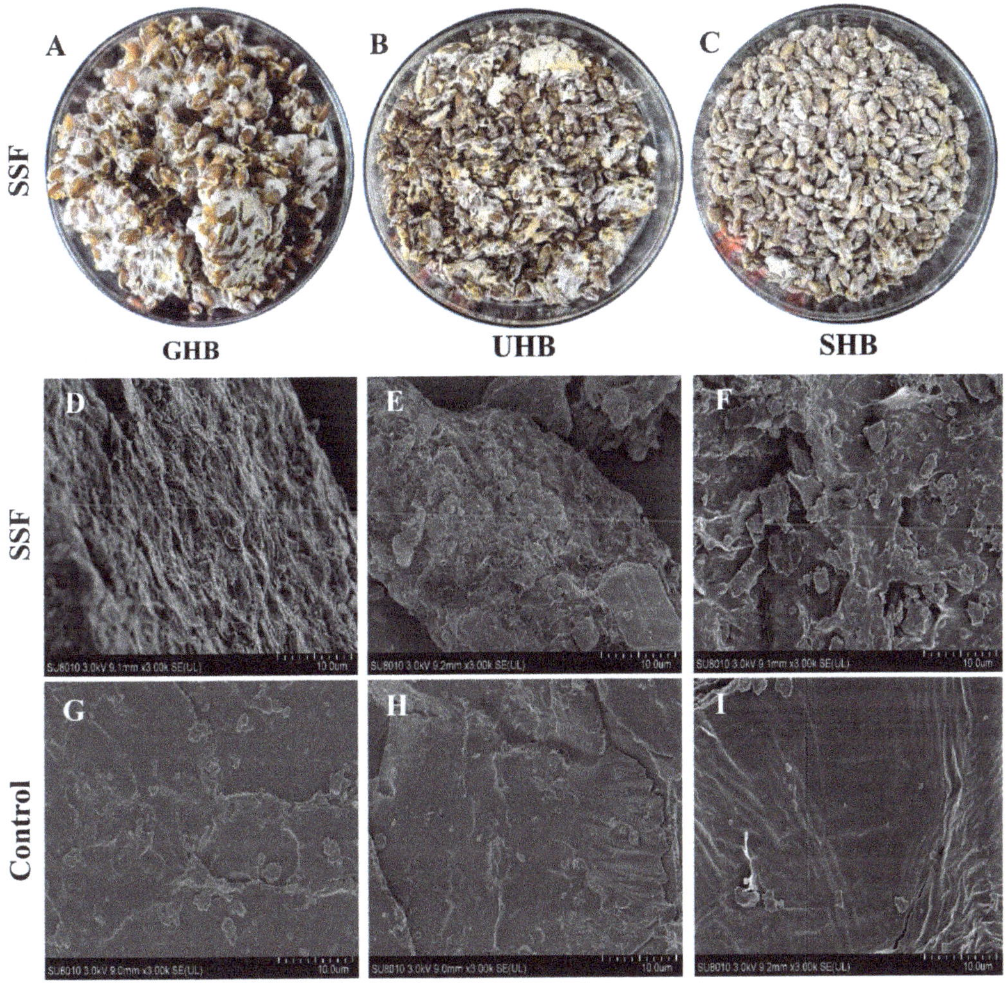


**Figure 1.** Appearance of solid-state fermentation highland barley. The letter (**A–C**) represents solid-state fermentation HB treated by germination (SSF-GHB), ultrasound (SSF-UHB), and soaking (SSF-SHB), respectively. Scanning electron microscopy of solid-state fermented and control highland barley. (**D–F**) indicate solid-state fermentation HB treated by germination (SSF-GHB), ultrasound (SSF-UHB), and soaking (SSF-SHB), respectively. (**G–I**) indicate germinated, sonicated, and soaked HB for control groups (GHB, UHB, SHB), respectively.

Due to the difficulty in separating the mycelial biomass from the grain, the indirect determination method was used with the ergosterol dosage to estimate the amount of mycelia present in the colonized HB. To quantify the changes in mycelial biomass during SSF, the correlation of ergosterol and mycelial biomass was performed. Clearly observed from Figure 2A,B, the quantitative correlation between mycelial biomass and ergosterol was performed ($y = 0.0044x + 0.0214$, $R^2 = 0.9913$), indicating that the ergosterol can indirectly quantify the mycelial biomass in SSF. The ergosterol values was determined in different colonized HB and was shown in Figure 2C. The levels of ergosterol in the SSF-GHB, SSF-UHB and SSF-SHB were 0.19 ± 0.01, 0.16 ± 0.00 and 0.08 ± 0.01 mg/g, respectively. However, the level of ergosterol in unfermented HB was below the limit of detection. The mycelial biomass of colonized HB was presented in Figure 2D. Among all fermented samples, the highest amount of mycelial biomass was observed in the SSF-GHB (37.79 ± 2.35 mg/g).

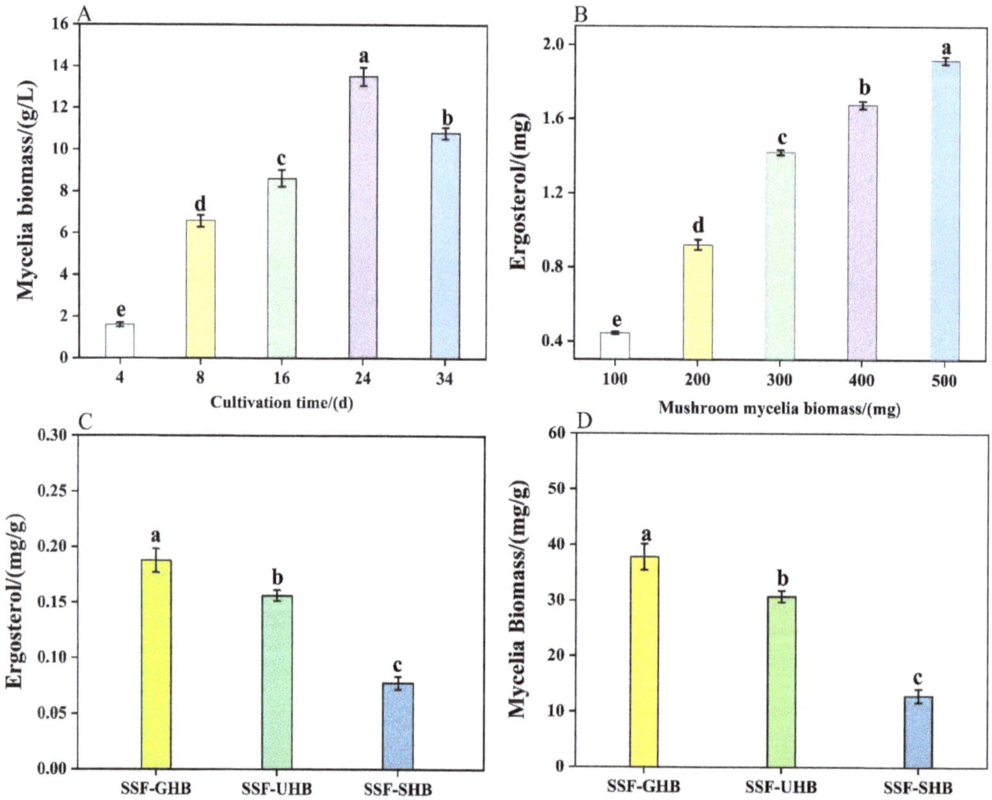


**Figure 2.** Ergosterol constituents of the highland barley colonized with *A. sinodeliciosus* var. Chaidam ZJU-TP-08 mycelia. (**A**): Mycelia biomass (g/L) produced by submerse cultivation during 34 d; (**B**): Ergosterol curve of mycelial at 24 d of cultivation; (**C**): Ergosterol determined (mg/g) of colonized Highland barley, the different colors of legends represent different group, (▭) SSF-GHB, (▭) SSF-UHB, (▭) SSF-SHB; (**D**): Indirect determination of mycelial biomass (mg/g) of colonized Highland barley, the different colors of legends represent different group, (▭) SSF-GHB, (▭) SSF-UHB, (▭) SSF-SHB. a–e: Values were determined in triplicate. Equal letters indicated that there were no significant difference at 5% ($p > 0.05$) in the parameter evaluated.

### 3.2. Alterations in Nutritional Compositions

As shown in Table 1, the effect of *A. sinodeliciosus* var. Chaidam ZJU-TP-08 after colonization on GHB, UHB and SHB was further examined, including the chemical (the moisture, ash, crude fat, reduced sugar, total sugar) and nutritional compositions (crude fiber, dietary fiber, crude protein, β-glucan and GABA). The contents of crude protein in the SSF-GHB, SSF-UHB and SSF-SHB were 13.02 ± 0.06, 12.97 ± 0.22 and 12.95 ± 0.08 g/100 g. Compared to the crude protein in unfermented HB (the control groups), the protein content in SSF-GHB, SSF-UHB and SSF-SHB were increased by 5.20%, 5.40% and 6.00% ($p < 0.05$), respectively. As shown in Table 2, the contents of total amino acids in the SSF-GHB, SSF-UHB and SSF-SHB were 11.90 ± 0.10, 11.75 ± 0.04 and 11.74 ± 0.17 g/100 g. Compared to the total amino acid content in unfermented GHB, UHB and SHB, the contents of total amino acids in SSF-GHB, SSF-UHB and SSF-SHB were increased by 5.40%, 5.60% and 5.10% ($p < 0.05$), respectively. Particularly, the levels of valine, phenylalanine, glutamic acid, alanine and aspartic acid were increased among all fermentation groups ($p < 0.05$).

It has been reported that fermentation strategy was one of the best food processing techniques to increase the nutrition level of grains, such as protein [28]. Herein, the total protein content results demonstrated the potential of *A. sinodeliciosus* var. Chaidam ZJU-TP-08 for enriching proteins with HB. The most accepted explanation for the increase in protein content is that the microorganisms using substrate as carbon and energy sources to produce fungal protein, fungi can depolymerize the cell wall components and assimilate nitrogenous compounds from the substrates, altering the solubility of the proteins and improving the degradability of plant biomasses [10,14]. Fermentation by *Pleurotus ostreatus* (*Avena sativa*) for 336 h could increase oat protein levels [29], which was similar to our results. Similarly, the content of protein were increased 3–20% in fermented rice, due to the metabolic capacity of macrofungi during the fermentation process [30]. In the present study, the content of glutamic acid and aspartic acid were abundant in the SSF-GHB, SSF-UHB and SSF-SHB, and these amino acids are responsible for the production of fresh taste and are important for improving the taste of HB flour. A similar observation was noted in the brown rice, corn and wheat flour colonized by *A. blazei*, *P. albidus* and *A. fuscosuccinea* mycelia [10].

Dietary fiber plays an important role in promoting human health. In this work, the contents of dietary fiber in the SSF-GHB, SSF-UHB and SSF-SHB were 3.80 ± 0.12, 3.23 ± 0.06 and 3.21 ± 0.11 g/100 g. Compared to the dietary fiber in the non-fermented HB groups (GHB, UHB and SHB), the contents of dietary fiber in the SSF-GHB, SSF-UHB and SSF-SHB were increased by 53.80%, 61.50% and 51.20% ($p < 0.05$), respectively. According to a previous study [31], it was proven that the secretion and action of enzymes such as β-glucosidase, cellulase and xylanase rendered the decrease in dietary fiber during fermentation, which break down the polysaccharides of dietary fiber into smaller carbohydrates that serve as a source of energy and carbon for the metabolic processes. However, another research reported that fiber components were partially hydrolyzed and released by fungal enzymes (cellulase, xylanase, etc.) during fermentation process [32]. Therefore, the possible reasons for the increase in dietary fiber content were that the degradation of cellulose and hemicellulose and the formation of loose structures caused more soluble polysaccharides to be released during the SSF process. Meanwhile, it has been demonstrated that the total fiber content of fermented red quinoa seeds with *Neurospora intermedia* was increased by 73%, mainly attributing to the fungal biomass components [33]. Therefore, we speculated that the increase in dietary fiber content was attributed to the biomass component of the fungi in this study.

**Table 1.** Nutritional constituents (dry matter) of the highland barley (the control groups and colonized groups with the mycelia of *A. sinodeliciosus* var. Chaidam ZJU-TP-08).

| Parameters | SSF | | | Control | | |
|---|---|---|---|---|---|---|
| | GHB | UHB | SHB | GHB | UHB | SHB |
| Moisture (g/100g) | 12.38 ± 0.72 [a] | 12.63 ± 1.11 [a] | 11.57 ± 0.50 [bc] | 11.11 ± 1.08 [bc] | 10.16 ± 0.52 [c] | 10.06 ± 0.92 [c] |
| Ash (g/100g) | 4.36 ± 0.14 [a] | 4.03 ± 0.10 [ab] | 4.03 ± 0.06 [ab] | 4.01 ± 0.07 [ab] | 3.69 ± 0.19 [b] | 3.50 ± 0.15 [b] |
| Fat (g/100g) | 1.33 ± 0.05 [d] | 1.43 ± 0.06 [cd] | 1.49 ± 0.02 [cd] | 1.52 ± 0.04 [bc] | 1.68 ± 0.11 [ab] | 1.70 ± 0.08 [a] |
| Reduce sugar (mg/g) | 11.17 ± 0.68 [a] | 7.22 ± 0.49 [b] | 6.32 ± 0.26 [b] | 6.94 ± 0.43 [b] | 4.27 ± 0.29 [c] | 3.89 ± 0.25 [c] |
| Total sugar (mg/g) | 49.10 ± 1.50 [a] | 40.25 ± 0.31 [b] | 33.12 ± 0.35 [c] | 18.85 ± 0.57 [e] | 18.66 ± 0.29 [e] | 22.07 ± 0.78 [d] |
| Crude fiber (g/100g) | 2.68 ± 0.05 [a] | 2.51 ± 0.01 [a] | 2.68 ± 0.28 [a] | 1.78 ± 0.16 [b] | 1.57 ± 0.09 [b] | 1.94 ± 0.22 [b] |
| Dietary fiber (g/100g) | 3.80 ± 0.12 [a] | 3.23 ± 0.06 [b] | 3.21 ± 0.11 [b] | 2.47 ± 0.11 [c] | 2.00 ± 0.12 [d] | 2.13 ± 0.20 [d] |
| Crude protein (g/100g) | 13.02 ± 0.06 [a] | 12.97 ± 0.22 [a] | 12.95 ± 0.08 [a] | 12.38 ± 0.10 [b] | 12.30 ± 0.05 [b] | 12.23 ± 0.04 [b] |
| Beta-glucan (mg/g) | 4.72 ± 0.05 [a] | 4.29 ± 0.13 [b] | 4.35 ± 0.21 [ab] | 3.80 ± 0.15 [c] | 3.56 ± 0.24 [c] | 3.67 ± 0.15 [c] |
| GABA (mg/g) | 0.88 ± 0.15 [a] | 0.78 ± 0.11 [a] | 0.77 ± 0.12 [a] | 0.87 ± 0.14 [a] | 0.79 ± 0.08 [a] | 0.83 ± 0.01 [a] |

[a], [b], [c], [d], [e]: Values were determined in triplicate. Equal letters in the same row indicated that there were no significant difference at 5% ($p > 0.05$) in the parameter evaluated.

**Table 2.** Amino acids of the highland barley (the control groups and colonized groups with the mycelia of *A. sinodeliciosus* var. Chaidam ZJU-TP-08).

| Amino Acids (g/100 g) | SSF | | | Control | | |
|---|---|---|---|---|---|---|
| | GHB | UHB | SHB | GHB | UHB | SHB |
| Essential amino | | | | | | |
| Lysine | 0.43 ± 0.00 [ab] | 0.47 ± 0.04 [a] | 0.44 ± 0.00 [ab] | 0.45 ± 0.02 [ab] | 0.40 ± 0.00 [b] | 0.44 ± 0.02 [ab] |
| Valine | 0.64 ± 0.04 [a] | 0.71 ± 0.06 [a] | 0.68 ± 0.01 [a] | 0.57 ± 0.19 [a] | 0.63 ± 0.01 [a] | 0.61 ± 0.03 [a] |
| Leucine | 0.89 ± 0.05 [a] | 0.89 ± 0.03 [a] | 0.94 ± 0.14 [a] | 0.92 ± 0.03 [a] | 0.87 ± 0.06 [a] | 0.84 ± 0.09 [a] |
| Isoleucine | 0.53 ± 0.03 [a] | 0.54 ± 0.01 [a] | 0.56 ± 0.01 [a] | 0.52 ± 0.03 [a] | 0.55 ± 0.00 [a] | 0.52 ± 0.01 [a] |
| Threonine | 0.47 ± 0.04 [a] | 0.47 ± 0.04 [a] | 0.42 ± 0.03 [a] | 0.47 ± 0.06 [a] | 0.46 ± 0.01 [a] | 0.47 ± 0.01 [a] |
| Methionine | 0.18 ± 0.03 [a] | 0.15 ± 0.00 [abc] | 0.17 ± 0.01 [ab] | 0.15 ± 0.00 [abc] | 0.14 ± 0.00 [c] | 0.14 ± 0.00 [bc] |
| Phenylalanine | 0.74 ± 0.02 [a] | 0.69 ± 0.05 [a] | 0.71 ± 0.13 [a] | 0.70 ± 0.02 [a] | 0.67 ± 0.05 [a] | 0.68 ± 0.04 [a] |
| Histidine | 0.34 ± 0.00 [a] | 0.34 ± 0.01 [a] | 0.34 ± 0.01 [a] | 0.34 ± 0.01 [a] | 0.32 ± 0.01 [a] | 0.31 ± 0.00 [a] |
| Non-essential | | | | | | |
| Arginine | 0.66 ± 0.01 [a] | 0.65 ± 0.01 [a] | 0.61 ± 0.04 [a] | 0.63 ± 0.02 [a] | 0.63 ± 0.02 [a] | 0.62 ± 0.02 [a] |
| Tyrosine | 0.61 ± 0.01 [a] | 0.61 ± 0.01 [a] | 0.56 ± 0.06 [a] | 0.63 ± 0.10 [a] | 0.60 ± 0.02 [a] | 0.62 ± 0.03 [a] |
| Serine | 0.30 ± 0.01 [a] | 0.28 ± 0.00 [ab] | 0.27 ± 0.01 [bc] | 0.26 ± 0.00 [bcd] | 0.25 ± 0.01 [d] | 0.26 ± 0.00 [cd] |
| Glutamic acid | 1.81 ± 0.16 [a] | 1.66 ± 0.07 [a] | 1.58 ± 0.19 [a] | 1.61 ± 0.05 [a] | 1.58 ± 0.07 [a] | 1.57 ± 0.13 [a] |
| Proline | 0.60 ± 0.01 [a] | 0.61 ± 0.02 [a] | 0.66 ± 0.06 [a] | 0.59 ± 0.02 [a] | 0.62 ± 0.08 [a] | 0.66 ± 0.10 [a] |
| Glycine | 0.70 ± 0.10 [a] | 0.67 ± 0.04 [a] | 0.69 ± 0.08 [a] | 0.64 ± 0.03 [a] | 0.63 ± 0.01 [a] | 0.63 ± 0.02 [a] |
| Alanine | 1.14 ± 0.11 [a] | 1.16 ± 0.08 [a] | 1.15 ± 0.20 [a] | 1.00 ± 0.11 [a] | 0.96 ± 0.05 [a] | 0.98 ± 0.06 [a] |
| Cystine | 0.33 ± 0.03 [a] | 0.33 ± 0.01 [a] | 0.35 ± 0.02 [a] | 0.32 ± 0.01 [a] | 0.32 ± 0.02 [a] | 0.31 ± 0.02 [a] |
| Aspartic acid | 1.64 ± 0.01 [a] | 1.52 ± 0.12 [a] | 1.62 ± 0.02 [a] | 1.50 ± 0.09 [a] | 1.51 ± 0.02 [a] | 1.51 ± 0.06 [a] |
| Total | 11.90 ± 0.10 [a] | 11.75 ± 0.04 [a] | 11.74 ± 0.17 [a] | 11.2 9± 0.19 [b] | 11.13 ± 0.02 [b] | 11.17 ± 0.09 [b] |

[a], [b], [c], [d]: Values were determined in triplicate. Equal letters in the same row indicated that there were no significant difference at 5% ($p > 0.05$) in the parameter evaluated.

HB has been shown to be rich in β-glucan with many benefits [11,12]. Therefore, the changes of SSF on β-glucan content were also evaluated. The contents of β-glucan in the

SSF-GHB, SSF-UHB and SSF-SHB were 4.72 ± 0.05, 4.29 ± 0.13 and 4.35 ± 0.21 mg/g, respectively. Compared to the β-glucan in the control groups (GHB, UHB and SHB), the contents of β-glucan in the SSF-GHB, SSF-UHB and SSF-SHB were increased by 24.21%, 20.51% and 18.53% ($p < 0.05$), respectively. The causes of elevated β-glucan can be attributed to two points of view. The increase in β-glucan content after fermentation was due to enhanced activity of enzymes secreted such as β-glucanases and carboxypeptidases, which caused degradation of total and insoluble β-glucan content into soluble β-glucan [34]. Furthermore, β-glucan was found in macrofungus mycelia [35]. We hypothesized that the increase in β-glucan after fermentation was possibly attributed to components of the fungal biomass.

GABA is a four-carbon non-protein free amino acid widely distributed in nature, which plays a role in regulating neuronal excitability throughout the nervous system [16]. The contents of GABA in the SSF-GHB, SSF-UHB and SSF-SHB were 0.88 ± 0.15, 0.78 ± 0.11 and 0.77 ± 0.12 mg/g ($p > 0.05$), respectively. The levels of GABA in the GHB, UHB and SHB (control groups) were 0.87 ± 0.14, 0.79 ± 0.08 and 0.83 ± 0.01 mg/g ($p > 0.05$), respectively. It was obvious that neither germination nor fermentation contributed significantly to the GABA content in this work. It has been reported that germination of grains could activate the corresponding enzymes to promote the biosynthesis of GABA, and brown rice germination promoted increased GABA content [16,17]. The exact cause for this change is still under examination in the future investigation.

### 3.3. Total Phenolic, Flavonoid Content and Antioxidant Activity

Phenolic compounds have been reported to help reduce the risk of various diseases and to promote the sensory characteristics of foods [17]. In this study, we evaluated the changes of SSF HB on phenolic compounds. Total phenols, total flavonoid and the total antioxidant activities upon different treatments of HB are presented in Table 3. For all fermentations, the levels of free, bound, and total phenolic compounds were superior to the unfermented substrate. Similarly, free, bound and total flavonoid exhibited the similar altered trends. The contents of total phenols in the SSF-GHB, SSF-UHB and SSF-SHB were 6.52 ± 0.22, 6.38 ± 0.47 and 5.58 ± 0.12 mg GAE/g, respectively. Compared to the control groups (GHB, UHB, and SHB), the contents of total phenols in the SSF-GHB, SSF-UHB and SSF-SHB were increased by 18.12%, 25.10% and 21.85%, respectively. The contents of total flavonoid in the SSF-GHB, SSF-UHB and SSF-SHB were 2.15 ± 0.07, 2.05 ± 0.04 and 1.87 ± 0.06 mg RE/g, respectively. Overall, the fermentation upon the three pretreated materials could improve the total phenolic and total flavonoid formation.

**Table 3.** Total phenols, total flavonoids and antioxidant activity of the highland barley (the control groups and colonized groups with the mycelia of *A. sinodeliciosus* var. Chaidam ZJU-TP-08).

| Biological Activity | | SSF | | | Control | | |
|---|---|---|---|---|---|---|---|
| | | GHB | UHB | SHB | GHB | UHB | SHB |
| Total phenols (mg GAE/g) | Free | 3.38 ± 0.06 $^{a}$ | 3.10 ± 0.26 $^{ab}$ | 2.60 ± 0.27 $^{bc}$ | 3.04 ± 0.12 $^{ab}$ | 2.46 ± 0.30 $^{c}$ | 2.21 ± 0.04 $^{c}$ |
| | Bound | 3.13 ± 0.08 $^{a}$ | 3.28 ± 0.21 $^{a}$ | 2.98 ± 0.15 $^{a}$ | 2.48 ± 0.10 $^{b}$ | 2.64 ± 0.04 $^{b}$ | 2.37 ± 0.03 $^{b}$ |
| | Total | 6.52 ± 0.22 $^{a}$ | 6.38 ± 0.47 $^{a}$ | 5.58 ± 0.12 $^{b}$ | 5.52 ± 0.16 $^{b}$ | 5.10 ± 0.35 $^{bc}$ | 4.58 ± 0.07 $^{c}$ |
| Total flavonoids (mg RE/g) | Free | 1.17 ± 0.04 $^{a}$ | 0.98 ± 0.07 $^{a}$ | 0.69 ± 0.09 $^{b}$ | 0.40 ± 0.03 $^{c}$ | 0.53 ± 0.21 $^{bc}$ | 0.48 ± 0.04 $^{bc}$ |
| | Bound | 0.94 ± 0.13 $^{b}$ | 0.86 ± 0.04 $^{bc}$ | 0.79 ± 0.03 $^{cd}$ | 1.09 ± 0.11 $^{a}$ | 0.71 ± 0.09 $^{de}$ | 0.67 ± 0.05 $^{e}$ |
| | Total | 2.15 ± 0.07 $^{a}$ | 2.05 ± 0.04 $^{b}$ | 1.87 ± 0.06 $^{c}$ | 1.49 ± 0.06 $^{c}$ | 1.24 ± 0.15 $^{d}$ | 1.15 ± 0.09 $^{d}$ |
| Total antioxidant capacity (μmol $Fe^{2+}$/g) | Free | 2.71 ± 0.06 $^{a}$ | 2.19 ± 0.13 $^{c}$ | 2.17 ± 0.09 $^{c}$ | 2.34 ± 0.05 $^{b}$ | 2.09 ± 0.06 $^{c}$ | 2.02 ± 0.04 $^{c}$ |
| | Bound | 4.99 ± 0.31 $^{a}$ | 4.61 ± 0.15 $^{b}$ | 3.49 ± 0.11 $^{d}$ | 3.93 ± 0.12 $^{c}$ | 3.46 ± 0.10 $^{d}$ | 3.13 ± 0.08 $^{d}$ |
| DPPH• radical scavenging (μmol TEAC/g) | Free | 1.19 ± 0.05 $^{a}$ | 1.07 ± 0.06 $^{b}$ | 0.75 ± 0.03 $^{d}$ | 0.92 ± 0.03 $^{c}$ | 0.99 ± 0.06 $^{bc}$ | 0.57 ± 0.03 $^{e}$ |
| | Bound | 0.95 ± 0.07 $^{a}$ | 0.85 ± 0.03 $^{b}$ | 0.74 ± 0.04 $^{c}$ | 0.51 ± 0.04 $^{d}$ | 0.50 ± 0.03 $^{d}$ | 0.50 ± 0.05 $^{d}$ |
| $ABTS^{+}$• radical scavenging (μmol TEAC/g) | Free | 2.63 ± 0.06 $^{a}$ | 2.40 ± 0.10 $^{b}$ | 2.19 ± 0.14 $^{c}$ | 1.99 ± 0.03 $^{d}$ | 2.19 ± 0.04 $^{c}$ | 2.47 ± 0.08 $^{ab}$ |
| | Bound | 2.24 ± 0.07 $^{a}$ | 2.22 ± 0.05 $^{a}$ | 2.13 ± 0.04 $^{a}$ | 1.43 ± 0.06 $^{b}$ | 1.25 ± 0.18 $^{b}$ | 1.35 ± 0.09 $^{b}$ |
| Hydroxyl radical scavenging (μmol TEAC/g) | Free | 1.22 ± 0.13 $^{a}$ | 1.31 ± 0.16 $^{a}$ | 0.99 ± 0.07 $^{b}$ | 0.92 ± 0.07 $^{b}$ | 0.74 ± 0.15 $^{c}$ | 0.68 ± 0.03 $^{c}$ |
| | Bound | 1.52 ± 0.09 $^{a}$ | 1.05 ± 0.06 $^{b}$ | 0.93 ± 0.05 $^{c}$ | 0.82 ± 0.05 $^{d}$ | 0.65 ± 0.06 $^{d}$ | 0.79 ± 0.04 $^{e}$ |

$^{a}$, $^{b}$, $^{c}$, $^{d}$, $^{e}$: Values were determined in triplicate. Equal letters in the same row indicated that there were no significant difference at 5% ($p > 0.05$) in the parameter evaluated.

SSF increased the content of total phenolic was attributed to three aspects. Firstly, the phenolic compounds in grains mainly existed in the form of bound phenols, which could be released by alkali, acid or microbial enzymes [26]. For example, macrofungi had been reported to be capable for production of enzymes such as amylase, xylanase, cellulase and other enzymes that promoted the release of bound phenolic compounds [36]. The enzyme β-glucosidase had been described as being able to hydrolyze phenolic glycosides to release free phenolic acids [37]. It had been proven that α-amylase, xylanase and β-glucosidase produced during SSF of wheat by *Aspergillus oryzae* and *Aspergillus awamori* were highly correlated with the release of the phenolic compounds [38]. According to the previous investigation, total phenolic contents of brown rice, corn and wheat were enhanced after colonization by *A. blazei*, *A. fuscosuccinea* and *P. albidus* [10]. Secondly, microbial metabolism could modify the bioactive substances in grains, leading to the synthesis of new substances such as phenolic compounds. In addition, the increase in phenolic content may be due to the breakdown of grain cell walls by microbially produced enzymes and the subsequent release of phenolic compounds [39]. It has been shown that fungi possess extracellular enzyme-releasing systems for oxidizing lignin systems, which degrade lignin and open the benzene ring, ultimately increasing the free phenolic content [40]. Macrofungal degradation of the grain cell wall could be one of the reasons for enhanced levels of phenolic compounds, however, considering the quite limited proportion of the lignin in grain cell wall of highland barley. Thus, this pathway may not be the main origin of total phenolic levels in fermented highland barley. The release of phenolic compounds by enzymes alone during fermentation was limited. It was said the pathway of polyketide metabolism has been shown in *Monascus* strains, and the biosynthesis of polyketids was capable of generating polyphenols, and this might cause the increase in phenolic content [41].

SSF had been increasingly employed to increase the content of phenolic compounds in certain food products, thus giving rise to enhance antioxidant activity. A single antioxidant method could not fully reflect the antioxidant capacity of the sample. Thus, four antioxidant methods were used to investigate the antioxidant capacities of the free and bound phenolic extracts in the present study, including T-AOC, DPPH•, $ABTS^{+}$• and hydroxyl radical scavenging activity. In this work, the levels of antioxidant in free and bound phenol extracted from SSF HB were significantly superior to those in the substrate without fermentation ($p < 0.05$), which was supported by previous findings [10,24,42,43]. Increased antioxidant properties of fermented grains may be inextricably linked to phenolic compounds. It has been demonstrated that soybean products fermented by SSF using *Trichoderma harzianum* showed intensive antioxidant activity than unfermented products, which was probably related to the markedly higher content of phenolic acids, flavonoids with more free hydroxyl groups achieved during SSF [44].

### *3.4. The Description and Characterization of Physicochemical Properties*

The physicochemical properties are largely responsible for the final textural and sensory qualities of baked, steamed and boiled process of HB flours. Therefore, the effect of SSF on physicochemical properties of HB was evaluated in the present work. The bulk density (BD), water-holding capacity (WHC), water solubility index (WSI) and pasting property of HB are presented in Table 4. The capacities of WSI in the SSF-GHB, SSF-UHB and SSF-SHB were 26.13 ± 1.88%, 27.36 ± 0.26% and 22.10 ± 1.05%, respectively. The capacities of WSI of all fermented HB flour samples were increased, compared to control groups. However, the levels of BD and WHC were insignificant difference to those of the control groups ($p > 0.05$). Previous research studies had also pointed out that fermentation increased the capacity of WHC in the samples, but there was no effect on BD index [27]. RVA was frequently used to study the pasting, gelatinization and other physicochemical properties of flours [45]. In this study, all fermented samples were significantly reduced in the peak viscosity (PV), tough viscosity (TV), breakdown viscosity (BV), final viscosity (FV), setback viscosity (SV) and peak temperature (PT), compared to the control groups ($p < 0.05$). The reduction in the pasting properties of the fermented samples might be partly attributed to the degradation of starch by α-amylase, which was released in fungal metabolism. It was reported that the solid state fermentation reduced the pasting properties of brown finger millet [14].

**Table 4.** Physical properties of the highland barley (the control groups and colonized groups with the mycelia of *A. sinodeliciosus* var. Chaidam ZJU-TP-08).

| Parameters | SSF | | | Control | | |
|---|---|---|---|---|---|---|
| | GHB | UHB | SHB | GHB | UHB | SHB |
| Functional properties | | | | | | |
| BD (g/cm$^3$) | 0.56 ± 0.00 $^a$ | 0.55 ± 0.01 $^a$ | 0.55 ± 0.01 $^a$ | 0.54 ± 0.00 $^a$ | 0.54 ± 0.00 $^a$ | 0.54 ± 0.01 $^a$ |
| WHC (g/g) | 5.20 ± 0.18 $^a$ | 5.23 ± 0.00 $^a$ | 5.13 ± 0.35 $^a$ | 5.20 ± 0.53 $^a$ | 5.49 ± 0.10 $^a$ | 5.66 ± 0.09 $^a$ |
| WSI (%) | 26.13 ± 1.88 $^{ab}$ | 27.36 ± 0.26 $^a$ | 22.10 ± 1.05 $^{abc}$ | 21.41 ± 4.54 $^{bc}$ | 18.91 ± 1.01 $^c$ | 10.92 ± 1.38 $^d$ |
| Pasting properties | | | | | | |
| PV (cP) | 293.00 ± 1.41 $^d$ | 140.50 ± 0.77 $^f$ | 209.50 ± 3.54 $^e$ | 652.00 ± 1.77 $^c$ | 829.00 ± 2.33 $^b$ | 1169.25 ± 1.06 $^a$ |
| TV (cP) | 285.00 ± 5.66 $^d$ | 140.00 ± 0.00 $^f$ | 205.00 ± 2.83 $^e$ | 608.00 ± 1.49 $^c$ | 775.00 ± 8.48 $^b$ | 1129.00 ± 5.66 $^a$ |
| BV (cP) | 5.50 ± 0.51 $^d$ | 0.50 ± 0.71 $^e$ | 6.50 ± 2.15 $^d$ | 44.00 ± 0.61 $^b$ | 54.00 ± 1.41 $^a$ | 41.00 ± 0.53 $^c$ |
| FV (cP) | 527.00 ± 0.71 $^d$ | 270.00 ± 2.02 $^f$ | 405.00 ± 20.51 $^e$ | 1003.00 ± 2.83 $^c$ | 1247.00 ± 17.68 $^b$ | 1698.00 ± 16.97 $^a$ |
| SV (cP) | 237.00 ± 2.73 $^d$ | 130.50 ± 2.09 $^f$ | 200.50 ± 17.68 $^e$ | 396.00 ± 4.24 $^c$ | 478.00 ± 26.16 $^b$ | 561.00 ± 11.31 $^a$ |
| PT (°C) | n.d. | n.d. | n.d. | 92.50 ± 2.12 $^a$ | 93.39 ± 1.51 $^a$ | 87.59 ± 1.90 $^b$ |
| Peak time (min) | 6.12 ± 0.11 $^b$ | 5.75 ± 0.09 $^c$ | 5.37 ± 0.04 $^d$ | 6.95 ± 0.07 $^a$ | 7.02 ± 0.02 $^a$ | 6.94 ± 0.01 $^a$ |

$^{a, b, c, d, e, f}$: Values were determined in triplicate. Equal letters in the same row indicate that there were no significant difference at 5% ($p > 0.05$) in the parameter evaluated. BD: bulk density, WHC: water-holding capacity, WSI: water solubility index, PV: peak viscosity, TV: trough viscosity, BV: breakdown viscosity, FV: final viscosity, SV: setback viscosity, PT: peak temperature.

### 3.5. Principal Component Analysis of Nutritional Constituents of Fermented and Unfermented HB

To highlight the key features of HB flour after SSF, principal component analysis (PCA) was performed, as the data show in Figure 3. Differences and similarities among the three SSF samples and their corresponding control groups in terms of eleven proximate compositions were visualized by PCA analysis, and the total variance (94.4%) was explained by the first two components, PC1 (83.2%) and PC2 (11.2%). SSF-GHB, SSF-UHB and SSF-SHB were located in the first and fourth quadrants, and all the proximate components (except for GABA) were distributed in the first and fourth quadrants, suggesting that SSF had a significant improvement in the nutrients, phenolic compounds and antioxidant properties of HB. However, GHB, UHB and SHB were located in the second and third quadrants, which showed insignificant correlation with the nutrients, phenolic components and antioxidant activities. In the previous study, principal component analysis (PCA) evaluated different solid-state fermentation media and the active components and antioxidant activities in mycelial fermentation products, indicating that solid-state fermentation could improve the active ingredients and antioxidant properties [46]. Similarly, our results of PCA analysis indicated that solid-state fermentation by *A. sinodeliciosus* var. Chaidam ZJU-TP-08 could enriched nutrients, phenolic components and antioxidant activities of HB.

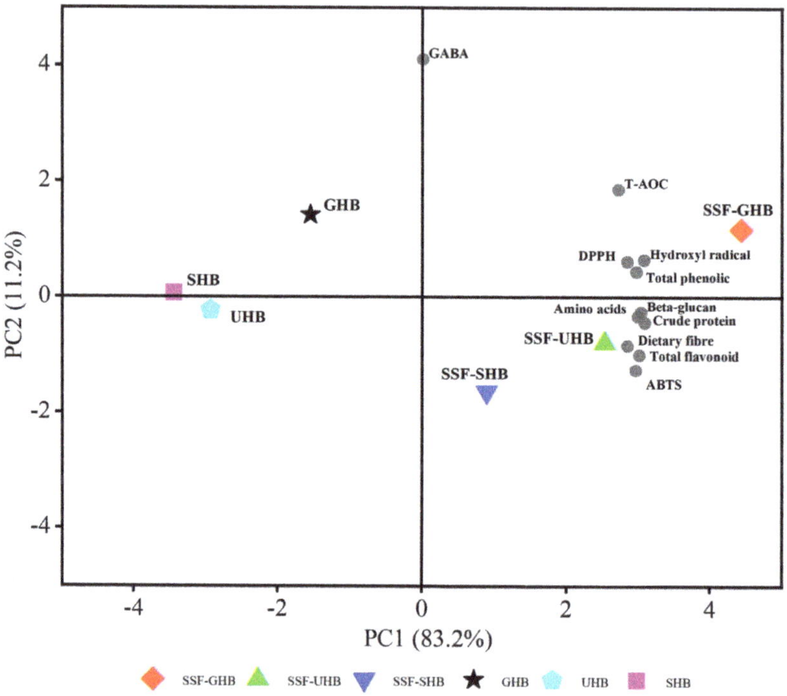


**Figure 3.** Nutritional constituents and biological activity of highland barley (the control groups and colonized groups with the mycelia of *A. sinodeliciosus* var. Chaidam ZJU-TP-08) were visualized by principal component analysis (PCA).

## 4. Conclusions

To sum up, the solid-state fermentation with *A. sinodeliciosus* var. Chaidam ZJU-TP-08 was an effective method to enhance the nutrients, phenolic components and antioxidant activities of HB. In all of the fermentation groups, SSF-GHB exhibited higher levels of ergosterol and β-glucan with $0.19 \pm 0.01$ and $4.72 \pm 0.05$ mg/g, respectively. In addition, compared to UHB, SSF-UHB increased levels of amino acids by 5.60%, dietary fiber by 61.50%, total phenols by 25.10% and total flavonoids by 65.32%, which were the most significant increases among all fermentation groups. For all fermentations, the levels of antioxidant activities were superior to those in the substrate without fermentation. Meanwhile, the WSI of all fermentations was increased. However, the pasting characteristics were decreased after solid-stated fermentation, indicating their potential as functional ingredients in the development of instant products and other food applications. In view of the results obtained, GHB, UHB and SHB colonized by *A. sinodeliciosus* var. Chaidam ZJU-TP-08 are potential processing approaches in functional food development, and further studies are recommended to elucidate the mechanisms involved in the biotransformation process.

**Author Contributions:** B.L.: Conceptualization, formal analysis, methodology, investigation, software, writing—original draft. H.L.: supervision, validation, writing—review and editing. Q.S.: validation, writing—review. Q.C.: conceptualization, supervision, investigation, formal analysis, funding acquisition, writing—review and editing. J.W.: conceptualization, methodology, supervision, writing—review and editing. All authors have read and agreed to the published version of the manuscript.

**Funding:** This study was funded by the Natural Science Foundation of China (31871904) and the Starry Night Science Fund of Zhejiang University Shanghai Institute for Advanced Study [(no. SN-ZJU-SIAS-004)].

**Institutional Review Board Statement:** This article does not contain any studies with human participants or animals performed by any of the authors.

**Informed Consent Statement:** All the authors agreed to submit this paper to your journal.

**Data Availability Statement:** The study did not report any data.

**Conflicts of Interest:** The authors declare no competing interests.

**Sample Availability:** Samples of solid state fermentation are available from the authors.

## Abbreviations

HB: Highland barley; GHB: Germinated Highland barley; UHB: Ultrasonic Highland Barley; SHB: Soaked Highland barley; SSF: solid-state fermentation; SSF-GHB: solid-state fermentation of GHB; SSF-UHB: solid-state fermentation of UHB; SSF-SHB: solid-state fermentation of SHB

## References


1. Kalac, P. A review of chemical composition and nutritional value of wild-growing and cultivated mushrooms. *J. Sci. Food Agric.* **2013**, *93*, 209–218. [CrossRef]
2. Sanchez, C. Reactive oxygen species and antioxidant properties from mushrooms. *Synth. Syst. Biotechnol.* **2017**, *2*, 13–22. [CrossRef] [PubMed]
3. He, W.S.; Cui, D.D.; Li, L.L.; Tong, L.-T.; Rui, J.; Li, H.; Zhang, H.; Liu, X. Cholesterol-reducing effect of ergosterol is modulated via inhibition of cholesterol absorption and promotion of cholesterol excretion. *J. Funct. Foods* **2019**, *57*, 488–496. [CrossRef]
4. Wang, Z.R.; Parra, L.A.; Callac, P.; Zhou, J.-L.; Fu, W.-J.; Dui, S.-H.; Hyde, K.D.; Zhao, R.-L. Edible species of *Agaricus* (Agaricaceae) from Xinjiang Province (Western China). *Phytotaxa* **2015**, *202*, 185–197. [CrossRef]
5. Kuang, H.; Jiao, Y.C.; Wang, W.; Wang, F.; Chen, Q. Characterization and antioxidant activities of intracellular polysaccharides from *Agaricus bitorquis* (QueL.) Sacc. Chaidam ZJU-CDMA-12. *Int. J. Biol. Macromol.* **2020**, *156*, 1112–1125. [CrossRef]
6. Lu, H.Y.; Jiao, Z.H.; Jiao, Y.C.; Wang, W.; Chen, Q. Phenolic Acids-Rich Fractions from *Agaricus bitorguis* (Quel.) Sacc. Chaidam ZJU-CDMA-12 Mycelia Modulate Hypoxic Stress on Hypoxia-Damaged PC12 Cells. *Molecules* **2020**, *25*, 4845. [CrossRef] [PubMed]
7. Lu, H.Y.; Lou, H.H.; Wei, T.Y.; Liu, Z.; Jiao, Y.; Chen, Q. Ultrasound enhanced production of mycelia and exopolysaccharide by *Agaricus bitorquis* (Quel.) Sacc. Chaidam. *Ultrason. Sonochem.* **2020**, *64*, 105040. [CrossRef]
8. Souilem, F.; Fernandes, A.; Calhelha, R.C.; Barreira, J.C.; Barros, L.; Skhiri, F.; Martins, A.; Ferreira, I.C. Wild mushrooms and their mycelia as sources of bioactive compounds: Antioxidant, anti-inflammatory and cytotoxic properties. *Food Chem.* **2017**, *230*, 40–48. [CrossRef]
9. Calvo-Lerma, J.; Asensio-Grau, A.; Garcia-Hernandez, J.; Heredia, A.; Andrés, A. Exploring the Impact of Solid-State Fermentation on Macronutrient Profile and Digestibility in Chia (*Salvia hispanica*) and Sesame (*Sesamum indicum*) Seeds. *Foods* **2022**, *11*, 410. [CrossRef]
10. Stoffel, F.; Santana, W.D.; Fontana, R.C.; Gregolon, J.G.N.; Kist, T.B.L.; De Siqueira, F.G.; Mendonça, S.; Camassola, M. Chemical features and bioactivity of grain flours colonized by macrofungi as a strategy for nutritional enrichment. *Food Chem.* **2019**, *297*, 124988. [CrossRef]
11. Li, Y.T.; Li, T.; Liu, R.H. Bioactive compounds of highland barley and their health benefits. *J. Cereal Sci.* **2022**, *103*, 103366. [CrossRef]
12. Guo, T.L.; Horvath, C.; Chen, L.; Zheng, B. Understanding the nutrient composition and nutritional functions of highland barley (Qingke): A review. *Trends Food Sci. Technol.* **2020**, *103*, 109–117. [CrossRef]
13. Xiang, X.; Tan, C.; Sun, X.J.; Zhao, Y.; Zhang, J.; Zhu, Y.; Bai, J.; Dong, Y.; Zhou, X. Effects of fermentation on structural characteristics and *in vitro* physiological activities of barley beta-glucan. *Carbohydr. Polym.* **2020**, *231*, 115685. [CrossRef]
14. Azeez, S.O.; Chinma, C.E.; Bassey, S.O.; Eze, U.R.; Makinde, A.F.; Sakariyah, A.A.; Okubanjo, S.S.; Danbaba, N.; Adebo, O.A. Impact of germination alone or in combination with solid-state fermentation on the physicochemical, antioxidant, in vitro digestibility, functional and thermal properties of brown finger millet flours. *LWT-Food Sci. Technol.* **2021**, *154*, 112734. [CrossRef]
15. Thakur, P.; Kumar, K.; Ahmed, N.; Chauhan, D.; Rizvi, Q.U.E.H.; Jan, S.; Singh, T.P.; Dhaliwal, H.S. Effect of soaking and germination treatments on nutritional, anti-nutritional, and bioactive properties of amaranth (*Amaranthus hypochondriacus* L.), quinoa (*Chenopodium quinoa* L.), and buckwheat (*Fagopyrum esculentum* L.). *Curr. Res. Food Sci.* **2021**, *4*, 917–925. [CrossRef]
16. Liu, S.; Wang, W.; Lu, H.; Shu, Q.; Zhang, Y.; Chen, Q. New perspectives on physiological, biochemical and bioactive components during germination of edible seeds: A review. *Trends Food Sci. Technol.* **2022**, *123*, 187–197. [CrossRef]
17. Oliveira, M.E.A.S.; Coimbra, P.P.S.; Galdeano, M.C.; Carvalho, C.W.P.; Takeiti, C.Y. How does germinated rice impact starch structure, products and nutrional evidences?—A review. *Trends Food Sci. Technol.* **2022**, *122*, 13–23. [CrossRef]
18. Bonto, A.P.; Tiozon, R.N.; Sreenivasulu, N.; Camacho, D.H. Impact of ultrasonic treatment on rice starch and grain functional properties: A review. *Ultrason. Sonochem.* **2021**, *71*, 105383. [CrossRef]

19. Lu, H.Y.; Liu, S.Y.; Zhang, S.L.; Chen, Q. Light Irradiation Coupled with Exogenous Metal Ions to Enhance Exopolysaccharide Synthesis from *Agaricus sinodeliciosus* ZJU-TP-08 in Liquid Fermentation. *J. Fungi* **2021**, *7*, 992. [CrossRef]
20. Nowak, R.; Nowacka-Jechalke, N.; Pietrzak, W.; Gawlik-Dziki, U. A new look at edible and medicinal mushrooms as a source of ergosterol and ergosterol peroxide-UHPLC-MS/MS analysis. *Food Chem.* **2022**, *369*, 130927. [CrossRef]
21. AOAC. Official Methods of Analysis. In *The Association of Official Analytical Chemists*, 18th ed.; North Fredrick Avenue Gaithersburg: Montgomery County, MD, USA, 2015.
22. Liu, B.; Yuan, D.X.; Li, Q.Y.; Zhou, X.; Wu, H.; Bao, Y.; Lu, H.; Luo, T.; Wang, J. Changes in Organic Acids, Phenolic Compounds, and Antioxidant Activities of Lemon Juice Fermented by *Issatchenkia terricola*. *Molecules* **2021**, *26*, 6712. [CrossRef] [PubMed]
23. Watchararparpaiboon, W.; Laohakunjit, N.; Kerdchoechuen, O. An Improved Process for High Quality and Nutrition of Brown Rice Production. *Food Sci. Technol. Int.* **2010**, *16*, 147–158. [CrossRef] [PubMed]
24. Bei, Q.; Liu, Y.; Wang, L.; Chen, G.; Wu, Z. Improving free, conjugated, and bound phenolic fractions in fermented oats (*Avena sativa* L.) with *Monascus anka* and their antioxidant activity. *J. Funct. Foods* **2017**, *32*, 185–194. [CrossRef]
25. Yang, X.J.; Dang, B.; Fan, M.T. Free and Bound Phenolic Compound Content and Antioxidant Activity of Different Cultivated Blue Highland Barley Varieties from the Qinghai-Tibet Plateau. *Molecules* **2018**, *23*, 879. [CrossRef]
26. Wu, H.; Liu, H.N.; Ma, A.M.; Zhou, J.-Z.; Xia, X.-D. Synergetic effects of *Lactobacillus plantarum* and *Rhizopus oryzae* on physicochemical, nutritional and antioxidant properties of whole-grain oats (*Avena sativa* L.) during solid-state fermentation. *LWT-Food Sci. Technol.* **2022**, *154*, 112687. [CrossRef]
27. Tang, Y.; Xiao, L.; Wu, X.Y.; Li, W.; Wu, T.; Zhang, P. Impact of germination pretreatment on the polyphenol profile, antioxidant activities, and physicochemical properties of three color cultivars of highland barley. *J. Cereal Sci.* **2021**, *97*, 103152. [CrossRef]
28. Adebo, J.A.; Njobeh, P.B.; Gbashi, S.; Oyedeji, A.B.; Ogundele, O.M.; Oyeyinka, S.A.; Adebo, O.A. Fermentation of Cereals and Legumes: Impact on Nutritional Constituents and Nutrient Bioavailability. *Fermentation* **2022**, *8*, 63. [CrossRef]
29. Espinosa-Paez, E.; Guadalupe Alanis-Guzman, M.; Hernandez-Luna, C.E.; Báez-González, J.G.; Amaya-Guerra, C.A.; Andrés-Grau, A.M. Increasing Antioxidant Activity and Protein Digestibility in *Phaseolus vulgaris* and *Avena sativa* by Fermentation with the *Pleurotus ostreatus* Fungus. *Molecules* **2017**, *22*, 2275. [CrossRef]
30. Suarti, B.; Sukarno; Ardiansyah; Budijanto, S. Bio-active compounds, their antioxidant activities, and the physicochemical and pasting properties of both pigmented and non-pigmented fermented de-husked rice flour. *AIMS Agric. Food* **2021**, *6*, 49–64. [CrossRef]
31. Ong, A.; Lee, C.L.K. Cooperative metabolism in mixed culture solid-state fermentation. *LWT-Food Sci. Technol.* **2021**, *152*, 112300. [CrossRef]
32. Starzynska-Janiszewska, A.; Baczkowicz, M.; Sabat, R.; Stodolak, B.; Witkowicz, R. Quinoa Tempe as a Value-Added Food: Sensory, Nutritional, and Bioactive Parameters of Products from White, Red, and Black Seeds. *Cereal Chem.* **2017**, *94*, 491–496. [CrossRef]
33. Gmoser, R.; Fristedt, R.; Larsson, K.; Undeland, I.; Taherzadeh, M.J.; Lennartsson, P.R. From stale bread and brewers spent grain to a new food source using edible filamentous fungi. *Bioengineered* **2020**, *11*, 582–598. [CrossRef]
34. Xiao, X.; Li, J.Y.; Xiong, H.; Tui, W.; Zhu, Y.; Zhang, J. Effect of Extrusion or Fermentation on Physicochemical and Digestive Properties of Barley Powder. *Front. Nutr.* **2022**, *8*, 794355. [CrossRef] [PubMed]
35. Wasser, S.P. Medicinal mushrooms as a source of antitumor and immunomodulating polysaccharides. *Appl. Microbiol. Biotechnol.* **2002**, *60*, 258–274. [CrossRef] [PubMed]
36. Hur, S.J.; Lee, S.Y.; Kim, Y.C.; Choi, I.; Kim, G.-B. Effect of fermentation on the antioxidant activity in plant-based foods. *Food Chem.* **2014**, *160*, 346–356. [CrossRef]
37. Vattem, D.A.; Shetty, K. Ellagic acid production and phenolic antioxidant activity in cranberry pomace (*Vaccinium macrocarpon*) mediated by *Lentinus edodes* using a solid-state system. *Process Biochem.* **2003**, *39*, 367–379. [CrossRef]
38. Bhanja, T.; Kumari, A.; Banerjee, R. Enrichment of phenolics and free radical scavenging property of wheat koji prepared with two filamentous fungi. *Bioresour. Technol.* **2009**, *100*, 2861–2866. [CrossRef]
39. Xiao, Y.; Rui, X.; Xing, G.L.; Wu, H.; Li, W.; Chen, X.; Jiang, M.; Dong, M. Solid state fermentation with *Cordyceps militaris* SN-18 enhanced antioxidant capacity and DNA damage protective effect of oats (*Auena sativa* L.). *J. Funct. Foods* **2015**, *16*, 58–73. [CrossRef]
40. Sanchez, C. Lignocellulosic residues: Biodegradation and bioconversion by fungi. *Biotechnol. Adv.* **2009**, *27*, 185–194. [CrossRef]
41. Patakova, P. Monascus secondary metabolites: Production and biological activity. *J. Ind. Microbiol. Biotechnol.* **2013**, *40*, 169–181. [CrossRef]
42. Liu, W.W.; Dun, M.Q.; Liu, X.Y.; Zhang, G.; Ling, J. Effects on total phenolic and flavonoid content, antioxidant properties, and angiotensin I-converting enzyme inhibitory activity of beans by solid-state fermentation with *Cordyceps militaris*. *Int. J. Food Prop.* **2022**, *25*, 477–491. [CrossRef]
43. Luo, J.Q.; Liu, S.Y.; Lu, H.Y.; Chen, Q.H.; Shi, Y. A comprehensive review ofmicroorganism-derived cyclic peptides: Bioactive functions and food safety applications. *Compr. Rev. Food Sci. Food Saf.* **2022**, *11*, 1–19. [CrossRef]

44. Singh, H.B.; Singh, B.N.; Singh, S.P.; Nautiyal, C.S. Solid-state cultivation of *Trichoderma harzianum* NBRI-1055 for modulating natural antioxidants in soybean seed matrix. *Bioresour. Technol.* **2010**, *101*, 6444–6453. [CrossRef]
45. Yuan, T.Z.; Liu, S.Y.; Reimer, M.; Isaak, C.; Ai, Y. Evaluation of pasting and gelling properties of commercial flours under high heating temperatures using Rapid Visco Analyzer 4800. *Food Chem.* **2021**, *344*, 128616. [CrossRef]
46. Hsu, J.Y.; Chen, M.H.; Lai, Y.S.; Chen, S.-D. Antioxidant Profile and Biosafety of White Truffle Mycelial Products Obtained by Solid-State Fermentation. *Molecules* **2022**, *27*, 109. [CrossRef]

Journal of *Fungi*

MDPI

*Article*

# Spermidine Synthase and Saccharopine Reductase Have Co-Expression Patterns Both in Basidiomycetes with Fusion Form and Ascomycetes with Separate Form

**Yayong Yang [1,2], Lei Shi [1,2], Xinyu Xu [1,2], Jin Wen [1,2], Tianyue Xie [1,2], Hui Li [3], Xiaoyu Li [1,2], Mengyu Chen [1,2], Xinyi Dou [1,2], Chengjin Yuan [1,2], Hanbing Song [1,2], Baogui Xie [2] and Yongxin Tao [1,2,*]**

[1] College of Horticulture, Fujian Agriculture and Forestry University, Fuzhou 350002, China
[2] Mycological Research Center, Fujian Agriculture and Forestry University, Fuzhou 350002, China
[3] Institute of Cash Crops, Hebei Academy of Agriculture and Forestry Sciences, Shijiazhuang 050051, China
[*] Correspondence: taoyongxinmuse@163.com; Tel.: +86-0591-83789281

**Abstract:** Gene fusion is a process through which two or more distinct genes are fused into a single chimeric gene. Unlike most harmful fusion genes in cancer cells, in this study, we first found that spermidine synthetase- (SPDS, catalyst of spermidine biosynthesis) and saccharopine reductase- (SR, catalyst of the penultimate step of lysine biosynthesis) encoding genes form a natural chimeric gene, *FfSpdsSr*, in *Flammulina filiformis*. Through the cloning of full-length ORFs in different strains and the analysis of alternative splicing in developmental stages, *FfSpdsSr* has only one copy and unique transcript encoding chimeric SPDS-SR in *F. filiformis*. By an orthologous gene search of *SpdsSr* in more than 80 fungi, we found that the chimeric *SpdsSr* exists in basidiomycetes, while the two separate *Spds* and *Sr* independently exist in ascomycetes, chytridiomycetes, and oomycetes. Further, the transcript level of *FfSpdsSr* was investigated in different developmental stages and under some common environmental factors and stresses by RT-qPCR. The results showed that *FfSpdsSr* mainly up-regulated in the elongation stage and pileus development of *F. filiformis*, as well as under blue light, high temperature, $H_2O_2$, and MeJA treatments. Moreover, a total of 15 sets of RNA-Seq data, including 218 samples of *Neurospora crassa*, were downloaded from the GEO database and used to analyze the expression correlation of *NcSpds* and *NcSr*. The results showed that the separate *NcSpds* and *NcSr* shared highly similar co-expression patterns in the samples with different strains and different nutritional and environmental condition treatments. The chimeric *SpdsSr* in basidiomycetes and the co-expression pattern of the *Spds* and *Sr* in *N. crassa* indicate the special link of spermidine and lysine in fungi, which may play an important role in the growth and development of fruiting body and in response to the multiple environmental factors and abiotic stresses.

**Keywords:** chimeric gene; lysine; spermidine; basidiomycetes; ascomycetes

check for updates

**Citation:** Yang, Y.; Shi, L.; Xu, X.; Wen, J.; Xie, T.; Li, H.; Li, X.; Chen, M.; Dou, X.; Yuan, C.; et al. Spermidine Synthase and Saccharopine Reductase Have Co-Expression Patterns Both in Basidiomycetes with Fusion Form and Ascomycetes with Separate Form. *J. Fungi* **2023**, *9*, 352. https://doi.org/10.3390/jof9030352

Academic Editors: Mingwen Zhao, Gen Zou and Jing Zhu

Received: 31 December 2022
Revised: 7 March 2023
Accepted: 8 March 2023
Published: 14 March 2023



## 1. Introduction

Gene fusion is a process by which the complete or partial sequences of two or more distinct genes are fused into a single chimeric gene or transcript as a result of DNA- or RNA-derived rearrangements [1]. Gene fusion could increase the diversity of gene functions and drive the evolution of organisms. However, natural gene fusion often leads to genomic disorders or cancer, leaving devastating consequences [2], because it can change the properties of precursor proteins and even perturb normal regulatory pathways so that the chimeric genes are often used as molecular diagnostic markers to detect and treat cancer in medicine [3,4].

As for the normal natural chimeric genes (rarely found), only the following cases have been reported in fungi. In *Metakinetoplastina protists*, the native fusion gene *SCAMK* formed through the insertion of a calmodulin-like III gene into the sucrose non-fermenting related kinase3 gene by an unequal crossover of homologous chromosomes in meiosis [5].

In *Penicillium brevicompactum*, Cytochrome P450 (MpaD) and a Hydrolase (MpaE) naturally fused to become the chimeric enzyme gene *MpaDE*, which is involved in the synthesis of mycophenolic acid (MPA) [6]. The *SPE3* gene (encoding spermidine synthase) and *LYS9* gene (encoding saccharopine dehydrogenase) were first found to be fused *SPE3-LYS9* in *Cryptococcus neoformans*, which encodes functional spermidine synthase and saccharopine dehydrogenase gene products [7]. However, in *Saccharomyces cerevisiae*, the spermidine synthase gene and saccharopine dehydrogenase gene were identified as independent *Spe3* and *LYS9*, respectively [8]. This indicates that the fusion of these two genes depends on the species. Whether the fusion behavior of these two different genes (*Spe* and *Lys*) is evolutionarily formed will need to be systematically analyzed in multiple species.

The function and effect of the formation of the normal natural chimeric genes, including SPE-LYS, has a lack of in-depth research. In *C. neoformans*, the spe3-LYS9 mutants (the SPE3-homologous region was replaced by the *NAT1* gene cassette) showed reduced capsule and melanin production and growth rate, and the SPE3-lys9 mutants (the LYS9-homologous region was replaced by the *NAT1* gene cassette) grew slowly and died upon lysine starvation, while the spe3-lys9 mutants (both the SPE3 and LYS9-homologous region were replaced by the *NAT1* gene cassette) were avirulent and unable to survive in vivo [7]. Due to the biosynthesis of lysine by higher fungi (basidomycetes and ascomycetes), it adopts the α-aminoadipate (AAA) pathway, in which the saccharopine dehydrogenase (also known as saccharopine reductase) catalyzes the penultimate step of the lysine biosynthesis [9–11]. This suggests that the chimeric SPE3-LYS9 gene may link spermidine and lysine biosynthesis, and their close connection is important for the growth and metabolism of *C. neoformans*.

Spermidine, as one of the main polyamines, is ubiquitous to all living organisms and involved in many fundamental cellular processes [12]. It has been reported that spermidine plays a role in regulating the growth, development, and secondary metabolism of fungi and participates in response to various abiotic stresses, such as high temperature, cold injury, drought, osmotic stress, oxidative stress and photooxidative damage, as well as jasmonic acid and abscisic acid induction [13]. Lysine, as a basic positively charged amino acid, plays an important physiological role in the organism and also functions in the growth, development, and response to stress in fungi [14]. The deficiency of lysine in fungi will not only damage protein synthesis but also seriously affect the immune system. In *Magnaporthe oryzae*, the deletion of the lysine synthase gene affects its growth, conidia germination, and pathogenicity in rice blast [15]. In *Flammulina filiformis*, higher levels of lysine content can facilitate stipe elongation and increase yield [16]. Thus, both spermidine and lysine are necessary for cell growth, proliferation, and differentiation, and they play an important role in maintaining normal physiological function and metabolism. Due to their important roles, there is a possibility that it is necessary to make spermidine and lysine act synchronously in many cases during the process of evolution, as some species fused their key biosynthesis catalytic enzyme together under natural conditions. Therefore, it is necessary to investigate the natural fusion phenomenon of spermidine synthase and saccharopine dehydrogenase in multiple species and summarize the conditions and circumstances when they are highly expressed, so as to better reveal the biological significance of their fusion.

In this study, we investigated more than 80 species from four common fungal groups (basidiomycetes, ascomycetes, chytridiomycetes and oomycetes) to explore the different forms of being (chimeric pattern or separate pattern) of spermidine synthase (SPDS) and saccharopine reductase (SR, the penultimate step of the lysine biosynthesis in fungi, equivalent to saccharopine dehydrogenase). Further, we selected *F. filiformis* (large agaric fungi and industrially produced edible fungi) as the representative species of basidiomycetes for research in this study because it has the typical morphological and developmental characteristics of fruiting body, it publishes the complete genome and transcriptome data of different developmental stages, and it has been widely studied in recent years. *Neurospora crassa*, as the representative species of ascomycetes, was also used for research in this study, because it has detailed genome maps and abundant expression profile data uploaded in the NCBI database. The structure, existence forms, and locus in the chromosome of *Spds* and *Sr* in *F. filiformis* and *N. crassa* were investigated and studied in detail. Moreover, the expression patterns of *SpdsSr* in *F. filiformis* and the separate *Spds* and *Sr* in *N. crassa* were analyzed to reveal the possible functional effects and mechanisms of the natural chimeric *SpdsSr* pattern in basidiomycetes and their separate but cooperative expression pattern in ascomycetes.

## 2. Materials and Methods

### 2.1. Strains and Culture Conditions

The *F. filiformis* heterokaryon strains FL19 (monokaryon L11 and monokaryon L22 were isolated by protoplasmic preparation) and NJ6 (monokaryon 6-3 and monokaryon 6-21 were isolated by protoplasmic preparation) were widely cultivated in the southeast of China. The six strains were obtained from the Fujian Edible Fungi Germplasm Resource Collection Center of China and routinely maintained on potato dextrose agar (PDA) at 25 °C. Mycelia of strains L11, L22, 6-3 and 6-21 were cultured in complete yeast medium (CYM) at 25 °C and was sampled for gene cloning and sequencing. Cultivation of the fruiting body of strain FL19 was performed as per the method described by Tao et al. [17]. The fruiting body was sampled at three major developmental stages: after inoculation, including primordia (PR); young fruiting body (YF); and elongation (EL). The pileus in the elongation stage (ELP) and stipe in the elongation stage (ELS) were also sampled. These fruiting body samples were used for the detection of gene relative expression.

Light and temperature are the most common environmental factors affecting fungi. In order to detect the expression of genes in response to light and temperature, on the mycelia of FL19 the following treatments were performed. For the light treatment, blue light with 3 $\mu$moL$\cdot$m$^{-2}\cdot$s$^{-1}$ intensity was used to irradiate the mycelia growing on CYM plate for 1 h and 3 h, according to the previous study of Li et al. [18], and the control was that under darkness. For the temperature treatment, four gradients—15 °C (cold stress), 20 °C (low-temperature stress), 25 °C (the optimum temperature, as control), and 30 °C (high-temperature stress)—were used to cultivate *F. filiformis* mycelia after inoculation on CYM plate, according to the study of Lv et al. [19]. Because fungi are often subjected to oxidative stress from growing substrates and mechanical damage from the external environment, the exogenous $H_2O_2$ (5 mM and 10 mM, according to Liu et al. [20]) as the representative of oxidative stress and exogenous MeJA (0.01 mM and 0.02 mM, according to Ren et al. [21]) as the representative of mechanical damage (due to the previous study that showed mechanical damage acts through JA signal [22]) were added into the CYM medium to cultivate *F. filiformis* mycelia. All samples were wrapped in foil, quick-frozen in liquid nitrogen, and stored in a −80 °C freezer for future use.

### 2.2. Bioinformatic Analysis of SPDS and SR in Filamentous Fungi

The sequences of the *Sr* gene (GenBank No.: DAA10591.1) and *Spds* gene (GenBank No.: DAA11489.1) in the model species *S. cerevisiae* were used to identify the corresponding orthologous genes by local BLAST in the *F. filiformis* L11 genome (GenBank Accession No. APIA00000000.1; BioProject: PRJNA191865), 6-3 genome (GenBank Accession No. GCA_011800155.1), and *N. crassa* OR74A genome (GenBank Accession No. GCA_000182925.2; BioProject: PRJNA13841).

The gene structures of *Spds* and *Sr* were visualized using GSDS 2.0 online software (http://gsds.gao-lab.org/, accessed on 5 September 2022), and their conserved domains were predicted by InterProScan online software (http://www.ebi.ac.uk/Tools/pfa/iprscan/, accessed on 7 September 2022). The phylogenetic tree was constructed by MEGAX software with the Neighbor-Joining method and examined by bootstrap testing with 1000 repeats [23].

### 2.3. Cloning and Sequencing of Chimeric *FfSpdsSr* Transcript in *F. filiformis*

Total RNA from *F. filiformis* mycelium and fruiting body samples was extracted using E.Z.N.A.® Plant RNA Kit (Omega Bio-Tek, Norcross, GA, USA), and cDNA was reverse transcribed using TransScript® One-Step gDNA Removal and cDNA Synthesis SuperMix Kit (TransGen Biotech, Beijing, China). According to the cDNA sequence of the chimeric *FfSpdsSr* in the *F. filiformis* L11 genome, its full-length ORF was obtained by PCR amplification and sequencing, using the primers designed at both ends of the ORF of *FfSpdsSr* (completely includes the full-length ORF). The two primer pairs FfSpds-qF/R and FfSr-qF/R, which were designed in both SPDS and SR conserved domains, respectively, were used to detect the relative expression level of the chimeric *FfSpdsSr* gene by real-time quantitative PCR (RT-qPCR). All the primers were designed by PrimerQuest Tool online software (https://sg.idtdna.com/PrimerQuest/Home/Index, accessed on 21 September 2022) and are listed in Supplementary Material Table S1. RT-qPCR was performed using a CFX96 Real-Time PCR Detection System (Bio-Rad, Hercules, CA, USA). RT-qPCR amplification was performed using PerfectStart® Uni RT&qPCR Kit (TransGen Biotech, Beijing, China), according to the method of Tao et al. [24]. According to the study of Yang et al. [25], *RNB*, *V-ATP*, and *β-TUB* were used as stable internal control genes for the normalization of the RT-qPCR in this study. The relative expression levels of the tested genes were determined according to the $2^{-\Delta\Delta Ct}$ method.

### 2.4. RNA-Seq Data Analyses

The expression correlation of *NcSpds* and *NcSr* in *N. crassa* was analyzed using multiple RNA-Seq data, which was downloaded from the Gene Expression Omnibus (GEO) database (https://www.ncbi.nlm.nih.gov/geo/, accessed on 25 September 2022). The total 15 RNA-Seq datasets were divided into four groups to conduct statistical analyses, according to the sample types: group 1, the different nutritional resources (GSE35227, GSE44100, GSE44673, GSE52316, GSE60986, GSE68517, GSE42692 and GSE51091); group 2, the different development stages (GSE41484); group 3, the different stresses (GSE53013, GSE52153, GSE53534 and GSE61949); and group 4, the different strains (GSE60004 and GSE45406), as well as the whole group, including all of the above samples. The RPKM values of *NcSpds* and *NcSr* were obtained from the above RNA-Seq data in each experimental sample. The correlation of RPKM values of *NcSpds* and *NcSr* genes was analyzed using Graphpad Prism 9 software.

### 2.5. Statistical Analysis

In this study, the results of RT-qPCR experiments were carried out with three independent biological replicates to ensure that the trends and relationships observed were reproducible. The error bars indicate the standard deviation (SD) from the mean of triplicate samples. The significance of the data was analyzed by using one-way ANOVA and multiple comparison tests.

## 3. Results

### 3.1. Different Gene Structures of Spermidine Synthase and Saccharopine Reductase in *F. filiformis* and in *N. crassa*

According to the lysine biosynthesis pathway in *S. cerevisiae*, the saccharopine dehydrogenase *LYS9* was used for local BLAST in the genome of *F. filiformis* L11, and its orthologous gene was identified in *F. filiformis* L11 with the ID number of *gene2942* (GenBank No.: OQ378313, Table S2). The length of the *gene2942* is 2498 bp, and it contains six exons and five introns (Figure 1A). Its full-length ORF with 2253 bp encodes a protein with 750 amino acids (aa). By predicting the conserved domain of the protein encoded by *gene2942*, we found that it contains two major conserved domains (Figure 1A): the polyamine biosynthesis domain (IPR030374) in the N-terminal and the saccharopine dehydrogenase domain in the C-terminal. The N-terminal polyamine biosynthesis domain is exactly the spermidine synthase tetramerization domain (IPR035246). The C-terminal saccharopine dehydrogenase domain includes the saccharopine dehydrogenase NADP binding domain (IPR005097) and the saccharopine dehydrogenase C-terminal domain (IPR032095), and it acts as saccharopine reductase in the penultimate step of the lysine biosynthesis pathway named by Liu et al. [26] in *F. filiformis*. It means that *gene2942* is a natural chimeric protein of spermidine synthase (SPDS) and saccharopine reductase (SR), named FfSPDS-SR. Additionally, its fourth exon, as the longest one (1433 bp) in which no termination codon appeared, acts as a bridge between *Spds* and *Sr* to form a natural fusion gene. Theoretically, FfSPDS-SR has the function of both spermidine synthase, which is responsible for catalyzing the biosynthesis of the spermidine, and saccharopine reductase, which is responsible for catalyzing the biosynthesis of the saccharopine as the penultimate intermediate in the biosynthesis of fungal lysine.

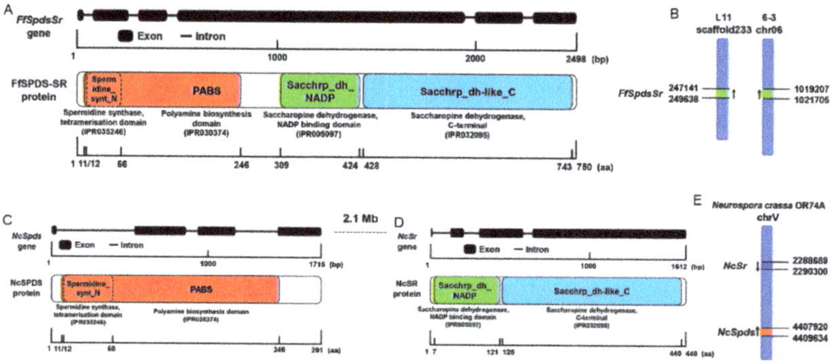


**Figure 1.** Gene structures and protein conserved domains of *FfSpdsSr* in *Flammulina filiformis* and *NcSpds* and *NcSr* in *Neurospora crassa*. (**A**) Gene structures and protein conserved domains of *FfSpdsSr* in *F. filiformis*. Thick lines represent exons and thin lines represent introns. (**B**) The location of the *FfSpdsSr* gene in the genome of *F. filiformis* strains L11 and 6-3. The green rectangles represent the region of the *FfSpdsSr* gene on the chromosome. The black arrows indicate the ORF direction of the *FfSpdsSr* gene. (**C**) Gene structures and protein conserved domains of *NcSpds* in *N. crassa*. Thick lines represent exons and thin lines represent introns. (**D**) Gene structures and protein conserved domains of *NcSr* in *N. crassa*. Thick lines represent exons and thin lines represent introns. (**E**) The blue rectangle and red rectangle represent the region of *NcSr* and *NcSpds* genes on the chromosome in *N. crassa*, respectively. The black arrows represent the ORF direction of *NcSpds* and *NcSr* genes.

To investigate whether there are other separate copies of *Spds* or *Sr*, we performed the local BLAST in the whole genome of *F. filiformis* strain L11 using the conserved domain region of *Spds* or *Sr*, respectively. The results showed that there is the only one copy that is chimeric *FfSpdsSr* (*gene2942*), located at the 247141-249638 bp of the scaffold233 in the genome of *F. filiformis* strain L11 (Figure 1B). The unique copy of chimeric *FfSpdsSr* also exists in the genomes of the other *F. filiformis* monokaryon strains, such as W23, L22, 6-3, and 6-21. In the genome of strain 6-3, obtained by the third-generation sequencing technology Pacbio [27], the unique chimeric *FfSpdsSr* was found to be located at the 1019207-1021705 bp of chromosome 06 (Figure 1B). This suggests that the unique and chimeric *SpdsSr* exists in *F. filiformis* universally.

In the same way, the conserved domain regions of SPDS and SR were used for local BLAST in the genome of *N. crassa* OR74A, respectively. Two separate orthologous proteins NCU06727 and NCU03748 were obtained. The length of the NCU06727 gene is 1715 bp and contains four exons and three introns (Figure 1C). Its full-length ORF with 876 bp encodes a protein with 291 aa, which contains only one main conserved domain: the polyamine biosynthesis domain (IPR030374). The exact domain of NCU06727 is the spermidine synthase tetramerization domain (IPR035246); thus, we renamed NCU06727 as NcSPDS (Figure 1C). The length of the NCU03748 gene is 1612 bp and contains four exons and three introns (Figure 1D). Its full-length ORF with 1347 bp encodes a protein with 448 aa, which contains only the saccharopine dehydrogenase domains. The detailed saccharopine dehydrogenase domains in NCU03748 include the saccharopine dehydrogenase NADP binding domain (IPR005097) and the saccharopine dehydrogenase C-terminal domain (IPR032095) (Figure 1D). Because that saccharopine dehydrogenase is in charge of the penultimate step of the fungal lysine biosynthesis and also called saccharopine reductase (SR) previously, we renamed NCU03748 as NcSR.

The polyamine biosynthesis domain (IPR030374) and the spermidine synthase tetramerization domain (IPR035246) are located at 11-246 aa and 12-66 aa in FfSPDS-SR and NcSPDS, respectively. Additionally, the saccharopine dehydrogenase NADP binding domain (IPR005097) and the saccharopine dehydrogenase C-terminal domain (IPR032095) are separated by 3 aa in both FfSPDS-SR and NcSR. The chimeric FfSPDS-SR protein is 11 more aa than NcSPDS plus NcSR, so these 11 aa region links SPDS and SR in chimeric FfSPDS-SR protein.

According to the genome of *N. crassa* OR74A, the *NcSr* gene was located at the 2288689-2290300 bp of the chromosome chrV (CM002240.1) (Figure 1E), while the *NcSpds* gene was located at the 4407920-4409634 bp of the chromosome chrV (CM002240.1) (Figure 1E). The *NcSpds* gene and *NcSr* gene are located on the same chromosome (chrV), but they are 2.1 Mb apart and have the opposite direction (Figure 1C,D). *NcSpds* and *NcSr* are two physically separate, relatively independent genes, with no forming of a chimeric gene.

### 3.2. The Chimeric FfSpdsSr Gene Presents in Heterokaryon of F. filiformis with Conserved and Unitary Transcript

In order to determine whether the alleles of the *FfSpdsSr* gene exist in different forms in two different nuclei of *F. filiformis* heterokaryon strain, we used two *F. filiformis* strains FL19 and NJ6, which are widely cultivated in the southeast of China. The heterokaryon FL19 separated two monokaryon L11 and monokaryon L22, while the heterokaryon NJ6 separated two monokaryon 6-3 and monokaryon 6-21. The genome sequences of the *FfSpdsSr* gene in L11, L22, 6-3, and 6-21 were performed multiple sequence alignment, the results showed that the length of the alleles of the *FfSpdsSr* gene in four strains is 2494-2498 bp, and the similarity was 99.3% among them (SNP site variation). Further, the full-length ORF of the *FfSpdsSr* gene was also performed PCR amplification and sequencing, using the primers designed at both ends of the ORF of *FfSpdsSr* (completely includes the ORF). The unique transcript (chimeric ORF of the chimeric *FfSpdsSr* gene) was confirmed by the single PCR band and the Sanger sequencing results in each of L11, L22, 6-3, and 6-21 strains (Figure 2A). Meanwhile, a single PCR band was also amplified for the full-length ORF of *FfV-ATP* and *Ffβ-TUB* in L11 and L22 as the positive control (Figure S1). The unique transcript sequence of *FfSpdsSr* is consistent with the results predicted by the genome, meaning that the exons, introns, and their sites of the *FfSpdsSr* gene are accurate. In addition, in order to explore whether the *FfSpdsSr* gene forms different alternative splicing variants during transcription through alternative splicing, we mapped the raw reads of the transcriptome from the different developmental stages on the *FfSpdsSr* gDNA sequence using ZOOM software. The mapping results showed that there are a total of five introns in the *FfSpdsSr* gene, which all showed a 5′ (GT) and 3′ (AG) splice site in four introns. Importantly, none of the reads were detected to be derived from adjacent exons and introns, meaning that all introns of *FfSpdsSr* were clipped out during transcription (Figure 2B). Together with the results of full-length ORF cloning and sequencing, this suggests that the gene *FfSpdsSr* has only one transcript: the fused CDS of *SpdsSr*.

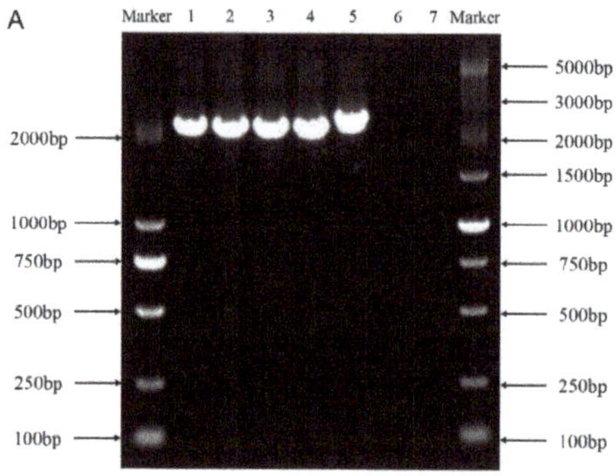


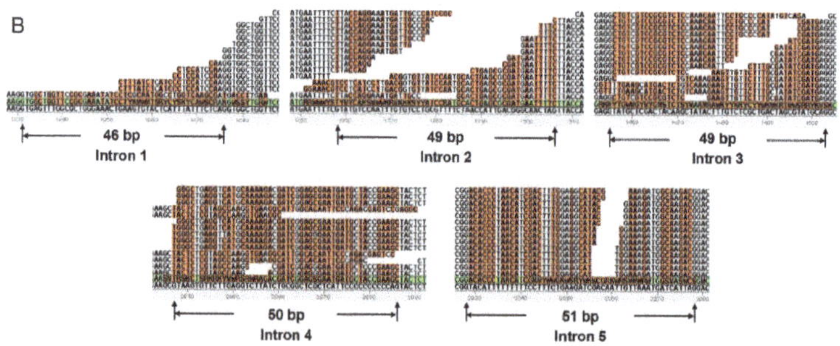


**Figure 2.** Validation of the full-length ORF and unitary transcript of *FfSpdsSr* gene in *F. filiformis*. (**A**) Full-length ORF amplification results of *FfSpdsSr* gene in *F. filiformis*. Numbers 1–4 represent the full-length ORF of chimeric *FfSpdsSr* amplified from cDNA of monokaryon L11 (isolated from dikaryon FL19), L22 (isolated from dikaryon FL19), 6-3 (isolated from dikaryon NJ6), and 6-21 (isolated from dikaryon NJ6) in *F filiformis*, respectively. Number 5 represents the full length of gDNA of *FfSpdsSr* in *F. filiformis* strain L11. Numbers 6–7 represent the PCR results using cDNA as template and the combined primers (the beginning of ORF of *NcSpds* as forward primer and the end of ORF of *NcSr* as reverse primer) in *N. crassa* strain, FGSC#4200 and FGSC#2489. The markers 2000 and 5000 are shown on the side of the electropherogram. (**B**) Alternative splicing analysis of *FfSpdsSr* in *F. filiformis*. The orange parts represent the intron region of *FfSpdsSr*.

Similarly, the beginning of the ORF of *NcSpds* as the forward primer and the end of the ORF of *NcSr* as the reverse primer were designed to test whether the chimeric *SpdsSr* transcript exists in *N. crassa*. The PCR, using cDNA as the template, was performed in the two strains (monokaryon FGSC#2489 and monokaryon FGSC#4200, which can mate into a heterokaryon strain) of *N. crassa*. There was no PCR band using the above primer pair in two *N. crassa* strains; however, a single PCR band was amplified for the full-length ORF of *NcV-ATP* and *Ncβ-TUB* as the positive control (Figure S1), confirming that *NcSpds* and *NcSr* do not form chimeric transcripts at the RNA level.

### 3.3. Gene Structure and Phylogeny Evolution of SPDS and SR in Fungi

In order to analyze whether chimeric *SpdsSr* is also present in fungi other than *F. filiformis*, we investigated 48 basidiomycetes, 27 ascomycetes, five chytridiomycetes, and two oomycetes; their complete genome sequences can be obtained in NCBI or JGI database. The conserved domain regions of SPDS and SR were used for local BLAST to obtain the orthologous gene of chimeric *SpdsSr* or the two separate *Spds* and *Sr* genes in the above species. The results showed that chimeric *SpdsSr* was found in all 48 basidiomycetes, while the separate *Spds* and *Sr* genes were found in all 27 ascomycetes, five chytridiomycetes, and two oomycetes. Furthermore, the phylogenetic trees were constructed based on the conserved SPDS and SR domain regions. Both the phylogenetic tree based on the SPDS domain (Figure 3A) and the phylogenetic tree based on the SR domain (Figure 3B) showed that 82 species are clustered into four groups: basidiomycetes, ascomycetes, chytridiomycetes, and oomycetes, which correspond to four phyla of fungi. This also suggests that the formation of chimeric *SpdsSr* is closely related to the evolution of fungi.

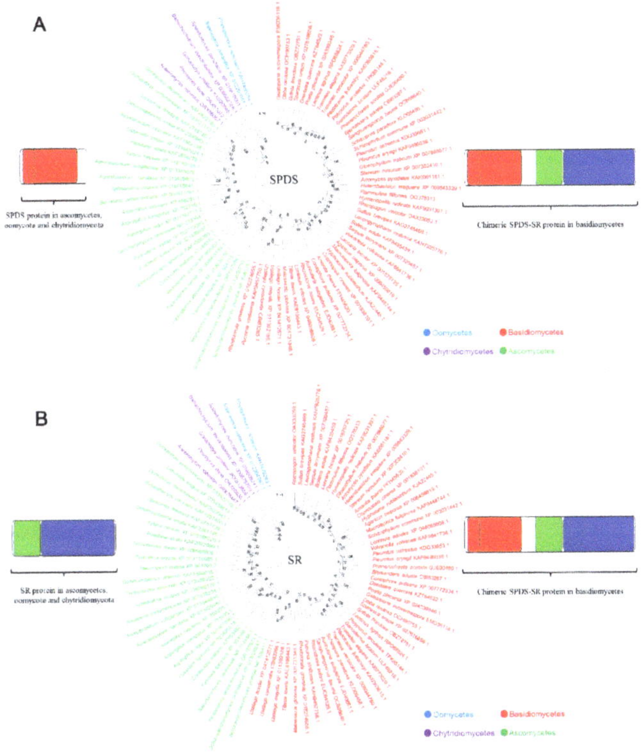


**Figure 3.** Phylogenetic analysis of SPDS and SR in fungi. (**A**) Phylogenetic trees based on SPDS domain in fungi, including basidiomycetes, ascomycetes, chytrid fungi, and oomycetes. (**B**) Phylogenetic trees based on SR domain in fungi, including basidiomycetes, ascomycetes, chytridiomycetes, and oomycetes.

### *3.4. Expression Patterns of the Chimeric Gene SpdsSr in F. filiformis*

In order to explore the roles of the chimeric *FfSpdsSr* gene in the growth and development of *F. filiformis* as well as in response to various environmental stresses, the transcription levels of *FfSpdsSr* were detected in different developmental stages, tissues, common environment factors (light and temperature), and stresses ($H_2O_2$ represents oxidative stress and exogenous MeJA represents mechanical damage signal). To verify whether the expression trend of 5′-end *FfSpds*-region and 3′-end *FfSr*-region of the chimeric gene *FfSpdsSr* was consistent, we designed the RT-qPCR primers FfSpds-qF/R and FfSr-qF/R on the *FfSpds* and *FfSr* conserved domain sequences, respectively. First of all, the expression trends of the *FfSpds*-region and the *FfSr*-region were highly consistent in all of the test samples because they belong to the unitary transcript of *FfSpdsSr*. During the three major developmental stages (PR, YF, EL), the transcription levels of *FfSpdsSr* showed an up-regulated trend from primordia to the elongation stage as the fruiting body grew (Figure 4A). For the stipe and pileus in the elongation stage, the transcription level of *FfSpdsSr* in the pileus (ELP) was about 4-fold higher than that in the stipe (ELS) (Figure 4B). Just as the blue light was usually used to shine *F. filiformis* in the factory, the two time points (1 h and 3 h, according to the previous study by Li et al. [18]) after blue light irradiation were used for the sample and detection. *FfSpdsSr* also up-regulated at 1 h and 3 h after blue light irradiation, and its relative expression level at BL-3h was higher than that at BL-1h (Figure 4C). According to the study of Lv et al. [19], four temperature gradients (15 °C, 20 °C, 25 °C, 30 °C) were used to detect the response expression of *FfSpdsSr*. *FfSpdsSr* showed no significant difference in expression under the 15 °C or 20 °C cultivations (low temperature) compared with the 25 °C cultivation (the optimum temperature), while it showed a 2.5-fold up-regulation under the 30 °C cultivation (high temperature) (Figure 4D). Besides the light and temperature, many cultivation substrates often bring oxidative stress and mechanical damage often occurs. During the growth of *F. filiformis*, we adopted the exogenous addition of $H_2O_2$ (5 mM and 10 mM, according to Liu et al. [20]) and MeJA (0.01 mM and 0.02 mM, according to Ren et al. [21]) as two cases of environmental stress to detect the response expression of *FfSpdsSr*. The RT-qPCR results showed that *FfSpdsSr* significantly up-regulated under 10 mM $H_2O_2$ treatment (Figure 4E) or under 0.01-0.02 mM MeJA treatment (Figure 4F).

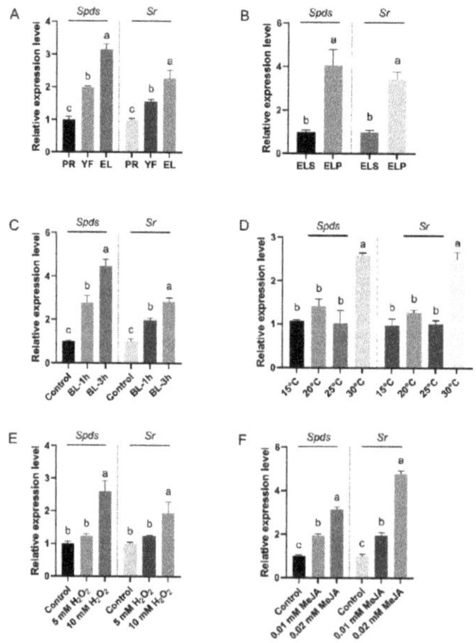

**Figure 4.** Relative expression levels of *FfSpdsSr* gene in different developmental stages, different tissues, and under different environmental factors in *F. filiformis*. (**A**) Relative expression levels of *FfSpdsSr* in different developmental stages of strain FL19. PR: primordia; YF: young fruiting body EL: elongation stage. (**B**) Relative expression levels of *FfSpdsSr* in different tissues of strain FL19. ELS: stipe in elongation stage; ELP: pileus in elongation stage. (**C**) Relative expression levels of *FfSpdsSr* under blue light conditions. BL-1h: 1 h under blue light irradiation, BL-3h: 3 h under blue light irradiation. (**D**) Relative expression levels of the *FfSpdsSr* gene of *F. filiformis* in strain FL19 under different temperature gradients, including 15 °C (cold stress), 20 °C (low-temperature stress), 25 °C (the optimum temperature, as control), and 30 °C (high-temperature stress). (**E**) Relative expression levels of *FfSpdsSr* gene of *F. filiformis* in strain FL19 under $H_2O_2$ condition. (**F**) Relative expression levels of *FfSpdsSr* gene of *F. filiformis* in strain FL19 under MeJA condition. (**A–F**) Spds represents the RT-qPCR results using the primers FfSpdsF/R designed in the SPDS domain, and Sr represents the RT-qPCR results using the primer FfSrF/R designed in the SR domain. All values are the means ± SD of three independent experiments, and the lowercase letters indicate significant differences among different samples ($p < 0.05$).

### *3.5. Similar Expression Patterns of NcSpds and NcSr in N. crassa*

Although *Spds* and *Sr* are two separate and independent genes in ascomycetes, we wondered if there was some correlation between expression and function. Since *N. crassa* is a model ascomycetes species, there have been numerous genome-wide expression profile data under a variety of different conditions; we downloaded, in total, 15 sets of RNA-Seq data from the GEO database, in order to analyze the expression correlation of *NcSpds* and *NcSr*. These 15 sets of RNA-Seq data were gathered from different experimenters around the world and included a total of 218 samples. Although different sequencing depths among RNA-Seq datasets led to a large range of changes in the absolute expression value for each gene, RPKM values of *NcSpds* and *NcSr* were calculated under the same internal standard and can be compared in each sample. Thus, we performed the correlation analysis based on the RPKM values of *NcSpds* and *NcSr* obtained from the above RNA-Seq datasets.

In group 1 (Figure 5A), there were a total of 54 samples involved in different nutritional resources, including avicel, sucrose, no carbon, xylan, and straw. In general, RPKM values of *NcSpds* and *NcSr* showed similar patterns under these nutritional conditions, and the correlation coefficient $r_s$ value was 0.9209 ($p < 0.0001$) (Figure 5B), which was larger than the threshold of the strong correlation coefficient, 0.8. In Figure 5B, most of the sample points clustered within the 95% confidence bands of the best-fit line. In addition, the RPKM peak of *NcSpds* and *NcSr* mainly appeared in the sample under the sucrose nutrient condition.

In group 2 (Figure 5C), there were a total of 13 samples involved in the different development stages, including perithecium wall, centrum parenchyma, asci, paraphyses, ostiole, etc. The RPKM values of *NcSpds* and *NcSr* showed slightly similar patterns under these development stages, and the correlation coefficient $r_p$ value was 0.7816 ($p = 0.0016$) (Figure 5D), which was slightly lower than the threshold of strong correlation coefficient, 0.8, but still higher than the threshold of acceptable correlation coefficient, 0.6. In addition, the RPKM peak of *NcSpds* and *NcSr* mainly appeared in the sample around sexual development.

In group 3 (Figure 5E), there were a total of 36 samples involved in the different stresses, including dimethylsulfoxide, staurosporine, light, dithiothreitol, tunicamycin, etc. In general, the RPKM values of *NcSpds* and *NcSr* showed highly similar patterns under these stresses, and the correlation coefficient $r_s$ value was 0.9305 ($p < 0.0001$) (Figure 5F), which was higher than the threshold of strong correlation coefficient 0.8. In Figure 5F, most sample points are clustered within the 95% confidence bands of the best-fit line. The RPKM peak of *NcSpds* and *NcSr* mainly appeared in the sample under dimethylsulfoxide stress.

In group 4 (Figure 5G), there were a total of 115 different mutant strain samples. The RPKM values of *NcSpds* and *NcSr* showed a slightly similar pattern under these strains, and the correlation coefficient $r_s$ value was 0.7367 ($p < 0.0001$) (Figure 5H), less than 0.8 but still higher than the threshold of acceptable correlation coefficient, 0.6. When we put all 218 samples together for analysis (Figure 5I), the absolute expression of *NcSpds* and *NcSr* showed a very similar trend, and the correlation coefficient $r_s$ value was 0.9214 ($p < 0.0001$) (Figure 5J), suggesting that *NcSpds* and *NcSr* were strongly correlated at the transcription level.

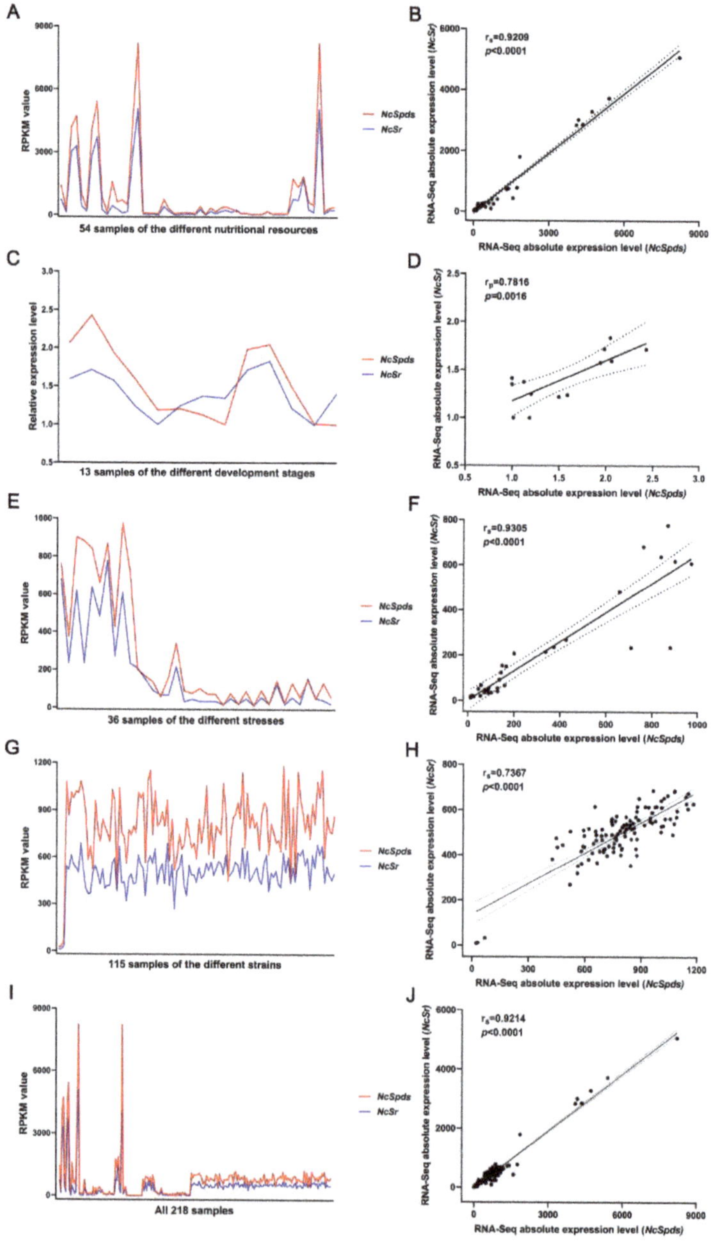


**Figure 5.** Correlation analyses of expression levels of *NcSpds* and *NcSr* in *N. crassa*. (**A**,**B**) Correlation analyses of expression levels of *NcSpds* and *NcSr* in the different nutritional resources, including avicel, sucrose, no carbon, xylan, barleystraw, cornstraw, rice straw, soybean straw, wheatstraw, pectin, orange peel powder (OPP), L-arabinose, D-glucose, and D-xylose (RNA-Seq data: GSE35227, GSE44100, GSE44673, GSE52316, GSE60986, GSE68517, GSE42692 and GSE51091). (**C**,**D**) Correlation analyses of expression levels of *NcSpds* and *NcSr* in the different development stages, including perithecium wall (PW), centrum parenchyma (CP), asci (AS), paraphyses (PP), and ostiole (OS) (RNA-Seq data GSE41484). (**E**,**F**) Correlation analyses of expression levels of *NcSpds* and *NcSr* under

the different stresses, including dimethylsulfoxide (DMSO), staurosporine (STS), dark, light for 15 min, light for 60 min, light for 120 min, light for 240 min, dithiothreitol (DTT, 5 mM), DTT (0.1 mM), $H_2O$, tunicamycin (TM, 40 μg/mL), and TM (5 μg/mL) (RNA-Seq data: GSE53013, GSE52153, GSE53534 and GSE61949). (**G**,**H**) Correlation analyses of expression levels of *NcSpds* and *NcSr* in the 115 different strains, including Δ3βG, CPL4, and CPL7 and the 112 strains collected in Louisiana [28] (RNA-Seq data: GSE60004 and GSE45406). (**I**,**J**) Correlation analyses of expression levels of *NcSpds* and *NcSr* in all 218 samples from RNA-Seq data. (**A**,**C**,**E**,**G**,**I**) The red lines and the blue lines represent the expression levels of *NcSpds* and *NcSr*, respectively. (**B**,**D**,**F**,**H**,**J**) Correlation analyses were performed on the expression levels of *NcSpds* and *NcSr*. The "$r_p$" was the Pearson correlation coefficient and the "$r_s$" was the Spearman correlation coefficient. The black dots represent the expression level of *NcSpds* and *NcSr* in each sample. The black dotted line represents the 95% confidence bands of the best-fit line.

## 4. Discussion

Gene fusion means that two or more independent genes are placed in one set of ORFs via chromosomal rearrangement and combination, which is under the same regulatory sequence, including promoter and terminator. The result of gene fusion is to form a chimeric gene, and the expression product of the chimeric gene is chimeric protein. In most cases, gene fusion can result in abnormal sequences, disordered expression of certain genes, and protein function deactivation, which can lead to or promote the development of tumors. In past studies, a natural chimeric gene was mostly found in cancer cells, so many chimeric genes have been established as important biomarkers and therapeutic targets in multiple cancer types [29]. To date, it has been rarely reported in normal plant, animal, and fungal cells. Thus, in this study, the discovery of the natural chimeric gene SPDS-SR in basidiomycetes such as *F. filiformis*, as well as the confirmation of its single copy and unique variant spliceosome, are of great biological significance for the study of gene evolution. This is because gene fusion events can serve as phylogenetic markers [30].

In the 27 ascomycetes investigated in this paper, SPDS and SR did not fuse but existed independently as two genes. The cis-regulatory sequences of the promoters of these two genes are also different. However, from 15 sets of expression profile data, the expression patterns of *Spds* and *Sr* showed a highly positive correlation. Due to these expression profile data being derived from high-throughput RNA-Seq under the different sample types, different treatments, different strains, and different experiment times, the positive correlation obtained by statistics is highly reliable. It also suggests that the transcription of *Spds* and *Sr* (although they are not fused) tended to synchronize in ascomycetes, especially in some special development periods and under some environmental factor treatments.

The similar expression patterns of *Spds* and *Sr* in ascomycetes may be due to the fact that they are regulated by the same transcription factor (although there are no reports about this). The fusion of the two genes may improve the regulation efficiency [6]. Another possibility is that the fungi require the combined action of polyamines and lysine during the special growth and development stages or in response to some environmental changes, as this study found that high expression of *NcSpds* and *NcSr* appeared during sexual development or under osmotic stress (DMSO treatment), etc. Coincidentally, *FfSpdsSr* also significantly up-regulated during the pileus' rapid development or under several environmental stressors, including blue light, high temperature, $H_2O_2$, and MeJA treatment. Therefore, co-expression of the two genes encoding synthetase is required in ascomycetes. While in basidiomycetes, these two genes directly form the one chimeric gene. The occurrence of *SpdsSr* forming a chimeric gene also suggests that spermidine and lysine biosynthesis are linked directly. Previous studies have proved the involvement of spermidine as an antioxidant against the potential abiotic-stress-derived oxidative damage [31]. Spermidine, as a type of signaling molecule, also play important roles in growth and development and the defense against biotic and abiotic stresses, while lysine is directly involved in many biological processes, such as constituent protein structure, stress response, secondary metabolites biosynthesis, etc. [32]. Both spermidine and lysine can promote growth, cell division, DNA replication, and cell differentiation through interacting with nucleic acids, proteins, and phospholipids to regulate the physical and chemical properties of the membrane, the structure and function of nucleic acids, and the activity of enzymes. It has been reported that the deficiency of polyamines and lysine can cause an imbalance of ROS homeostasis and a disorder of physiological function.

Gene fusion of related functions can improve the efficiency of biochemical reactions and signal transduction, as well as promote the coordinated regulation of their expression [33]. Based on the analysis of the cis-elements of the promoter of the *SpdsSr* gene in basidiomycetes and *Spds* and *Sr* genes in ascomycetes and the summary of the existing expression profile data, it can be seen that *SpdsSr*, *Spds*, and *Sr* are mainly found in the rapid development period of fruiting bodies (such as the rapid elongation stage of *F. filiformis*); in the tissue parts of sexual reproductive organs (such as the pileus development and the basidiospore formation); and when multiple environmental factors, including blue light, high-temperature stress, oxidative stress (such as $H_2O_2$ treatment), and hormonal changes (such as MeJA treatment), are up-regulated. This suggests that *SpdsSr* (or the combination of spermidine and lysine) may be mainly involved in the growth and development process of fungi and play important roles in coping with environmental stress. This is also supported by the fact that the fusion of SPE3-LYS9 in *C. neoformans* is important for virulence and survival in vivo, and that the spe3-lys9 mutants were avirulent and unable to survive in vivo [7].

From an evolutionary point of view, ascomycetes seem to be more junior and basidiomycetes more advanced. Basidiomycetes (especially large edible fungi) have the characteristics of fruiting bodies, such as a stipe and pileus. Meanwhile, during the development of fruiting bodies, basidiomycetes are more susceptible to changes in environmental factors such as temperature and light. These environmental factors also regulate the changes in the shapes and colors of basidiomycetes from time to time, so that fungi can better adapt to the living environment. The functional convergent evolution of the fused *SpdsSr* suggests the evolutionary advantages of basidiomycetes in utilizing spermidine and lysine. At present, there is no sufficient evidence as to whether *SpdsSr* is to meet the growth and development and cope with the changes of environmental factors. However, the phenomenon of the natural fusion form and co-expression pattern in basidiomycetes should indeed attract the interest of fungal researchers. It is also worth attention to further reveal the biological significance of this fusion phenomenon from heredity and evolution.

## 5. Conclusions

The spermidine synthetase and saccharopine reductase encoding genes form a natural chimeric gene, *FfSpdsSr,* in *F. filiformis*, which is also found to present in the other 48 tested basidiomycetes, while the two separate *Spds* and *Sr* genes independently present in *N. crassa* and are also found to present in the other tested 27 ascomycetes, five chytridiomycetes, and two oomycetes. In *F. filiformis*, the full-length ORF PCR results and no alternative splicing variant of *FfSpdsSr* confirms that the unique chimeric *FfSpdsSr* transcript functions in *F. filiformis*. *FfSpdsSr* mainly up-regulated in the elongation stage and pileus development of *F. filiformis*, as well as under the blue light, high-temperature, $H_2O_2$, and MeJA treatments. Interestingly, the separate *NcSpds* and *NcSr* in *N. crassa* were found to share highly similar co-expression patterns in the total 15 RNA-Seq datasets, including 218 samples with different strains and different nutritional and environmental condition treatments. This implies that the fusion of *Spds* and *Sr* may have great biological significance and meet evolutionary needs.

**Supplementary Materials:** The following supporting information can be downloaded at: https://www.mdpi.com/article/10.3390/jof9030352/s1, Figure S1: PCR of the full-length ORF of internal control genes in *F. filiformis* and *N. crassa*; Table S1: The primers used in this study; Table S2: The gene accession number used in this study.

**Author Contributions:** Conceptualization, Y.T. and B.X.; methodology, Y.Y., X.X., L.S., J.W. and T.X.; software, X.L., M.C., X.D. and C.Y.; resources, H.L., X.L. and H.S.; data curation, Y.Y., T.X., L.S. and H.S.; writing—original draft preparation, Y.Y., X.X. and J.W.; writing—review and editing, Y.T., B.X., Y.Y. and X.X.; visualization, Y.Y. and Y.T.; supervision, Y.T. and B.X. All authors have read and agreed to the published version of the manuscript.

**Funding:** This work was supported by grants from the Natural Science Foundation of Fujian Province of China (2019J01380), the National Natural Science Foundation of China (31902088), the Natural Science Foundation for Distinguished Young Scholar of Fujian Agriculture and Forestry University of China (xjq201919), the Excellent Master Thesis Cultivation Fund of Fujian Agriculture and Forestry University (Yayong Yang: 1122YS01011), the National College students Innovation and Entrepreneurship Plan Training Project (Zhuohan Jing: 202210389031), the College Students Innovation and Entrepreneurship Plan Training Project of Fujian Agriculture and Forestry University (Xiaoyu Li: 202210389213), the Seed Industry Innovation and Industrialization Project of Fujian Province of China (zycxny2021012), the Key Research and Development Planning Project of Hebei Province (21326315D), and the HAAFS Science and Technology Innovation Special Project (2022KJCXZX-JZS-9).

**Institutional Review Board Statement:** This study did not involve humans or animals.

**Informed Consent Statement:** Not applicable.

**Data Availability Statement:** All experimental data in this study will be made available upon reasonable request from readers. Publicly available datasets were analyzed in this study. Data is available at NCBI GEO: GSE35227, GSE44100, GSE44673, GSE52316, GSE60986, GSE68517, GSE42692, GSE51091, GSE41484, GSE53013, GSE52153, GSE53534, GSE61949, GSE60004 and GSE45406. This data can be found here: https://www.ncbi.nlm.nih.gov/geo/ (accessed on 25 September 2022).

**Conflicts of Interest:** The authors declare no conflict of interest.

## Abbreviations

| | |
|---|---|
| SPDS | Spermidine synthetase |
| SR | Saccharopine reductase |
| AAA | $\alpha$-Aminoadipate |
| CYM | Complete yeast medium |
| PDA | Potato dextrose agar |
| PR | Primordia |
| YF | Young fruiting body |
| EL | Elongation stage |
| ELP | Pileus in elongation stage |
| ELS | Stipe in elongation stage |
| JA | Jasmonic acid |
| MeJA | Methyl jasmonate |
| RT-qPCR | Real-Time quantitative PCR |
| SD | Standard deviation |
| BL-1h | 1 h under blue light irradiation |
| BL-3h | 3 h under blue light irradiation |
| GEO | Gene Expression Omnibus |

## References

1. Williford, A.; Betrán, E. Gene Fusion. *eLS* **2013**. [CrossRef]
2. Rogers, R.L.; Hartl, D.L. Chimeric Genes as a Source of Rapid Evolution in *Drosophila melanogaster*. *Mol. Biol. Evol.* **2012**, *29*, 517–529. [CrossRef]
3. Ma, L.; Yang, S.; Zhao, W.; Tang, Z.; Zhang, T.; Li, K. Identification and analysis of pig chimeric mRNAs using RNA sequencing data. *BMC Genom.* **2012**, *13*, 429. [CrossRef]

4. Dai, X.; Theobard, R.; Cheng, H.; Xing, M.; Zhang, J. Fusion genes: A promising tool combating against cancer. *Biochim. Biophys. Acta Rev. Cancer* **2018**, *1869*, 149–160. [CrossRef]
5. Chen, F.; Zhang, L.; Lin, Z.; Cheng, Z.-M.M. Identification of a novel fused gene family implicates convergent evolution in eukaryotic calcium signaling. *BMC Genom.* **2018**, *19*, 306. [CrossRef] [PubMed]
6. Hansen, B.G.; Mnich, E.; Nielsen, K.F.; Nielsen, J.B.; Nielsen, M.T.; Mortensen, U.H.; Larsen, T.O.; Patil, K.R. Involvement of a Natural Fusion of a Cytochrome P450 and a Hydrolase in Mycophenolic Acid Biosynthesis. *Appl. Environ. Microbiol.* **2012**, *78*, 4908–4913. [CrossRef] [PubMed]
7. Kingsbury, J.M.; Yang, Z.; Ganous, T.M.; Cox, G.M.; McCusker, J.H. Novel Chimeric Spermidine Synthase-Saccharopine Dehydrogenase Gene (*SPE3-LYS9*) in the Human Pathogen *Cryptococcus neoformans*. *Eukaryot. Cell* **2004**, *3*, 752–763. [CrossRef]
8. Winston, M.K.; Bhattacharjee, J.K. Biosynthetic and regulatory role of lys9 mutants of *Saccharomyces cerevisiae*. *Curr. Genet.* **1987**, *11*, 393–398. [CrossRef]
9. Zabriskie, T.M.; Jackson, M.D. Lysine biosynthesis and metabolism in fungi. *Nat. Prod. Rep.* **2000**, *17*, 85–97. [CrossRef]
10. Xu, H.; Andi, B.; Qian, J.; West, A.H.; Cook, P.F. The α-Aminoadipate Pathway for Lysine Biosynthesis in Fungi. *Cell Biochem. Biophys.* **2006**, *46*, 43–64. [CrossRef]
11. Song, Z.; He, M.; Zhao, R.; Qi, L.; Chen, G.; Yin, W.-B.; Li, W. Molecular Evolution of Lysine Biosynthesis in Agaricomycetes. *J. Fungi* **2021**, *8*, 37. [CrossRef]
12. Rocha, R.O.; Wilson, R.A. Essential, deadly, enigmatic: Polyamine metabolism and roles in fungal cells. *Fungal Biol. Rev.* **2019**, *33*, 47–57. [CrossRef]
13. Valdés-Santiago, L.; Ruiz-Herrera, J. Stress and polyamine metabolism in fungi. *Front. Chem.* **2014**, *1*, e42. [CrossRef]
14. Galili, G. New Insights into the Regulation and Functional Significance of Lysine Metabolism in Plants. *Annu. Rev. Plant Biol.* **2002**, *53*, 27–43. [CrossRef]
15. Bai, P.; Park, C.H.; Shirsekar, G.; Songkumarn, P.; Bellizzi, M.; Wang, G.L. Role of lysine residues of the *Magnaporthe oryzae* effector AvrPiz-t in effector- and PAMP-triggered immunity. *Mol. Plant Pathol.* **2019**, *20*, 599–608. [CrossRef]
16. Liu, F.; Wang, W.; Chen, B.-Z.; Xie, B.-G. Homocitrate Synthase Expression and Lysine Content in Fruiting Body of Different Developmental Stages in *Flammulina velutipes*. *Curr. Microbiol.* **2015**, *70*, 821–828. [CrossRef] [PubMed]
17. Tao, Y.; Chen, R.; Yan, J.; Long, Y.; Tong, Z.; Song, H.; Xie, B. A hydrophobin gene, Hyd9, plays an important role in the formation of aerial hyphae and primordia in *Flammulina filiformis*. *Gene* **2019**, *706*, 84–90. [CrossRef] [PubMed]
18. Li, J.; Xu, C.; Jing, Z.; Li, X.; Li, H.; Chen, Y.; Shao, Y.; Cai, J.; Wang, B.; Xie, B.; et al. Blue light and its receptor white collar complex (FfWCC) regulates mycelial growth and fruiting body development in *Flammulina filiformis*. *Sci. Hortic.* **2023**, *309*, 111623. [CrossRef]
19. Lv, X. Annotation of Pheromone Signaling Pathway and Functional Study of Gene ste12 in *Flammulina filiformis*. Master's Thesis, Shandong Agricultural University, Taian, China, 2022. [CrossRef]
20. Liu, Y.; Ma, X.; Long, Y.; Yao, S.; Wei, C.; Han, X.; Gan, B.; Yan, J.; Xie, B. Effects of β-1,6-Glucan Synthase Gene (*FfGS6*) Overexpression on Stress Response and Fruit Body Development in *Flammulina filiformis*. *Genes* **2022**, *13*, 1753. [CrossRef]
21. Ren, A.; Qin, L.; Shi, L.; Dong, X.; Mu, D.S.; Li, Y.X.; Zhao, M.W. Methyl jasmonate induces ganoderic acid biosynthesis in the basidiomycetous fungus *Ganoderma lucidum*. *Bioresour. Technol.* **2010**, *101*, 6785–6790. [CrossRef]
22. Schulze, A.; Zimmer, M.; Mielke, S.; Stellmach, H.; Melnyk, C.W.; Hause, B.; Gasperini, D. Wound-Induced Shoot-to-Root Relocation of JA-Ile Precursors Coordinates Arabidopsis Growth. *Mol. Plant* **2019**, *12*, 1383–1394. [CrossRef] [PubMed]
23. Kumar, S.; Stecher, G.; Li, M.; Knyaz, C.; Tamura, K. MEGA X: Molecular Evolutionary Genetics Analysis across Computing Platforms. *Mol. Biol. Evol.* **2018**, *35*, 1547–1549. [CrossRef]
24. Tao, Y.; van Peer, A.F.; Huang, Q.; Shao, Y.; Zhang, L.; Xie, B.; Jiang, Y.; Zhu, J.; Xie, B. Identification of novel and robust internal control genes from *Volvariella volvacea* that are suitable for RT-qPCR in filamentous fungi. *Sci. Rep.* **2016**, *6*, 29236. [CrossRef] [PubMed]
25. Yang, Y.; Xu, X.; Jing, Z.; Ye, J.; Li, H.; Li, X.; Shi, L.; Chen, M.; Wang, T.; Xie, B.; et al. Genome-Wide Screening and Stability Verification of the Robust Internal Control Genes for RT-qPCR in Filamentous Fungi. *J. Fungi* **2022**, *8*, 952. [CrossRef] [PubMed]
26. Liu, J.; Wang, R.; Zhang, D.; Shang, X.; Tan, Q. Analysis of genes related to lysine biosynthesis based on whole genome of *Flammulina velutipes*. *Microbiol. China* **2016**, *43*, 2225–2233. [CrossRef]
27. Xu, W.; Huang, R.; Liu, Y.; Tong, Z.; Han, X.; Xie, L.; Xie, B. Genome sequencing and assembly strategy analyses of *Flammulina filiformis*. *Mycosystema* **2018**, *37*, 1578–1585. [CrossRef]
28. Akey, J.M.; Palma-Guerrero, J.; Hall, C.R.; Kowbel, D.; Welch, J.; Taylor, J.W.; Brem, R.B.; Glass, N.L. Genome Wide Association Identifies Novel Loci Involved in Fungal Communication. *PLOS Genet.* **2013**, *9*, e1003669. [CrossRef]
29. Kim, P.; Zhou, X. FusionGDB: Fusion gene annotation DataBase. *Nucleic Acids Res.* **2019**, *47*, D994–D1004. [CrossRef]
30. Leonard, G.; Richards, T.A. Genome-scale comparative analysis of gene fusions, gene fissions, and the fungal tree of life. *Proc. Natl. Acad. Sci. USA* **2012**, *109*, 21402–21407. [CrossRef]
31. Groppa, M.D.; Benavides, M.P. Polyamines and abiotic stress: Recent advances. *Amino Acids* **2008**, *34*, 35–45. [CrossRef]

32. Quezada, H.; Aranda, C.; DeLuna, A.; Hernández, H.; Calcagno, M.L.; Marín-Hernández, A.; González, A. Specialization of the paralogue LYS21 determines lysine biosynthesis under respiratory metabolism in *Saccharomyces cerevisiae*. *Microbiology* **2008**, *154*, 1656–1667. [CrossRef] [PubMed]
33. Yanai, I.; Wolf, Y.I.; Koonin, E.V. Evolution of gene fusions: Horizontal transfer versus independent events. *Genome Biol.* **2002**, *3*, research0024.1. [CrossRef]

Journal of *Fungi*

MDPI

*Article*

# Phylogenesis of the Functional 1-Aminocyclopropane-1-Carboxylate Oxidase of Fungi and Plants

**Yanan Li, Man Qi, Qi Zhang, Zhixu Xu, Yan Zhang, Yuqian Gao, Yuancheng Qi, Liyou Qiu * and Mingdao Wang ***

Key Laboratory of Enzyme Engineering of Agricultural Microbiology, Ministry of Agriculture and Rural Affairs, College of Life Sciences, Henan Agricultural University, Zhengzhou 450002, China

* Correspondence: qliyou@henau.edu.cn (L.Q.); wangmingdao@henau.edu.cn (M.W.)

**Abstract:** The 1-aminocyclopropane-1-carboxylic acid (ACC) pathway that synthesizes ethylene is shared in seed plants, fungi and probably other organisms. However, the evolutionary relationship of the key enzyme ACC oxidase (ACO) in the pathway among organisms remains unknown. Herein, we cloned, expressed and characterized five ACOs from the straw mushroom (*Volvariella volvacea*) and the oyster mushroom (*Pleurotus ostreatus*): VvACO1-4 and PoACO. The five mushroom ACOs and the previously identified AbACO of the button mushroom contained all three conserved residues that bound to Fe(II) in plant ACOs. They also had variable residues that were conserved and bound to ascorbate and bicarbonate in plant ACOs and harbored only 1–2 of the five conserved ACO motifs in plant ACOs. Particularly, VvACO2 and AbACO had only one ACO motif 2. Additionally, VvACO4 shared 44.23% sequence identity with the cyanobacterium *Hapalosiphon* putative functional ACO. Phylogenetic analysis showed that the functional ACOs of monocotyledonous and dicotyledonous plants co-occurred in Type I, Type II and Type III, while putative functional gymnosperm ACOs also appeared in Type III. The putative functional bacterial ACO, functional fungi and slime mold ACOs were clustered in ancestral Type IV. These results indicate that ACO motif 2, ACC and Fe(II) are essential for ACO activity. The ACOs of the other organisms may come from the horizontal transfer of fungal ACOs, which were found ordinarily in basidiomycetes. It is mostly the first case for the horizontal gene transfers from fungi to seed plants. The horizontal transfer of ACOs from fungi to plants probably facilitates the fungal-plant symbioses, plant–land colonization and further evolution to form seeds.

**Keywords:** 1-aminocyclopropane-1-carboxylic acid oxidase; basidiomycete; sequence motif analysis; horizontal gene transfer



**Citation:** Li, Y.; Qi, M.; Zhang, Q.; Xu, Z.; Zhang, Y.; Gao, Y.; Qi, Y.; Qiu, L.; Wang, M. Phylogenesis of the Functional 1-Aminocyclopropane-1-Carboxylate Oxidase of Fungi and Plants. *J. Fungi* **2023**, *9*, 55. https://doi.org/10.3390/jof9010055

Academic Editors: Mingwen Zhao, Gen Zou and Jing Zhu

Received: 23 November 2022
Revised: 20 December 2022
Accepted: 27 December 2022
Published: 29 December 2022



## 1. Introduction

A number of bacteria, fungi, slime molds, algae, and most groups of land plants synthesize, perceive and respond to ethylene, which regulates a variety of physiological activities, including growth, development, reproduction, senescence, and stress responses [1]. However, the ethylene biosynthetic pathways of these organisms are different, including the ethylene-forming enzyme (EFE) pathway, 2-keto-4-methylthiobutyric acid (KMBA) pathway, 1-aminocyclopropane-1-carboxylic acid (ACC) pathway and others. The key enzymes of the ACC pathway are ACC synthase (ACS) and ACC oxidase (ACO). ACS converts methionine-derived S-adenosyl-L-methionine to ACC, which is oxidized to ethylene by ACC oxidase [2,3].

The EFE pathway has been explored for the biosynthesis of ethylene by *Pseudomonas syringae* [4–8], *Ralstonia solanacearum* [9,10] and cyanobacterium *Synechocystis* [11]. Whereas the ethylene biosynthesis pathway in cyanobacterium *Hapalosiphon* is probably the ACC pathway since supplementation with ACC increases ethylene production [12]; however, the ACO of *Hapalosiphon* has not been identified.

Fungal ethylene biosynthesis pathways are quite diverse. The EFE pathway is involved in ethylene production by *Fusarium oxysporum* [13], *Penicillium cyclopium* [14] and *P. digitatum* [15]. The other ascomycetes: *Aspergillus terreus* [16], *Botrytis cinerea* [17,18], and *Fusarium oxysporum* [19], have the KMBA pathway for synthesizing ethylene. Similar to seed plants, the basidiomycetes *Agaricus bisporus* [20,21] and slime mold *Dictyostelium mucoroides* [22] synthesize ethylene via the ACC pathway.

The ethylene biosynthesis pathway in seed plants is a well-known ACC pathway, but the pathway should be absent in non-seed plants [23]. Although several red algae, green algae and mosses can convert ACC to ethylene, such as the red algae *Pterocladiella capillacea* [24], green algae *Haematococcus pluviales* [25], *Acetabularia mediterranea* [26], *Ulva (Enteromorpha) intestinalis* [27], *Spirogyra pratensis* [28] and moss *Funaria hygrometrica* [29]; however, a liverwort *Riella helicophylla* and fern *Regnellidium diphyllum* [30,31] do not convert [$^{14}$C]-labeled ACC into ethylene. ACO genes are not present in the genomes of species of red algae [32,33], ferns, or bryophytes [34,35] and have not been reported in species of green algae. Thus, the non-seed plants produce ethylene presumably by an unknown non-ACC-dependent pathway [36].

ACO belongs to the 2-oxoglutarate-dependent dioxygenase (2OGD) superfamily of nonheme iron-containing proteins [37]. All 2OGDs have a typical 2-His-1-carboxylate motif required for Fe(II) binding. Dilley et al. [38] proposed the ACO reaction mechanism by using *Malus domestica* ACO1 as a model enzyme. Fe(II) participates in binding to ACC. Several positively charged amino-acid residues from ACO C-terminal $\alpha$-helix 11, including Arg175, Arg244, Ser246, Lys158, Lys292, Arg299 and Phe300, form a "nest" binding with the other two substrates, ascorbate and bicarbonate. Ascorbate and bicarbonate coordinate the activation the ACO reaction. In particular, ascorbate provides a binding site to ACO for oxygen and donates electrons [38].

The 2OGD superfamily is widely distributed in microorganisms, fungi and mammals and involves many functions [39]; therefore, many proteins have sequence similarities to ACOs. It is difficult to directly extract ACO from biological tissues, and functional characterization of ACO requires the use of recombinant protein. Therefore, only a few functional ACO enzymes have been reported [40]. The phylogenetic analysis of the functional and putative ACOs divides the seed plant ACOs into three types, Type I, Type II, and Type III, and groups non-seed plant ACOs into one cluster of "ancient", suggesting that the seed plant ACOs diverged from a shared non-seed plant ancestral ACO or 2ODG [40]. Nevertheless, the evolutionary relationship of the functional ACOs between plants and fungi remains unknown.

In previous studies, we found that the compost and casing soil of *Agaricus bisporus* contained ACC [41], and inhibitors that inhibited ACS and ACO in plants also inhibited button mushroom ethylene production. Two ACS genes and one ACO gene were cloned from the *Agaricus bisporus* genome and reduced gene expression decreased ethylene synthesis in *Agaricus bisporus*, similar to plants [42]. Ethylene inhibited mycelial growth and primordium formation but induced post-harvest mushroom maturation and senescence [20,43,44]. In this study, we cloned four and one ACO genes from the straw mushroom (*Volvariella volvacea*) and the oyster mushroom (*Pleurotus ostreatus*), respectively. The heterologously expressed proteins of the five ACO genes showed ACO activities. The evolutionary relationships of the ACOs in fungi and plants were explored.

## 2. Materials and Methods

### 2.1. Strains and Plasmids

*Pleurotus ostreatus* Heikang 650, and *Volvariella volvacea* V23 were provided by Henan Province Edible Fungi Germplasm Resource Bank. *E. coli* BL21, *Pichia pastoris* GS115, pET-28a and pPIC9K were purchased from BioVector NTCC Inc. (Beijing, China).

### 2.2. Bioinformatics Analysis

The ACO protein sequences were aligned using ClustalX 2.0. The phylogenetic tree was constructed by MEGA, version 11.0, using the maximum likelihood method with the bootstrap method (1000 replicates), Poisson model and complete deletion. Motifs were identified by MEME 5.4.1 with the following parameters: "five motifs should MEME find" mode, and other parameters were left as default. An NCBI Conserved Domain Search was used to predict the function of the motifs.

### 2.3. RT-PCR

The mycelia of the straw mushrooms and oyster mushrooms cultured in PDA plates were collected, the total RNA of the mycelia was extracted by the STE method [45], and cDNA was synthesized using HiScript® III RT SuperMix for qPCR (+gDNA wiper) (Vazyme Biotech Co., Ltd., Nanjing, China) following the manufacturers' instructions. The four ACO genes in the straw mushrooms (designated *VvACO1-4*) and the one ACO gene in the oyster mushrooms (designated *PoACO*) were cloned by PCR. The PCR system contained 100 ng cDNA, 2 μL primer (10 μM), 25 μL 2× Tolo Fast Pfu Premix, and $ddH_2O$ to reach a total volume of 50 μL. The primers used are listed in Table 1. PCR was performed at 95 °C for 5 min, 32 cycles of 98 °C for 5 s, 57 °C for 30 s, and 72 °C for 30 s, followed by a 72 °C extension for 10 min. The PCR product was purified and then sequenced by Tsingke Biotechnology Co., Ltd. (Zhengzhou, China).

**Table 1.** List of primers used in the present study.

| Name | Nucleotide Sequence (5′–3′) | Application |
|---|---|---|
| PoACO-F | ATGCCTACGAAGGCATCTACA | Cloning of PoACO |
| PoACO-R | TTAGTTGAAATGCTTAATAACAGTTCCACGAATCC | |
| pPIC9K-ACO-F | AAGGCGAATTAATTCGCGGCCGCATGCCGACCAAAGCAAGCA | Construction for pPIC9K-PoACO |
| pPIC9K-ACO-R | GCTGAAGCTTACGTAGAATTCTTAATGATGATGATGATGATGAA-TCACTGTACCACG | |
| VvACO3-F | GGATCCATGTCCCATCTCAATTCTGCAGCC | Cloning of VvACOs and Construction for the VvACOs expression vectors |
| VvACO3-R | AAGCTTTTAATAGGCACCGACTTGCCCG | |
| VvACO1-F | GGAATTCCATATGATGGCGGCTCCTAAATCTACC | |
| VvACO1-R | GCTCTAGATTAATACCTCCCCTTCTTCTCGTC | |
| VvACO2-F | CGGGATCCATGGCTGGCTTTGCCACGT | |
| VvACO2-R | GCTCTAGATCAAGCCTCAATGCGCTCAG | |
| VvACO4-F | CGGGATCCATGTTTATCTTAAGAACTCGGCTTC | |
| VvACO4-R | GCTCTAGATTACGAGTAGGTCACTGCCAGG | |

### 2.4. Construction of Mushroom ACO Expression Vectors

The four ACO genes of the straw mushrooms were digested with two enzymes (*Bam*HI, *Hin*dIII, *Eco*RI or *Xba*I) and ligated with the plasmid pCold TF or pET-28a to obtain prokaryotic expression vectors. The ACO gene of the oyster mushrooms was digested with *Eco*RI and *Not*I and then ligated with the plasmid pPIC9K to obtain a eukaryotic expression vector.

### 2.5. Mushroom ACO Expression and Purification

Four ACO prokaryotic expression vectors of the straw mushrooms were transformed into *E. coli* BL21, which was cultured in LB medium at 37 °C and 220 rpm to an $OD_{600}$ of 0.4 and then induced by 0.5 mM IPTG at 16 °C and 180 rpm for 12 h. The target proteins were purified using a HisTrap™ HP gravity flow column (GE Healthcare, Uppsala, Sweden), eluted with 200 mM–250 mM imidazole, and the eluate was detected by SDS-PAGE.

The oyster mushroom ACO eukaryotic expression vector was transformed into *Pichia pastoris* GS115 using the electroporation method. The transformants were grown and induced to express proteins according to methods described elsewhere [46]. The culture supernatant was collected and concentrated in a 10 kDa ultrafiltration tube. The concentrated solution was determined by SDS-PAGE.

### 2.6. ACO Enzyme Activity Assay

The enzyme activities of the expressed ACO proteins were assayed using a previously described method [21] with a slight modification. Briefly: a 5.5 mL headspace vial contained a 1.8 mL mixture composed of 150 mM $NaHCO_3$, 30 mM ascorbic acid, 1.0 mM ACC, 0.1 mM $FeSO_4$, and 0.1 mg enzyme protein in 100 mM Tris-HCl buffer at pH 7.2. After shaking incubation at 30 °C for 1 h, 1 mL gas was withdrawn for ethylene determination by gas chromatography (GC-2010 plus, Shimadzu, Kyoto, Japan) [47].

## 3. Results

### 3.1. Cloning, Expression and Enzyme Activities of ACO Proteins from Straw Mushrooms and Oyster Mushrooms

There are four ACO genes annotated in the straw mushroom *V. volvacea* V23 genome [48] and designated *VvACO*1-4 (Table 2). One ACO gene was annotated in the oyster mushroom *P. ostreatus* PC15 genome [49] and named *PoACO* (Table 2). The primers were designed according to the ACO gene sequences (Table 1), and four and one ACO genes were cloned from the mRNA of the straw mushroom *V. volvacea* V23 and *P. ostreatus* Heikang 650, respectively. After sequencing and alignment, the sequences of the five genes were completely consistent with the genome data.

**Table 2.** Selected ACO sequences used in the phylogenetic analyses.

| Species | Protein Name | Protein ID | Type | Protein (aa) | Reference |
|---|---|---|---|---|---|
| *Volvariella volvacea* | VvACO4 | JGI 111930 | 4 | 354 | This study |
| | VvACO3 | JGI 116615 | 4 | 374 | This study |
| | VvACO2 | JGI 118606 | 4 | 381 | This study |
| | VvACO1 | JGI 111142 | 4 | 353 | This study |
| *Pleurotus ostreatus* | PoACO | KDQ32580 | 4 | 324 | This study |
| *Agaricus bisporus* | AbACO | JGI 195789 | 4 | 368 | [20] |
| *Dictyostelium mucoroides* | DmACO | BAF64840 | 4 | 368 | [22] |
| *Arabidopsis thaliana* | AtACO1 | AT2G19590.1 | 2 | 310 | [50] |
| | AtACO2 | AT1G62380.1 | 1 | 320 | [51] |
| | AtACO5 | AT1G77330.1 | 3 | 307 | [50] |
| Apple (*Malus domestica*) | MdACO1 | MDP0000195885 | 1 | 314 | [52] |
| | MdACO6 | MDP0000025650 | 3 | 298 | [53] |
| | MdACO7 | MDP0000200896 | 2 | 305 | [54] |
| Tomato (*Solanum lycopersicum*) | SlACO1 | Solyc07g049530.2.1 | 1 | 315 | [55] |
| | SlACO5 | Solyc07g026650.2.1 | 2 | 301 | [56] |
| | SlACO7 | Solyc06g060070.2.1 | 3 | 314 | [57] |
| Maize (*Zea mays*) | ZmACO20 | Zm00008a017510_T01 | 1 | 323 | [58] |
| | ZmACO15 | Zm00008a037502_T01 | 3 | 314 | [22] |
| Rice (*Oryza sativa*) | OsACO1 | LOC_Os09g27820.1 | 1 | 322 | [59] |
| | OsACO6 | LOC_Os06g37590.1 | 2 | 293 | [60] |
| | OsACO7 | LOC_Os01g39860.1 | 2 | 312 | [22] |
| | OsACO4 | LOC_Os11g08380.1 | 3 | 309 | [22] |
| Douglas fir (*Pseudotsuga menziesii*) | PmACO | ABF20554 | 4 | 320 | [61] |
| Loblolly pine (*Pinus taeda*) | PtACO1 | ADD65762 | 4 | 333 | [62,63] |
| | PtACO2 | ADD65761 | 4 | 333 | [62,63] |
| | PtACO3 | ADD65760 | 4 | 323 | [62,63] |

The predicted molecular weights of VvACO1-4 were 39.97 kD, 42.08 kD, 42.04 kD and 39.44 kD, respectively. We first used the pET-28a plasmid to express the four proteins, but only VvACO3 was successfully expressed (Figure 1C). The other three proteins were expressed using the pCold TF plasmid. The molecular weights of VvACO1, VvACO2 and VvACO4 expressed by pCold TF vectors were approximately 90, 92, and 90 kDa, respectively, due to the molecular weight of TF (trigger factor from *E. coli*) being approximately 50 kDa (Figure 1A,B,D).

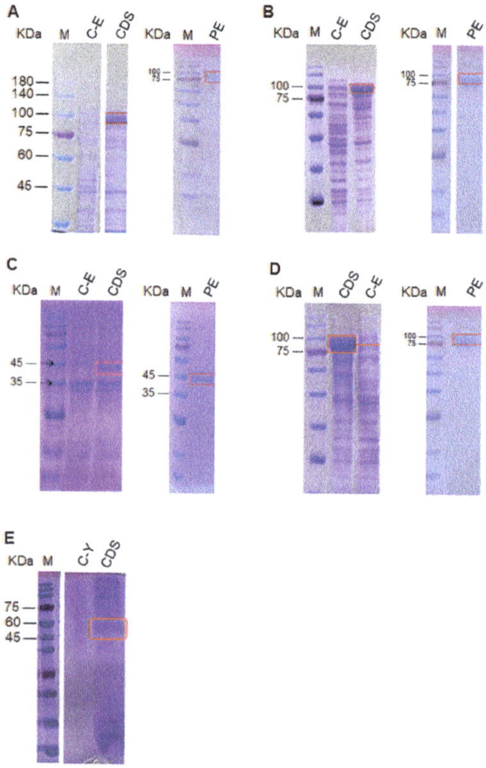


**Figure 1.** Expression and purification of the ACO proteins from straw mushrooms and oyster mushrooms. (**A**), VvACO1; (**B**), VvACO2; (**C**), VvACO3; (**D**), VvACO4; (**E**), PoACO. M, protein marker; C-E, control *E. coli* cell extract; CDS, cell disruption supernatant; C-Y, control yeast cell extract; PE, purified enzyme.

PoACO protein had an expected molecular weight of 36.95 kDa. We used pET-28a and pCold TF vectors to express PoACO but could not obtain free active protein. Then, we successfully obtained free active protein using the plasmid pPIC9K expressed in *Pichia* yeast. The molecular weight of the PoACO protein was approximately 50 kDa (Figure 1E), probably caused by glycosylation modification.

The ability of the purified proteins to oxidize ACC to produce ethylene was determined. All five proteins were able to oxidize ACC to produce ethylene, among which VvACO2 had the highest specific enzyme activity (Figure 2).

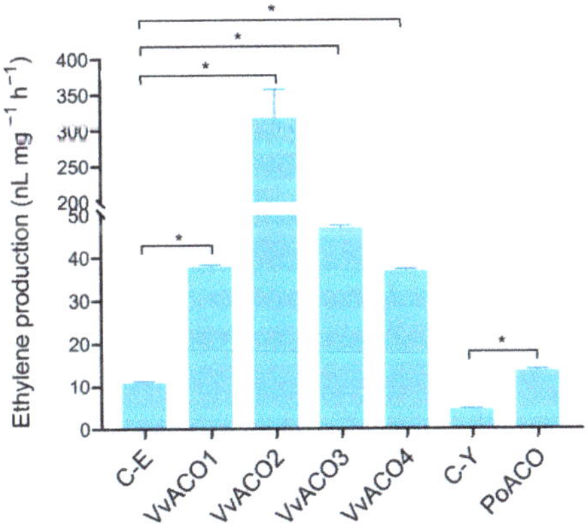


**Figure 2.** Specific activities of the heterologously expressed ACO enzymes from straw mushroom and oyster mushroom. The data shown were the mean of three independent experiments ± standard deviation (SD). They were analyzed by one-way analysis of variance (ANOVA) followed by the Tukey–Kramer multiple-comparison post hoc test. * Indicates a significant difference at $p < 0.05$.

### 3.2. Residue Analysis of Functional Fungi, Slime Mold and Plant ACOs

Fungal functional ACOs included AbACO from button mushrooms [20,21], VvACO1-4 from straw mushrooms and PoACO from oyster mushrooms identified in this study. The functional ACO of slime mold was DmACO from *D. mucoroides* [22]. These seven functional ACOs were sequence aligned with the four representative plant-functional ACOs AtACO2 (Type I), AtACO1 (Type II), and AtACO5 and OsACO4 (Type III). Except for the sequence identity of PoACO and AbACO, which reached 41.36%, the sequence identity among the ACO proteins of other fungi and slime molds was 19.4–31.3%. The ACO proteins of these fungi and slime molds were only 18.0–27.3% identical in sequence to those of plants. Six of the eight conserved amino acid residues of the Fe(II) ascorbate family of dioxygenases [21] were harboured in all 11 ACO proteins, and the other two residues were variable. The two variable residues can occur in both plant ACOs, five fungal ACOs and slime mold ACOs. The three conserved Fe(II) binding residues in plant ACOs [64] were all involved in the six conserved residues, whereas the proposed ascorbate binding residues [38] and bicarbonate binding residues [65] in plant ACOs were all highly variable in fungal and slime mold ACOs. In addition, an entirely conserved RXS motif proposed to be involved in binding the carboxylate of ACC in plant ACOs [66] was also variable in XS in fungal ACOs (Figure 3).

### 3.3. Phylogenetic Analysis of Microbial and Plant ACOs

The identified functional ACOs of fungi, slime molds and some angiosperms and possibly functional ACOs of several bacteria and gymnosperms (Table 2) were selected to perform phylogenetic analysis to study their evolutionary relationships. These 27 ACOs should be divided into four types. Type I, Type II and Type III all contained the ACOs of monocotyledonous and dicotyledonous plants, while ACOs of gymnosperms were all grouped into Type III. Type IV contained only the ACOs of fungi, slime molds and bacteria (Figure 4).

Five conserved motifs were discovered in the 27 ACOs, which all belong to ACO motifs. The seed plant ACOs all contained these five conserved motifs, except OsACO6 and OsACO7, which had four conserved motifs. The ACOs of bacteria, fungi and slime moulds had only 1–2 conserved motifs, especially VvACO2, and AbACO had only one ACO motif 2 (PIRNAIVVNIGDQJEVJSNGRYKSVWHRV) (Figure 5).

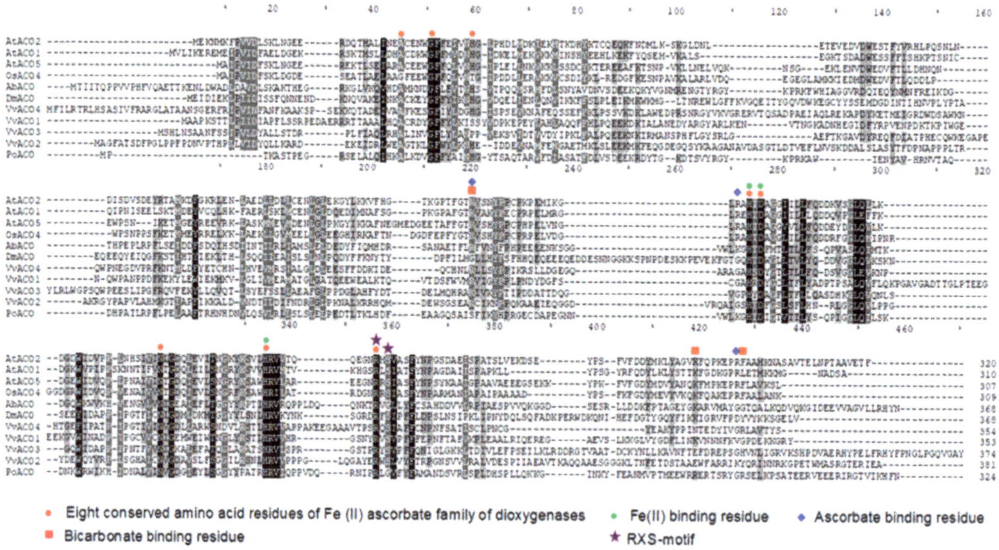


**Figure 3.** Selected functional ACO protein sequence alignment of mushrooms, slime molds and plants. Protein sequences were aligned using ClustalX 2.0. The important and conserved amino acid residues were marked in accordance with the legend shown.

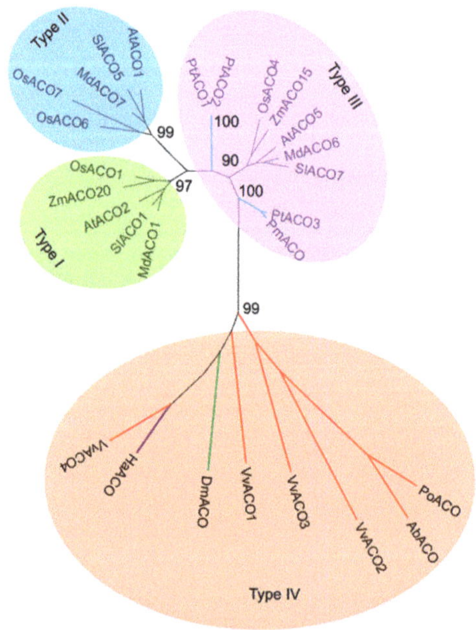


**Figure 4.** Unrooted phylogenetic tree of 27 ACO proteins generated by MEGA, version 11.0, with the maximum likelihood method. Numbers near branches show percentage bootstrap support. Red clades indicate fungal ACO proteins; blue clades indicate ACO proteins of gymnosperms; green clades indicate the ACO proteins of slime mold; purple clades indicate bacterial ACO proteins; black clades indicate the ACO proteins of angiosperms. The protein sequence information is listed in Table 2.

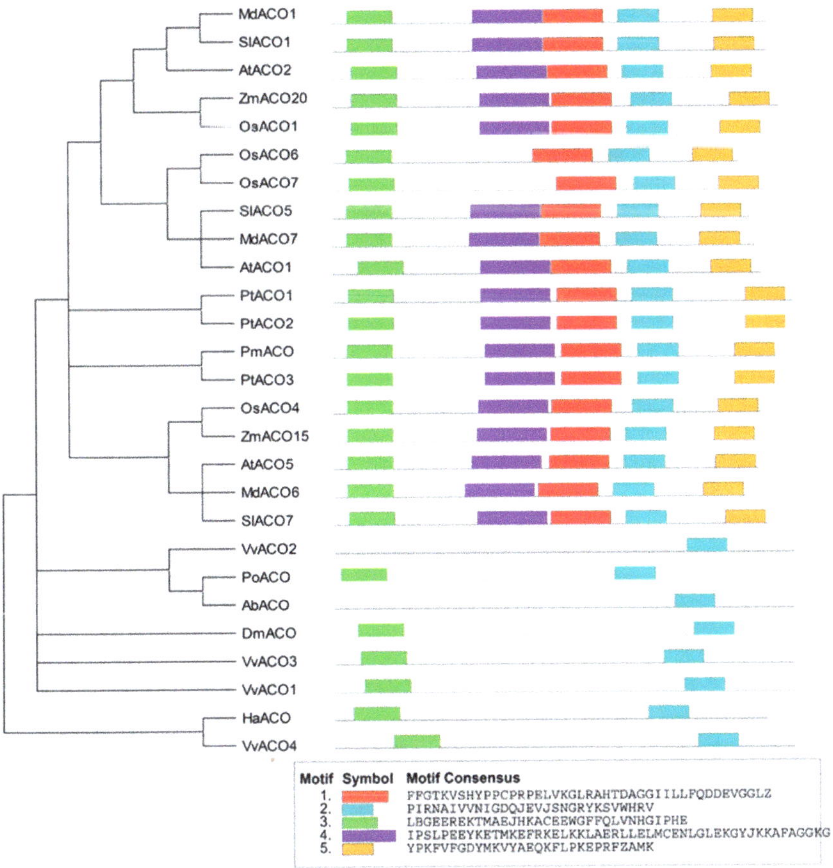


**Figure 5.** The phylogenetic relationships and conserved motif structure of the ACO protein sequences from the functional ACOs of fungi, slime molds and angiosperms plants, and the putative functional ACOs of cyanobacteria and gymnosperms. The phylogenetic trees were constructed MEGA version 11.0 with the maximum likelihood method. The conserved motif structure was discovered by MEME. The protein sequence information listed in Table 2.

## 4. Discussion

Five putative ACO genes from straw mushrooms and oyster mushrooms were cloned and expressed, and all showed ACO activities, such as the button mushroom AbACO and slime mold DmACO. The amino acid sequence identity between AbACO and plant ACOs is less than 25%, and the conserved ascorbate binding residues, bicarbonate binding residues and RXS in plant ACOs are not conserved in AbACO. The three conserved residues that bind Fe(II) in plant ACOs are conserved in AbACO. Mutation of the three conserved residues that bind Fe(II) reduces the specific activity of AbACO by 24–38%. However, mutating the G in the RXG of AbACO to S increased the specific activity of the enzyme two-fold and dramatically reduced the bicarbonate dependence. It is suggested that AbACO activity requires ACC, Fe(II) and bicarbonate but does not require ascorbate [21]. Similar to AbACO, 2–3 residues out of the three conserved and binding ascorbate residues in plant ACOs were not conserved in fungi and slime molds. Of the three conserved residues that bind bicarbonate in plant ACOs, one and 2–3 were not conserved in AtACO1 and the ACOs of fungi and slime molds. The RXS motif conserved in plant ACOs was replaced with RXG in PoACO, as in AbACO. The ACOs of straw mushroom, oyster mushroom and slime mold had 1–2 conserved ACO motifs, whereas the ACOs of plants had 4–5. In particular, VvACO2 and AbACO had only one ACO motif 2, indicating that ACO motif 2, ACC and Fe(II) should be essentially required for ACO activity.

Despite the fact that several ethylene receptors and ethylene signal transduction components of seed plants may originate from green algae [28], the ACOs in seed plants should not come from green algae. Houben and Van de Poel conducted a phylogenetic tree analysis of the ACOs of angiosperm plants along with putative ACOs of gymnosperms and other non-seed plants (without fungi), dividing these ACOs into Type I, Type II, Type III and the other group of the ancestral non-seed plant node of putative ACOs. Type I and Type II include ACOs of common angiosperm plants, and Type III includes ACOs of common angiosperm plants and putative ACOs of some gymnosperms. They proposed that the three types of ACOs diverged in parallel from a shared non-seed plant ancestral ACO [40]. However, neither putative *ACO* genes from gymnosperms nor non-seed plants were functionally characterized in vitro. The putative ACOs of the gymnosperms and other non-seed plants were almost all not ACOs but other 2OGDs.

The putative ACO protein PmACO of *Pseudotsuga menziesii* was able to bind to the *Arabidopsis* ACO antibody. Its expression was upregulated by mechanical wounding, consistent with the wound-induced increase in ethylene levels [62]. Three putative ACO genes, *PtACO*1-3, were cloned from *Pinus taeda*, and their promoters responded to indole-3-acetic acid (IAA), wounding, and gravitropic reorientation [63,64]. Cyanobacterium *Hapalosiphon* may have the ACC pathway to synthesize ethylene [12]. Two isopenicillin N synthase family oxygenases were annotated in the genome of *Hapalosiphon* sp. MRB220. One of them was 44.23% identical to VvACO4 in the deduced amino acid sequence and very likely to be ACO, named HaACO (GenBank WP_053457617.1). In this study, phylogenetic analysis of the identified functional ACOs of fungi, slime molds and some angiosperms, and possibly functional ACOs of several bacteria and gymnosperms, showed that the ACOs can be divided into four types. The ACOs of seed plants were grouped into Type I, Type II and Type III, similar to the above report [40]. Nevertheless, the ancestral Type IV contained only fungi, slime molds and bacteria. This suggests that the ACOs of seed plants may originate from microorganisms.

This study and a previous report [40] showed that the ACOs of monocotyledonous and dicotyledonous plants co-occur in Type I, Type II and Type III. Gymnosperm ACOs also appeared in Type III, implying the existence of horizontal gene transfer of ACOs in seed plants. In this study, the ACOs of bacteria, fungi and slime molds were clustered in Type IV, indicating that there was also horizontal gene transfer among them. ACOs are rarely found in bacteria but ordinarily in basidiomycetes, and the sequence identity of cyanobacterial HaACO and VvACO4 is as high as 44.23%, indicating that HaACO is likely to be horizontally transferred from fungi. Fungal ACOs contained only 1–2 conserved ACO motifs and were quite primitive and simple compared to plant ACOs. In addition, ACOs were not found in non-seed plants [34]. Therefore, ACOs in seed plants should be horizontally transferred from fungi. Horizontal gene transfers from fungi to seed plants have not been discovered previously [66]. The horizontal gene transfers from fungi to seed plants found in this study may be the first to be reported.

Ascomycota and Basidiomycota diverged about 500 million years ago—a little time before plants started their colonization of land [67]. The horizontal transfer of ACOs from fungi to plants probably facilitates fungal–plant symbioses, plant–land colonization and further evolution to form seeds.

## 5. Conclusions

We cloned, expressed and characterized five ACOs from basidiomycetes straw mushrooms and oyster mushrooms. The five ACOs, plus the previously identified AbACO of button mushroom, contained all three conserved residues that bind to Fe(II) in plant ACOs but did not contain the entire residues that conserved binding to ascorbate and bicarbonate in plant ACOs and only had 1–2 of the five conserved ACO motifs in plant ACOs. Interestingly, VvACO4 shared 44.23% sequence identity with the putative cyanobacterial HaACO. Evolutionary analysis of the functional ACOs of fungi, slime molds and angiosperm plants and the putative functional ACOs of cyanobacteria and gymnosperms can classify these ACOs into four types. The ACOs of monocotyledonous and dicotyledonous plants cooccurred in Type I, Type II and Type III. Gymnosperm ACOs also appeared in Type III. The ACOs of bacteria, fungi, and slime molds were clustered in Type IV. These results indicate that ACO motif 2, ACC and Fe(II) are essential for ACO activity. Horizontal gene transfer of ACOs exists between monocotyledonous and dicotyledonous plants, between seed plants and gymnosperms and between fungi and bacteria. The ACO genes of the other organisms may come from the horizontal transfer of fungal ACO genes.

**Author Contributions:** Conceptualization, L.Q. and M.W.; methodology, Y.L., M.Q., Q.Z., Z.X. and Y.Z.; formal analysis, Y.G. and Y.Q.; writing—original draft preparation, Y.L. and M.W.; writing—review and editing, L.Q.; funding acquisition, Y.L., L.Q. and M.W. All authors have read and agreed to the published version of the manuscript.

**Funding:** This research was funded by the Science and Technology Department of Henan Province, grant numbers 202102110044 and 222102110302, and the National Natural Science Foundation of China, grant number 32102455.

**Institutional Review Board Statement:** Not applicable.

**Informed Consent Statement:** Not applicable.

**Data Availability Statement:** Not applicable.

**Conflicts of Interest:** The authors declare no conflict of interest.

## References

1. Poel, B.V.D.; Cooper, E.D.; Delwiche, C.F.; Chang, C. An evolutionary perspective on the plant hormone ethylene. In *Ethylene in Plants*; Wen, C.K., Ed.; Springer: Dordrecht, The Netherlands, 2015; pp. 109–134.
2. Kende, H. Ethylene biosynthesis. *Annu. Rev. Plant Physiol. Plant Mol. Biol.* **1993**, *44*, 283–307. [CrossRef]
3. Ali, S.; Kim, W.-C. Plant growth promotion under water: Decrease of waterlogging-induced ACC and ethylene levels by ACC deaminase-producing bacteria. *Front. Microbiol.* **2018**, *9*, 1096. [CrossRef] [PubMed]
4. Fukuda, H.; Ogawa, T.; Ishihara, K.; Fujii, T.; Nagahama, K.; Omata, T.; Inoue, Y.; Tanase, S.; Morino, Y. Molecular cloning in *Escherichia coli*, expression, and nucleotide sequence of the gene for the ethylene-forming enzyme of *Pseudomonas syringae* pv. phaseolicola PK2. *Biochem. Biophys. Res. Commun.* **1992**, *188*, 826–832. [CrossRef]
5. Tao, L.; Dong, H.J.; Chen, X.; Chen, S.F.; Wang, T.H. Expression of ethylene-forming enzyme (EFE) of *Pseudomonas syringae* pv. glycinea in *Trichoderma viride*. *Appl. Microbiol. Biotechnol.* **2008**, *80*, 573. [CrossRef]
6. Martinez, S.; Fellner, M.; Herr, C.Q.; Ritchie, A.; Hu, J.; Hausinger, R.P. Structures and mechanisms of the non-heme Fe(II)-and 2-oxoglutarate-dependent ethylene-forming enzyme: Substrate binding creates a twist. *J. Am. Chem. Soc.* **2017**, *139*, 11980–11988. [CrossRef]
7. Zhang, Z.; Smart, T.J.; Choi, H.; Hardy, F.; Lohans, C.T.; Abboud, M.I.; Richardson, M.S.W.; Paton, R.S.; McDonough, M.A.; Schofield, C.J. Structural and stereoelectronic insights into oxygenase-catalyzed formation of ethylene from 2-oxoglutarate. *Proc. Natl. Acad. Sci. USA* **2017**, *114*, 4667–4672. [CrossRef]
8. Li, M.; Martinez, S.; Hausinger, R.P.; Emerson, J.P. Thermodynamics of iron (II) and substrate binding to the ethylene-forming enzyme. *Biochemistry* **2018**, *57*, 5696–5705. [CrossRef]
9. Weingart, H.; Völksch, B.; Ullrich, M.S. Comparison of ethylene production by *Pseudomonas syringae* and *Ralstonia solanacearum*. *Phytopathology* **1999**, *89*, 360–365. [CrossRef]
10. Valls, M.; Genin, S.; Boucher, C. Integrated regulation of the type III secretion system and other virulence determinants in *Ralstonia solanacearum*. *PLoS Pathog.* **2006**, *2*, e82. [CrossRef]
11. Zavřel, T.; Knoop, H.; Steuer, R.; Jones, P.R.; Červený, J.; Trtílek, M. A quantitative evaluation of ethylene production in the recombinant cyanobacterium *Synechocystis* sp. PCC 6803 harboring the ethylene-forming enzyme by membrane inlet mass spectrometry. *Bioresour Technol.* **2016**, *202*, 142–151. [CrossRef]
12. Huang, T.C.; Chow, T.J. Ethylene production by blue-green algae. *Bot. Bull. Acad. Sin.* **1984**, *25*, 81–86.
13. Hottiger, T.; Boller, T. Ethylene biosynthesis in *Fusarium oxysporum* f. sp. tulipae proceeds from glutamate/2-oxoglutarate and requires oxygen and ferrous ions in vivo. *Arch. Microbiol.* **1991**, *157*, 18–22. [CrossRef]
14. Pazout, J.; Pazoutova, S. Ethylene is synthesised by vegetative mycelium in surface cultures of *Penicillium cyclopium* Westling. *Can. J. Microbiol.* **1989**, *35*, 384–387. [CrossRef]

15. Fukudaa, H.; Fujiia, T.; Ogawaa, T. Preparation of a cell-free ethylene- forming system from *Penicillium digitatum*. *Agric. Biol. Chem.* **1986**, *50*, 977–981. [CrossRef]
16. Akhtar, M.J.; Arshad, M.; Khalid, A.; Mahmood, M.H. Substrate-dependent biosynthesis of ethylene by rhizosphere soil fungi and its influence on etiolated pea seedlings. *Pedobiologia* **2005**, *49*, 211–219. [CrossRef]
17. Chagué, V.; Elad, Y.; Barakat, R.; Tudzynski, P.; Sharon, A. Ethylene biosynthesis in *Botrytis cinerea*. *FEMS Microbiol. Ecol.* **2002**, *40*, 143–149. [CrossRef]
18. Cristescu, S.M.; De Martinis, D.; Hekkert, S.L.; Parker, D.H.; Harren, F.J.M. Ethylene production by *Botrytis cinerea* in vitro and in tomatoes. *Appl. Environ. Microbiol.* **2002**, *68*, 5342–5350. [CrossRef]
19. Tzeng, D.; DeVay, J. Ethylene production and toxi genicity of methionine and its derivatives with riboflavin in cultures of *Verticillium*, *Fusarium* and *Colletotrichum* species exposed to light. *Physiol. Plant* **1984**, *62*, 545–552. [CrossRef]
20. Zhang, C.; Huang, T.; Shen, C.; Wang, X.; Qi, Y.; Shen, J.; Song, A.; Qiu, L.; Ai, Y. Downregulation of ethylene production increases mycelial growth and primordia formation in the button culinary-medicinal mushroom, *Agaricus bisporus* (Agaricomycetes). *Int. J. Med. Mushrooms* **2016**, *18*, 1131–1140. [CrossRef]
21. Meng, D.; Shen, L.; Yang, R.; Zhang, X.; Sheng, J. Identification and active site analysis of the 1-aminocyclopropane-1-carboxylic acid oxidase catalysing the synthesis of ethylene in *Agaricus bisporus*. *Biochim. Biophys. Acta Gen. Subj.* **2014**, *1840*, 120–128. [CrossRef]
22. Amagai, A.; Maeda, Y. The ethylene action in the development of cellular slime molds: An analogy to higher plants. *Protoplasma* **1992**, *167*, 159–168. [CrossRef]
23. Li, D.; Mou, W.; Van de Poel, B.; Chang, C. Something old, something new: Conservation of the ethylene precursor 1-amino-cyclopropane-1-carboxylic acid as a signaling molecule. *Curr Opin Plant Biol.* **2022**, *65*, 102116. [CrossRef] [PubMed]
24. Garcia-Jimenez, P.; Robaina, R.R. Effects of ethylene on tetrasporogenesis in *Pterocladiella capillacea* (Rhodophyta). *J. Phycol.* **2012**, *48*, 710–715. [CrossRef] [PubMed]
25. Maillard, P.; Thepenier, C.; Gudin, C. Determination of an ethylene biosynthesis pathway in the unicellular green alga, *Haematococcus pluvialis*. Relationship between growth and ethylene production. *J. Appl. Phycol.* **1993**, *5*, 93–98. [CrossRef]
26. Driessche, T.V.; Kevers, C.; Collet, M.; Gaspar, T. *Acetabularia mediterranea* and ethylene: Production in relation with development, circadian rhythms in emission, and response to external application. *J. Plant Physiol.* **1988**, *133*, 635–639. [CrossRef]
27. Plettner, I.N.A.; Steinke, M.; Malin, G. Ethene (ethylene) production in the marine macroalga *Ulva* (*Enteromorpha*) *intestinalis* L. (*Chlorophyta, Ulvophyceae*): Effect of light-stress and co-production with dimethyl sulphide. *Plant Cell Environ.* **2005**, *28*, 1136–1145. [CrossRef]
28. Ju, C.; Van de Poel, B.; Cooper, E.D.; Thierer, J.H.; Gibbons, T.R.; Delwiche, C.F.; Chang, C. Conservation of ethylene as a plant hormone over 450 million years of evolution. *Nat. Plants* **2015**, *1*, 14004. [CrossRef]
29. Rohwer, F.; Bopp, M. Ethylene synthesis in moss protonema. *J. Plant Physiol.* **1985**, *117*, 331–338. [CrossRef]
30. Chernys, J.; Kende, H. Ethylene biosynthesis in *Regnellidium diphyllum* and *Marsilea quadrifolia*. *Planta* **1996**, *200*, 113–118. [CrossRef]
31. Osborne, D.J.; Walters, J.; Milborrow, B.V.; Norville, A.; Stange, L.M.C. Evidence for a non-ACC ethylene biosynthesis pathway in lower plants. *Phytochemistry* **1996**, *42*, 51–60. [CrossRef]
32. Uji, T.; Matsuda, R.; Takechi, K.; Takano, H.; Mizuta, H.; Takio, S. Ethylene regulation of sexual reproduction in the marine red alga *Pyropia yezoensis* (Rhodophyta). *J. Appl. Phycol.* **2016**, *28*, 3501–3509. [CrossRef]
33. Endo, H.; Mizuta, H.; Uji, T. α-aminoisobutyric acid mimics the effect of 1-aminocyclopropane-1-carboxylic acid to promote sexual reproduction in the marine red alga *Pyropia yezoensis* (Rhodophyta). *J. Appl. Phycol.* **2021**, *33*, 1081–1087. [CrossRef]
34. Li, F.-W.; Brouwer, P.; Carretero-Paulet, L.; Cheng, S.; de Vries, J.; Delaux, P.-M.; Eily, A.; Koppers, N.; Kuo, L.Y.; Li, Z.; et al. Fern genomes elucidate land plant evolution and cyanobacterial symbioses. *Nat. Plants* **2018**, *4*, 460–472. [CrossRef] [PubMed]
35. Katayose, A.; Kanda, A.; Kubo, Y.; Takahashi, T.; Motose, H. Distinct functions of ethylene and ACC in the basal land plant *Marchantia polymorpha*. *Plant Cell Physiol.* **2021**, *62*, 858–871. [CrossRef] [PubMed]
36. Van de Poel, B.; Chang, C. Is losing ethylene a losing game? *Mol. Plant* **2022**, *15*, 788–790. [CrossRef]
37. Kawai, Y.; Ono, E.; Mizutani, M. Evolution and diversity of the 2-oxoglutarate-dependent dioxygenase superfamily in plants. *Plant J.* **2014**, *78*, 328–343. [CrossRef]
38. Dilley, D.R.; Wang, Z.; Kadirjan-Kalbach, D.K.; Ververidis, F.; Beaudry, R.; Padmanabhan, K. 1-aminocyclopropane-1-carboxylic acid oxidase reaction mechanism and putative post-translational activities of the ACCO protein. *AoB Plants* **2013**, *5*, plt031. [CrossRef]
39. Aravind, L.; Koonin, E.V. The DNA-repair protein AlkB, EGL-9, and leprecan define new families of 2-oxoglutarate- and iron-dependent dioxygenases. *Genome Biol.* **2001**, *2*, research0007.1. [CrossRef]
40. Houben, M.; Van de Poel, B. 1-Aminocyclopropane-1-carboxylic acid oxidase (ACO): The enzyme that makes the plant hormone ethylene. *Front. Plant Sci.* **2019**, *10*, 695. [CrossRef]
41. Zhang, D.; Qi, Y.; Gao, Y.; Shen, J.; Qiu, L. Preliminary exploring on the mechanism of casing soil stimulating the primordium formation of *Agaricus bisporus*. *Edible Fungi* **2010**, *32*, 9–11.
42. Eun, H.D.; Ali, S.; Jung, H.; Kim, K.; Kim, W.C. Profiling of ACC synthase gene (ACS11) expression in *Arabidopsis* induced by abiotic stresses. *Appl. Biol. Chem.* **2019**, *62*, 1–11. [CrossRef]

43. Chen, S.; Qiu, C.; Huang, T.; Zhou, W.; Qi, Y.; Gao, Y.; Shen, J.; Qiu, L. Effect of 1-aminocyclopropane-1-carboxylic acid deaminase producing bacteria on the hyphal growth and primordium initiation of *Agaricus bisporus*. *Fungal Ecol.* **2013**, *6*, 110–118. [CrossRef]
44. Li, T.; Zhang, J.; Gao, X.; Chen, J.; Zheng, Y.; Gao, Y.; Qiu, L. The molecular mechanism for the ethylene regulation of postharvest button mushrooms maturation and senescence. *Postharvest Biol. Technol.* **2019**, *156*, 110930. [CrossRef]
45. Shui, P.R.; Zheng, X.B.; Lin, J.F.; Guo, L.Q. A rapid and efficient method for isolating high quality total RNA from edible fungi. *Acta Edulis Fungi* **2008**, *15*, 32–36.
46. Furqan, B.R.N. Heterologous expression and characterization of thermostable lipase (Lk1) in *Pichia pastoris* GS115. *Biocatal Agric Biotechnol.* **2020**, *23*, 101448. [CrossRef]
47. Zhang, C.; Shang, D.; Zhang, Y.; Gao, X.; Liu, D.; Gao, Y.; Li, Y.; Qi, Y.; Qiu, L. Two hybrid histidine kinases involved in the ethylene regulation of the mycelial growth and postharvest fruiting body maturation and senescence of *Agaricus bisporus*. *Microbiol. Spectrum* **2022**, *10*, 02411–02422. [CrossRef]
48. Bao, D.; Gong, M.; Zheng, H.; Chen, M.; Zhang, L.; Wang, H.; Jiang, J.; Wu, L.; Zhu, Y.; Zhu, G.; et al. Sequencing and comparative analysis of the straw mushroom (*Volvariella volvacea*) genome. *PLoS ONE* **2013**, *8*, e58294. [CrossRef]
49. Riley, R.; Salamov, A.A.; Brown, D.W.; Nagy, L.G.; Floudas, D.; Held, B.W.; Levasseur, A.; Lombard, V.; Morin, E.; Otillar, R.; et al. Extensive sampling of basidiomycete genomes demonstrates inadequacy of the white-rot/brown-rot paradigm for wood decay fungi. *Proc. Natl. Acad. Sci. USA* **2014**, *111*, 9923–9928. [CrossRef]
50. Vandenbussche, F.; Vriezen, W.H.; Smalle, J.; Laarhoven, L.J.J.; Harren, F.J.M.; Van Der Straeten, D. Ethylene and auxin control the Arabidopsis response to decreased light intensity. *Plant Physiol.* **2003**, *133*, 517–527. [CrossRef]
51. Raz, V.; Ecker, J.R. Regulation of differential growth in the apical hook of *Arabidopsis*. *Development* **1999**, *126*, 3661–3668. [CrossRef]
52. Dong, J.G.; Olson, D.; Silverstone, A.; Yang, S.F. Sequence of a cDNA coding for a 1-aminocyclopropane-1-carboxylate oxidase homolog from apple fruit. *Plant Physiol.* **1992**, *98*, 1530–1531. [CrossRef] [PubMed]
53. Clouse, R.M.; Carraro, N. A novel phylogeny and morphological reconstruction of the PIN genes and first phylogeny of the ACC-oxidases (ACOs). *Front. Plant Sci.* **2014**, *5*, 296. [CrossRef] [PubMed]
54. Mellidou, I.; Buts, K.; Hatoum, D.; Ho, Q.T.; Johnston, J.W.; Watkins, C.B.; Schaffer, R.J.; Gapper, N.E.; Giovannoni, J.J.; Rudell, D.R.; et al. Transcriptomic events associated with internal browning of apple during postharvest storage. *BMC Plant Biol.* **2014**, *14*, 328. [CrossRef] [PubMed]
55. Hamilton, A.J.; Bouzayen, M.; Grierson, D. Identification of a tomato gene for the ethylene-forming enzyme by expression in yeast. *Proc. Natl. Acad. Sci. USA* **1991**, *88*, 7434–7437. [CrossRef] [PubMed]
56. Sell, S.; Hehl, R. A fifth member of the tomato 1-aminocyclopropane-1-carboxylic acid (ACC) oxidase gene family harbours a leucine zipper and is anaerobically induced. *J. DNA Seq. Mapp.* **2005**, *16*, 80–82. [CrossRef]
57. Zhang, Z.; Mao, C.; Shi, Z.; Kou, X. The amino acid metabolic and carbohydrate metabolic pathway play important roles during salt-stress response in tomato. *Front. Plant Sci.* **2017**, *8*, 1231. [CrossRef]
58. Gallie, D.R.; Young, T.E. The ethylene biosynthetic and perception machinery is differentially expressed during endosperm and embryo development in maize. *Mol. Genet. Genomics* **2004**, *271*, 267–281. [CrossRef]
59. Chae, H.S.; Cho, Y.G.; Park, M.Y.; Lee, M.C.; Eun, M.Y.; Kang, B.G.; Kim, W.T. Hormonal cross-talk between auxin and ethylene differentially regulates the expression of two members of the 1-aminocyclopropane-1-carboxylate oxidase gene family in rice (*Oryza sativa* L.). *Plant Cell Physiol.* **2000**, *41*, 354–362. [CrossRef]
60. Iwai, T.; Miyasaka, A.; Seo, S.; Ohashi, Y. Contribution of ethylene biosynthesis for resistance to blast fungus infection in young rice plants. *Plant Physiol.* **2006**, *142*, 1202–1215. [CrossRef]
61. Hudgins, J.W.; Ralph, S.G.; Franceschi, V.R.; Bohlmann, J. Ethylene in induced conifer defense: cDNA cloning, protein expression, and cellular and subcellular localization of 1-aminocyclopropane-1-carboxylate oxidase in resin duct and phenolic parenchyma cells. *Planta* **2006**, *224*, 865–877. [CrossRef]
62. Yuan, S.; Wang, Y.; Dean, J.F. ACC oxidase genes expressed in the wood-forming tissues of loblolly pine (*Pinus taeda* L.) include a pair of nearly identical paralogs (NIPs). *Gene* **2010**, *453*, 24–36. [CrossRef] [PubMed]
63. Yuan, S.; Dean, J.F. Differential responses of the promoters from nearly identical paralogs of loblolly pine (*Pinus taeda* L.) ACC oxidase to biotic and abiotic stresses in transgenic *Arabidopsis thaliana*. *Planta* **2010**, *232*, 873–886. [CrossRef] [PubMed]
64. Shaw, J.; Chou, Y.; Chang, R.; Yang, S.F. Characterization of the ferrous ion binding sites of apple 1-aminocyclopropane-1-carboxylate oxidase by site-directed mutagenesis. *Biochem. Biophys. Res. Commun.* **1996**, *700*, 697–700. [CrossRef]
65. Zhang, Z.; Ren, J.-S.; Clifton, I.J.; Schofield, C.J. Crystal structure and mechanistic implications of 1-aminocyclopropane-1-carboxylic acid oxidase—The ethylene-forming enzyme. *Chem. Biol.* **2004**, *11*, 1383–1394. [CrossRef]
66. Richards, T.A.; Soanes, D.M.; Foster, P.G.; Leonard, G.; Thornton, C.R.; Talbot, N.J. Phylogenomic analysis demonstrates a pattern of rare and ancient horizontal gene transfer between plants and fungi. *Plant Cell* **2009**, *21*, 1897–1911. [CrossRef] [PubMed]
67. Eastwood, D.C. Evolution of fungal wood decay. In *Deterioration and Protection of Sustainable Biomaterials, ACS Symposium Series*; Schultz, T.P., Goodell, B., Nicholas, D.D., Eds.; Amercian Chemical Society: Washington, DC, USA, 2014; pp. 93–112.

MDPI
St. Alban-Anlage 66
4052 Basel
Switzerland
www.mdpi.com

*Journal of Fungi* Editorial Office
E-mail: jof@mdpi.com
www.mdpi.com/journal/jof

www.ingramcontent.com/pod-product-compliance
Lightning Source LLC
Chambersburg PA
CBHW061604210926
51850CB00009B/2203